Telecom Power Systems

Telecom Power Systems

Dorin O. Neacșu

CRC Press
Taylor & Francis Group
Boca Raton London New York

CRC Press is an imprint of the
Taylor & Francis Group, an **informa** business

CRC Press
Taylor & Francis Group
6000 Broken Sound Parkway NW, Suite 300
Boca Raton, FL 33487-2742

International Standard Book Number-13: 978-1-138-09930-2 (Hardback)

Visit the Taylor & Francis Web site at
http://www.taylorandfrancis.com

and the CRC Press Web site at
http://www.crcpress.com

Contents

Preface

Servers, computers, laptops, tablets, cell phones, smartphones, gaming consoles, radio mobile communication walkie-talkies, telecom data centres, computing centres, satellite communication systems, aviation communication systems, marine communications, radar systems, cell towers, wireless broadband access, fixed-line applications, the Internet backbone and so on are all ultimately supplied with high current at low voltages.

Power electronics deals with electronic processing of energy. This makes it a support technology for a variety of industrial sectors. Since each sector brings its own peculiarities, specific topologies, design practices, software control or usage habits, it is well deserved to have dedicated textbooks for each major application field.

This book addresses topics specific to the application of power electronics to telecom systems. It is written at an introductory level and aims to help any student, engineer or business graduate getting into the specifics of telecom power systems. The telecommunications field is very modern, with a good employment prospect for young graduates. They will find the book interesting if they either work directly with telecom power or need a fair understanding of the power system architecture and requirements while working with any telecommunications system.

The manuscript starts off with the architecture of telecom power systems and continues with detailed descriptions of each module.

Diode- and transistor-based rectification are considered to take energy from a national grid and build towards a high-voltage direct current (DC) bus. Next, isolated DC/DC converters can set multiple other DC bus voltages. Together, the rectifier and the high-power isolated DC converter make up the so-called 'telecom rectifier'. This is usually the power supply at the cabinet or equipment level.

Other topologies of isolated DC/DC converters are better used to supply an entire rack of a cabinet; that is, a subsystem of all the equipment or of the cabinet. The lower-level voltage enters digital circuits called mainboards or motherboards where multiple point-of-load converters provide advanced processors with multiple low voltages, each at a very high current.

The voltage supplied to the high-performance processors through a point-of-load converter needs to be well regulated and allow fine adjustments. This implies a good knowledge of control systems theory, which is presented in the book. Here, modern requirements add software control of these point-of-load converters through the power management bus, a dedicated serial communication.

Auxiliary equipment requirements like uninterruptible power supplies, storage energy systems or charging systems are also explained, along with particular classifications or suggestions for usage.

The presentation of each telecom power system is completed by a very large number of practical examples able to build the reader's confidence in acquired knowledge. Since power electronics is a mature technology with historical roots, each notion is introduced to the reader with its history.

A combination of basic power electronics knowledge and specific industrial requirements was used to compose this work. Specific terminology and industry standards are explained in layman's terms. Since telecom power is full of terms specific to the field (e.g., telecom rectifier, brick converter, POL converter, LDO converter and so on), the book also serves as a tool for developing engineering language in a young specialist.

Due to specifics of power electronics applications in telecommunications, the industry developed a series of particular standards and practices which influence thinking and common design. Hence, a generic course in power electronics would not cover all these aspects. On the other hand, a telecom industry engineer would find it more difficult to take a conventional power electronics course and seek further details of its application to his field. This book fills in this gap and provides quick information to any electrical or electronics engineer wanting to master a discussion of telecom power systems.

Similar information is sometimes available through industrial seminars at various events, which are hard and expensive to follow for a recent graduate. Moreover, industrial seminars may sometimes be overcrowded with product details rather than application details. Since the field of the telecom industry offers so many new jobs, this is expected to become a must-have book for any entry-level professional.

The novelty and dynamics of the application topic make possible the adoption of such a textbook within many engineering technology colleges with a course of study in telecommunications engineering. Currently, teachers use conventional power electronics textbooks since no targeted book is available.

The manuscript represents an improved and evolved version of a lecture course offered within the Technical University of Iași, Romania, for the last four years (Spring sessions of 2014, '15, '16 and '17) with around 75–85 students per year, split into two different tracks: in Romanian or English, all registered with an undergraduate program in telecommunications engineering. Numerous similar undergraduate or graduate programs in telecommunications engineering have been launched within the last 10 years in the United States, Europe and Asia.

Westford, Massachusetts, USA, and Iași, Romania

MATLAB® and Simulink® the registered trademark of The MathWorks, Inc. For product information, please contact:

The MathWorks, Inc.
3 Apple Hill Drive
Natick, MA 01760-2098 USA Tel: 508 647 7000
Fax: 508-647-7001
E-mail: info@mathworks.com
Web: www.mathworks.com

Author

The author has broad experience in the application of advanced power electronics concepts in industry coming from a 30-year career alternating academic and industrial R&D positions.

Professor Dorin O. Neacşu received MSc and PhD degrees in electronics from the Technical University of Iasi, Iasi, Romania, in 1988 and 1994, respectively, and an MSc degree in engineering management from the Gordon Institute for Leadership, Tufts University, Medford, Massachusetts, United States, in 2005.

He was involved with TAGCM-SUT, Iasi, from 1988 to 1990, and with the faculty in the Department of Electronics, Technical University of Iasi, between 1990 and 1999. During this time, he held visiting positions at Université du Quebec a Trois Rivières, Canada, and General Motors/Delphi, Indianapolis, Indiana, United States. Following 1999, he was involved with U.S. industry as an electrical engineer, consultant, product manager and project manager, and with U.S. academic activities at the University of New Orleans, Massachusetts Institute of Technology and United Technologies Research Center. Since 2012, he has been an associate professor with the Technical University of Iasi, Romania, and a repeat visiting associate professor with Northeastern University, Boston, Massachusetts, United States. He has maintained a continuous stream of R&D publications since 1992 within various professional organizations around the world, has organized eight professional education seminars (tutorials) at IEEE conferences and holds three U.S. patents. He has published a book entitled *Switching Power Converters – Medium and High Power* (Boca Raton, Florida, United States: CRC Press/Taylor & Francis, 2006) with a second edition from December 2013. Other ISBN books or college textbooks have been published in the United States, Canada and Romania.

Dr. Dorin O. Neacşu received the 2015 'Constantin Budeanu' Award of the Romanian Academy of Sciences, the highest Romanian research recognition in electrical engineering. He is an associate editor of the *IEEE Transactions on Power Electronics* and *IEEE Industrial Electronics Society's Magazine,* a reviewer for IEEE transactions and conferences and a member of various IEEE committees.

More about the author at:

- ORCID = 0000-0003-0572-1838
- ThomsonReuters ResearcherID = 5276-2011
- https://www.linkedin.com/in/dorin-o-neacsu-82a842

1

Power System Architecture for Telecommunications Applications

1.1 Context

The current development of telecommunications systems brings up new requirements for power supply systems [1]. The delivery and processing of electrical energy from the transmission and distribution network (known as the 'power grid' in North America) to low-voltage electric power distribution, then to each piece of telecommunications equipment, is made through electronic power circuits. Modern electronic power circuits need to satisfy certain specific design requirements concerning high-quality voltage, without harmonics or electrical noise, as well as with increased efficiency, in order to reduce electrical energy loss. The cost of electrical energy during an operation rises above the initial cost of telecom power system equipment in just three years. Hence, operation at high efficiency within the energy processing system is more important than reduction of initial cost of equipment.

Another contemporary design requirement relates to improvement of power density within electronic power equipment. This means *more watts processed within lower-dimension equipment*.

Meeting customer expectations requires technology development in power system architecture, optimization of reliability aspects, protection for extreme operation conditions, standardization and modularity of the electronic power equipment, development and usage of specific power semiconductors, electronic power circuit topology optimization, minimization of thermal effects and improvement of packaging systems.

Since development of this field has already occurred over some time, ideas have matured into several standard architecture structures. Figure 1.1 illustrates the most-used electricity delivery method for telecom equipment. The solution is given here just as an example, and various similar solutions are possible.

The energy is taken from the low-voltage AC electrical power distribution and converted to DC voltage. A semi-regulated DC voltage bus is hence set forth at around 360–410 V DC. More recent grid-related requirements force the rectifier stage to operate with the unity power factor and low harmonics of the current, usually under a total of 5%. This conversion of energy from grid to a fixed DC bus is achieved with equipment usually called a *telecom rectifier*, which may be able to power an entire cabinet.

Moreover, the telecom rectifier may also have a secondary energy path from an energy storage unit able to secure uninterruptible power delivery when the national grid voltage fails. Options for this feature are discussed in Chapter 14, along with an example for powering an isolated cell tower.

In order to develop the concepts required for the alternative energy source, the theory of inverter operation is covered and inverters are discussed briefly in Chapters 11 and 12. Inverters are power converters able to produce AC voltages from a DC source, and they become the centrepieces of uninterruptible power supplies because most energy storage systems are actually DC sources.

The high-voltage DC bus in Figure 1.1 is next converted with galvanic isolation into a lower-voltage DC bus, most typically set at 48 V DC. The selection of this voltage level has a historical reason, as telephony systems used to work with 48 V DC.

The 48 V DC level was found to be high enough for operation of the telephone on long-distance telephone lines and low enough not to cause serious danger if somebody touched the telephone wires. *Telephone exchange centres* started by using 48 V DC. Even when newer-generation equipment allowed for *automatic exchangers*, these were still designed to make use of existing 48 V DC battery storage systems, already available with telecom service providers. This legacy design practice continued until the recent development of digital equipment. Hence, the 48 V DC intermediary bus can still be found in telecom power systems.

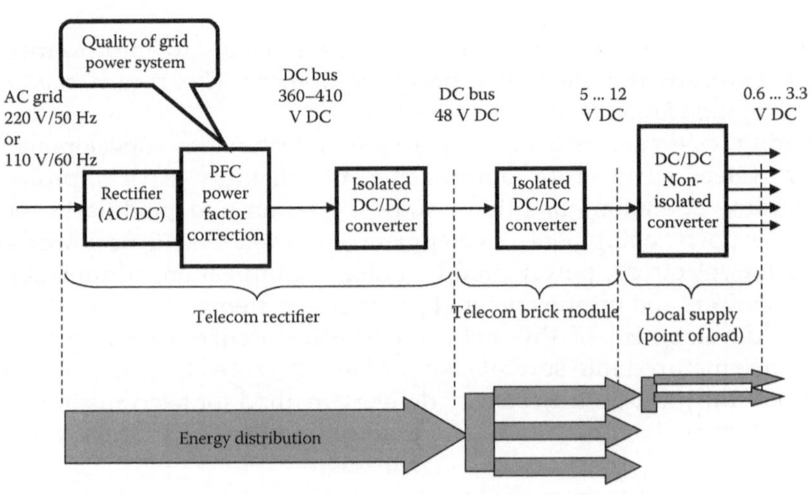

FIGURE 1.1
Example of power system architecture for a telecommunications application.

For completeness of information, the positive grounded system (also known as the −48 V DC system) has a telephone legacy. The negative voltage on the long-distance line was a better option than positive voltage in order to prevent electrochemical reactions from destroying the copper cable quickly. On the contrary, using a positive grounded system takes advantage of a technology called *cathodic protection,* discovered by Sir Humphry Davy for the British Navy in 1824. First developed to keep the copper hulls of British naval ships from corroding, this technology has been applied to protecting everything from oil rigs to gas pipelines to telephony cabinets. By keeping the cabinet frame predominantly at a more positive voltage, corrosion is reduced and the life of the equipment is increased.

Back to Figure 1.1. The 48 V DC is further sent to various subsystems, like equipment racks within a cabinet. For each electronic load, a second isolated converter provides lower voltage levels, like 12, 5 or 3.3 V DC.

Finally, such voltages supply electronic boards, sometimes called *motherboards* or *mainboards* since they house the processors, where sections of digital circuits ask for a large variety of voltages and currents at a very high regulation; that is, with very low ripple. Since all these digital circuits share the same ground, these local converters are usually built without galvanic isolation. The local non-isolated converters are also called *point-of-load* (PoL) converters.

Even though it represents a 10-year-old technology, an ancillary IBM server can be considered herein as an example [2]. Its design specification included seven voltages on the circuit board, accounting for a total of 1,200 W installed power. They included:

1.3 V DC @ 180 A (240 W)

1.5 V DC @ 120 A (180 W)

1.8 V DC @ 60 A (110 W)

2.5 V DC @ 80 A (200 W)

3.3 V DC @ 60 A (200 W)

5.0 V DC @ 20 A (100 W)

12 V DC @ 15 A (180 W)

The supply voltage to the entire board was 12 V DC, with a low-power input available at 18 V DC for fans. The thermal requirements of the chassis demanded 85% efficiency. Furthermore, there was a requirement for software-controlled power management, a precursor of today's power management bus (*PMBus*).

1.2 Example of Architecture for a Telecommunications Application

The components of the system in Figure 1.1 allow an understanding of the power flow and performance requirements or expectations for each subsystem. While Figure 1.1 shows the block diagram of an electrical system, Figure 1.2 sketches the actual telecom equipment, location of the power system subsystems and the power flow, along with efficiency expectations for each subsystem. As can be seen from Figure 1.2, there is energy loss on each power subsystem. As a notable example, 1 W saved at the server component level (ultimate client) brings a savings of 2.84 W in total consumption [19]. Hence, system efficiency optimization starts from the point-of-load converters, where energy management trends to thrive.

With the development of telecom equipment, the sum of all the losses becomes an important part of global energy systems. It is thus estimated that 5% of worldwide energy consumption is produced by computation equipment including telecommunications [3]. The telecom sector is responsible for approximately 1.4 trillion tons of CO_2 emissions – that is more than entire worldwide aviation emissions! Loss reduction within the electrical supply of telecom equipment is of high importance. In order to achieve such an energy loss reduction, understanding the operation of and specific requirements for

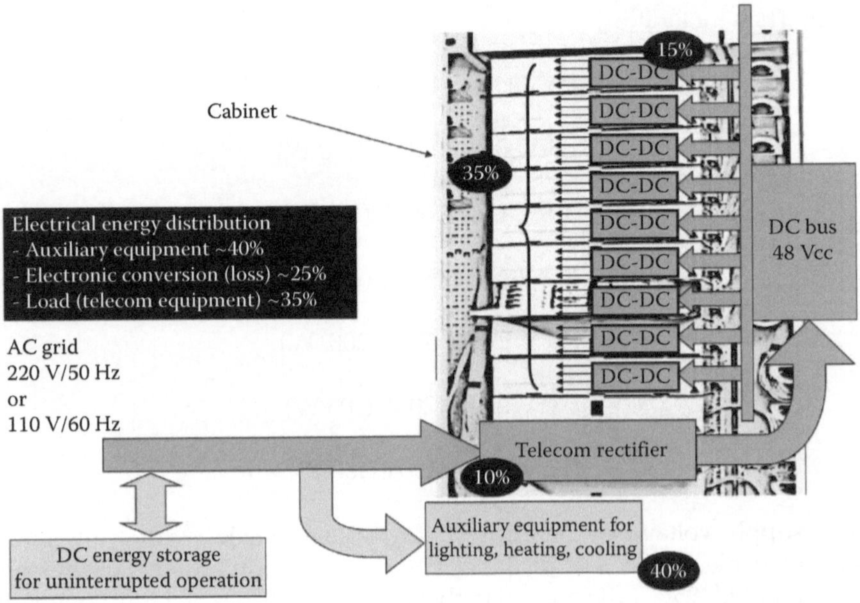

FIGURE 1.2
Power flow through the subsystems from the electrical architecture of Figure 1.1.

each subsystem of telecom equipment is important. A quick overview of the functions and duties for each module from Figure 1.2 follows.

1.2.1 Role of and Requirements for Rectifier

A rectifier represents an electronic power converter capable of transforming electrical energy from AC to DC form. Since the actual connection to a distribution network is governed by standards, depending on the maximum power level, the rectifier can be designed with diodes or a combination of rectifier diodes and transistors working in commutation. The various solutions are sketched in Figure 1.3.

At lower power levels, a simple diode rectifier is used. Whenever the power level rises, a solution achieves the unity power factor when using a controlled rectifier. This implies a switched transistor. Unity power factor means synchronization of the current's alternative waveform with the supplying voltage for reduction of reactive power circulation. A more perfected solution for power transfer, with improved control of the power factor and harmonics, is using a fully controlled bridge of transistors. Above 1 kW, the electrical energy supply may be preferred with a three-phase system involving three-phase rectifiers.

While Figure 1.3 introduces some of the options for the rectifier subsystem, the shown circuits are not the only solutions possible. Chapters 9 and 10 are devoted to the study, design and analysis of all options on the matter.

1.2.2 Role of and Requirements for High-Power Isolated Direct Current/Direct Current Converter

This converter subsystem ensures electrical energy conversion from a DC source to a DC load which requires another supply voltage. Moreover, the converter provides conversion with galvanic isolation, which means the

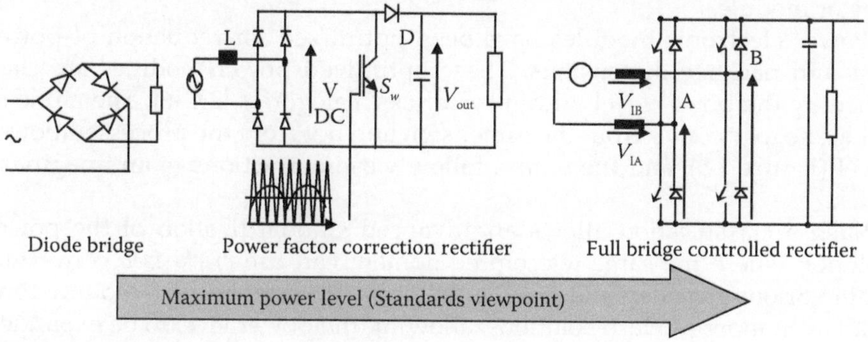

FIGURE 1.3
Possible circuits for AC/DC rectification.

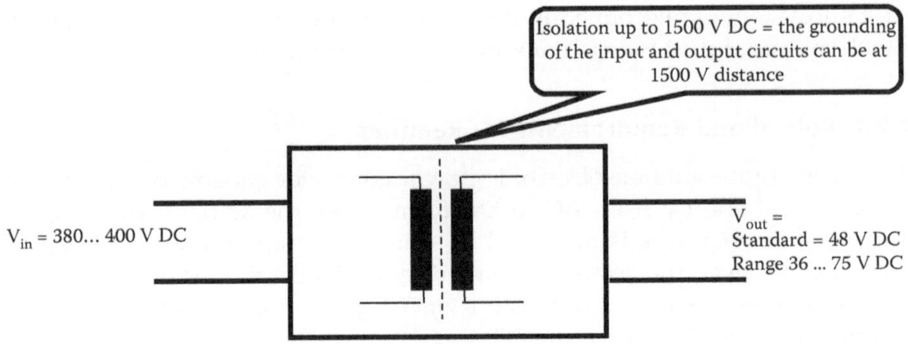

FIGURE 1.4
Functional schematic of an isolated DC/DC converter.

input and output circuits do not share the same grounding. Usually, galvanic isolation is achieved with electrical energy transfer through a transformer (more details in Chapter 6). An example of a functional schematic of an isolated DC/DC converter is shown in Figure 1.4.

Whatever circuitry is considered within this DC/DC converter, it is important to provide energy transfer with maximum efficiency. Certain architecture solutions for power systems include this high-power isolated DC/DC converter module within the rectifier, and both constitute the telecom rectifier. Various circuits for a high-power isolated DC/DC converter are therefore possible, and these are discussed in Chapters 6 and 7.

1.2.3 Role of and Requirements for Isolated Converter ('Brick' Converter)

Due to frequent usage of the high-voltage isolated converter structure within telecom power systems, this converter is built as a hybrid power semiconductor module.

Power electronic modules have been optimized for reduction of power loss and package dimensions. These optimized power modules are classified by the power level within the bricks' categories [2,4–6]. The name of each category comes from the dimension and not from the processed power level (Figure 1.5), and the names follow various fractions of an imaginary brick.

Such a classification allows an advanced standardization of the power solution where the same telecom equipment can use *brick-sized* converters from various suppliers independently. They can even be replaced any time with even more modern solutions, allowing the power level to be expanded according to the ever-more-demanding loads while the system packaging or wiring generally does not suffer changes.

(a)

(b)

FIGURE 1.5
Example of 1/4 brick (b) and 1/8 brick (a) converters.

The most-used dimensional bricks are:

- Converters of size 1/8 brick are usually optimized to deliver around 180 W at a voltage of 12, 5 or 3.3 V. Typical dimensions are 0.90 × 2.30 in., (22 × 56 mm). They are usually built with a *flyback convertor* or *synchronous forward converter* topology. Both topologies are addressed in Chapter 6.
- Converters of size 1/4 brick are optimized to deliver around 400 W at a voltage of 12, 5 or 3.3 V. Their typical dimensions are 1.45 × 2.30 in., (37 × 56 mm). They are usually built with a converter topology based on phase control named *phase-shift converter*, which will be detailed in Chapter 7.

While the previous definitions and data are provided as generic information on performance, telecom power modules have evolved over the years, and they continue to incorporate more and more technology advancements.

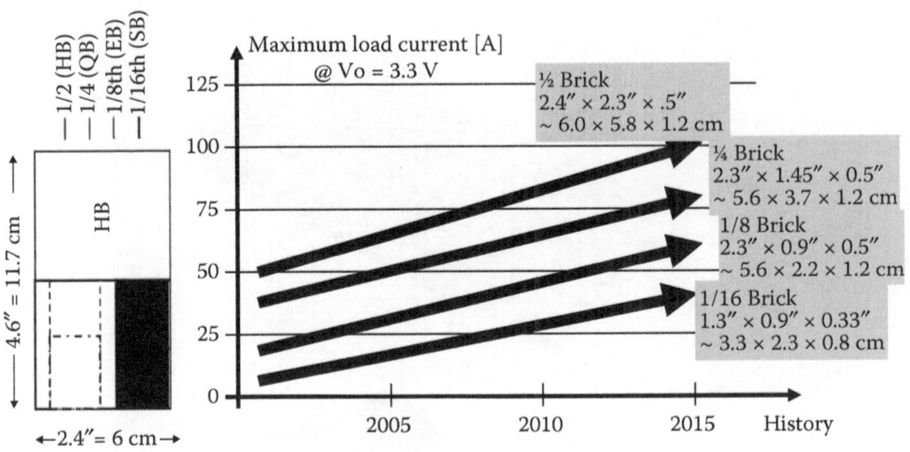

FIGURE 1.6
Evolution of technology over the years for the same dimensional bricks.

Such a trend leads to packaging more processing power within the same module dimension. In time, technology development allowed the processing of higher power for a given voltage transfer and for a similar size of the package for power conversion. An example of conversion from 48 V DC down to 3.3 V DC is shown graphically in Figure 1.6.

1.2.4 Role of and Requirements for Point-of-Load Direct Current/Direct Current Converters

Digital telecommunications circuits require well-regulated voltages at low DC levels, such as 5, 3.3, 1.8, 1.2 and 0.6 V and similar. It is most likely that within the same printed circuit board or even within the same integrated processor circuit, a large variety of DC voltages referenced to the same ground will be used. In order to achieve very well-regulated DC levels, local DC converters for distribution of DC voltages are used, with an additional role in accurate voltage stabilization or regulation.

The regulation requirements for the supply voltage of microprocessors are actually quite demanding. For instance, the MPC7448 RISC Microprocessor Hardware Specifications [7] define a fixed ±50-mV ripple allowance for all supply voltages in the range 1.0 V … 1.2 V DC, and ±5% for voltages in the range 1.5 V … 2.5 V DC. Even lower voltage ripple requirements are set in the *Design Guide & Applications Manual for Maxi, Mini, Micro Family DC-DC Converters and Accessory Modules* [8], where high- and low-frequency ripple is to be limited to 10 mV peak to peak for a possible range of output voltages of 0.6–1.35 V DC, set with a digital word on 5- or 7-bit voltage identification down (VID).

The number and variety of point-of-load power supplies within a processor board, also known as a motherboard, is continuously increasing. It is estimated that by the 2020s, the power-per-board requirement in data servers will have grown towards 5 kW. This is over a tenfold increase since the early 1980s, when computer electronic boards emerged.

Point-of-load power supplies can be implemented with switched-mode power supplies or with linear voltage regulators. Switched-mode power supplies follow the buck/boost converter principles discussed in Chapter 3. They feature a high efficiency, up to 96%. However, the buck or boost power converter topologies need to add a compensation law and protection circuitry to make up for a complete power supply. Control algorithms and their implementation are discussed in Chapters 4 and 5.

Linear regulators, also known as voltage stabilizers, traditionally have the drawback of a low efficiency due to the large voltage drop on the controlling series transistor. Efficiencies are in the range ~40%–70% without special filtering requirements. This issue is addressed with *low-drop-out* (LDO) voltage regulators, which work with a low collector-emitter voltage drop and mostly involve a *PNP*series transistor as a regulator. Details of this technology are addressed in Chapter 4.

Numerous control integrated circuits (ICs) are available for either converter topology.

The most important feature of digital point-of-load converters relates to *dynamic voltage and frequency scaling* (DVFS) techniques. DVFS is a technique for altering the voltage and/or frequency of a computing, telecom or embedded system based on performance and power requirements. Since the core processors and other digital peripherals in such systems are made in CMOS-integrated circuit technology, DVFS techniques can be understood from this perspective.

The voltage at which the circuit needs to be operated for stable operation can be lowered along the frequency increase, which leads to energy savings. More and more commercial processors support DVFS technology for saving power. The limitations of DVFS are:

- It harms the performance and may increase execution time or lead to missed deadlines.
- It requires a programmable clock generator and a power supply which allows energy overhead.
- Voltage transitions may require time on the order of tens of microseconds.
- Due to an increase in leakage energy at higher voltages and the trend of using multicore processors instead of increasing clock frequency, the returns from DVFS are diminishing.

Multiple DVFS techniques are reported in the literature, and these depend mostly on the software implementation. Chapter 8 reviews the most important trends of DVFS technology. For the power supply designer, it is important to ensure power supply capability for dynamic change of the DC voltage level. This is the premise of the PMBus feature.

For instance, a processor used at maximum capacity can reduce its supply voltage down to 1.0 V instead of the nominal 1.2 V DC used at lower usage in order to generate less heat loss. A good example within this class of converters is the AMD Hammer family of microprocessors. Such an operation mode allows the utilization of a DC converter called a voltage regulation module (VRM) or Voltage Regulator Down (VRD by Intel Spec) [7].

With the digital power technology of point-of-load converters moving into the commodity market, a standardization effort has followed to reduce costs and ensure interchangeability of similar products. This is why leading power supply companies (CUI Inc, Ericsson Power Modules and Murata), founded the Architects of Modern Power® (AMP) Group [9] in 2014.

Similar to the brick classification system used for isolated bus converters, a classification of power converters with the same sizes, matching feature sets and compatible configuration files has been defined:

- picoAMP™, intended for applications with load current between 6 and 18 Amp with land-grid-array (LGA) configurations
- microAMP™, intended for applications with load current between 20 and 25 Amp in horizontal and vertical printed-circuit-board configurations
- megaAMP™, intended for applications with load current between 40 and 50 Amp in horizontal and vertical configurations
- gigaAMP™, intended for applications with load current to 60 Amp in LGA configurations
- teraAMP™, intended for applications with load current between 90 and 120 Amp in horizontal and vertical configurations

This standardization may change since this field is very dynamic, with this classification of products set after 2014 only.

1.3 What Is Specific to Power Converters? Switched-Mode Operation

In order to save energy, power semiconductor devices operate in switched mode instead of the active region, as used to be the case with analogue electronics. Different design requirements are further imposed by the efficiency and size of packaging. Linear-mode power supplies (also called linear regulators or linear stabilizers) do not offer efficiency improvement opportunities

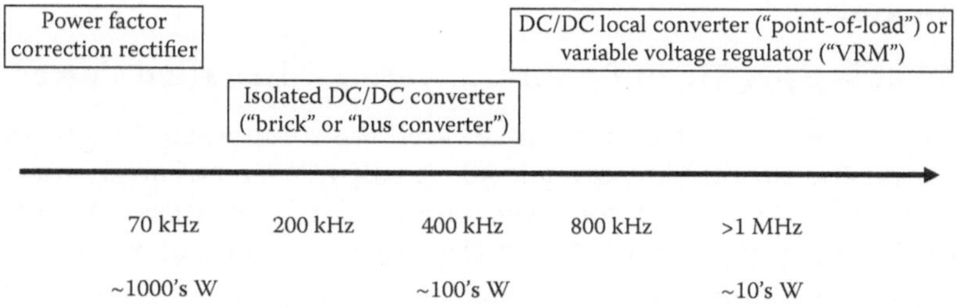

FIGURE 1.7
Switched-mode power converter solutions at various power and frequency levels.

since the efficiency consistently depends on the ratio between the input and output voltages, which is generally given for a certain application. Conversely, switched-mode power supplies allow performance improvements in both efficiency and packaging size. Hence, they are being adopted more and more to replace linear-mode power supplies.

A modern power converter system is characterized by switched-mode operation of power semiconductor devices. Using a higher switching frequency would help with reduction of filtering requirements and bring improvements in converter size with smaller magnetic devices (Figure 1.7). The switching frequency is, however, limited by the induced switching loss. The maximum achievable switching frequency is usually inversely proportional to the installed power. Power converters can process larger power at lower switching frequencies, while converters delivering lower levels of power can operate at higher switching frequencies. Figure 1.8 illustrates this graphically.

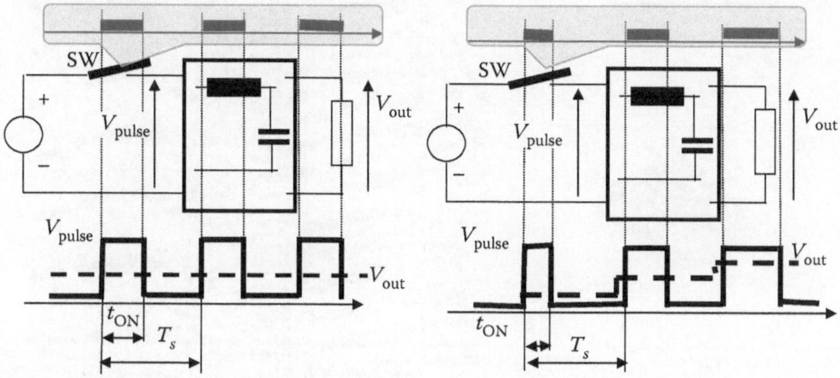

FIGURE 1.8
Operation at higher switching frequencies is advantageous to reduce the requirements for the power filter which depicts the low-frequency components.

1.4 What Is Specific to Power Converters? Installed Power

The selection of the electronic circuit as well the component design depends strongly on the processed power level. A full-wave rectifier circuit can be used within an integrated converter circuit able to process 2 W within a 2 mm × 2 mm printed circuit board (Figure 1.9). A similar rectifier circuit topology can be found within a six-floor-tall building set with multiple power semiconductor devices linked to each other to create a single-switch function and using the complex cooling structures of microcontroller-controlled pumps, all within a 500-MW *high-voltage DC* (HVDC) *link* system (Figure 1.9). Either case represents a different implementation of the same circuit schematic at a different power level. Although this is an extreme example, it is important to keep in mind that the design decision comes from the level of power processed through the power supply. Hence, converters are quoted with their *installed power.*

1.5 System Architecture Concepts

The power electronic equipment used within a power system has a somewhat simpler role: to take energy from an AC universal grid and deliver the energy to one or multiple DC loads [10–12].

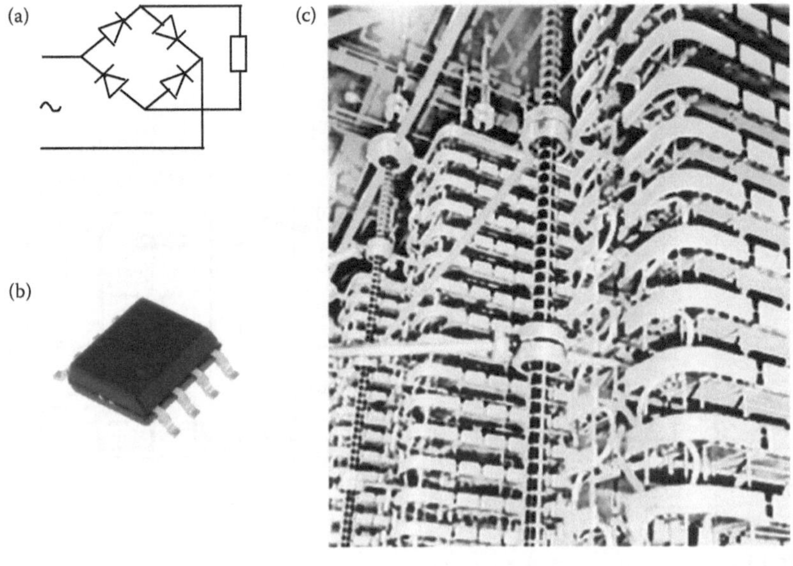

FIGURE 1.9
Different construction of power converters for various power levels.

In certain situations, the primary energy source is a battery or any other similar storage system able to allow uninterrupted operation when the AC grid disappears. Energy is stored as DC voltage at a voltage as high as possible, on the order of hundreds of volts. It is always necessary to charge these storage cells from the universal AC grid whenever the grid is present. Such a charging converter uses a rectifier followed by a DC/DC converter. Sometimes the same converter is used in reversed mode for a DC/AC conversion to supply other AC loads when the grid is missing. Chapter 14 will discuss various energy storage systems and present converter solutions for uninterruptible power delivery.

The power system architecture achieving this simple task of supplying loads from the universal AC system can be structured as a *centralized delivery system* or *decentralized delivery system* [9,10,13].

The centralized delivery system is usually built around a single isolated converter, and it produces multiple output voltages from a single DC voltage supply [14]. A good example is the flyback converter shown in Figure 1.10, with design and operation presented in detail in Chapter 6. The converter in this example produces five different DC output voltages from a single DC supply voltage through a centralized or unique controller.

The traditional solution for distributed power architecture consists of energy conversion from the 48 V DC telecom bus to various loads through various isolated DC/DC converters and through separate distributed control systems (Figure 1.11).

The novel solution for distributed power architecture (Figure 1.12) benefits from a local intermediary 12 V DC bus, followed by energy conversion to various DC loads through non-isolated direct DC converters. As mentioned before, such direct converters are also called point-of-load systems.

This novel or modern solution was imposed by the important decrease in the cost of point-of-load converters which came with the advent of integrated

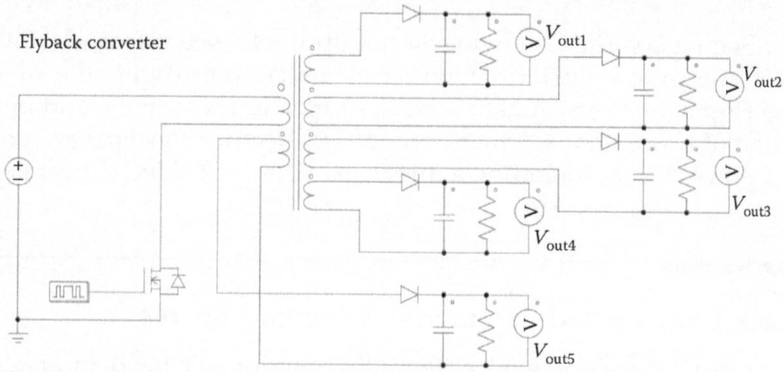

FIGURE 1.10
Power converter delivering five output voltages from a single source through a centralized controller.

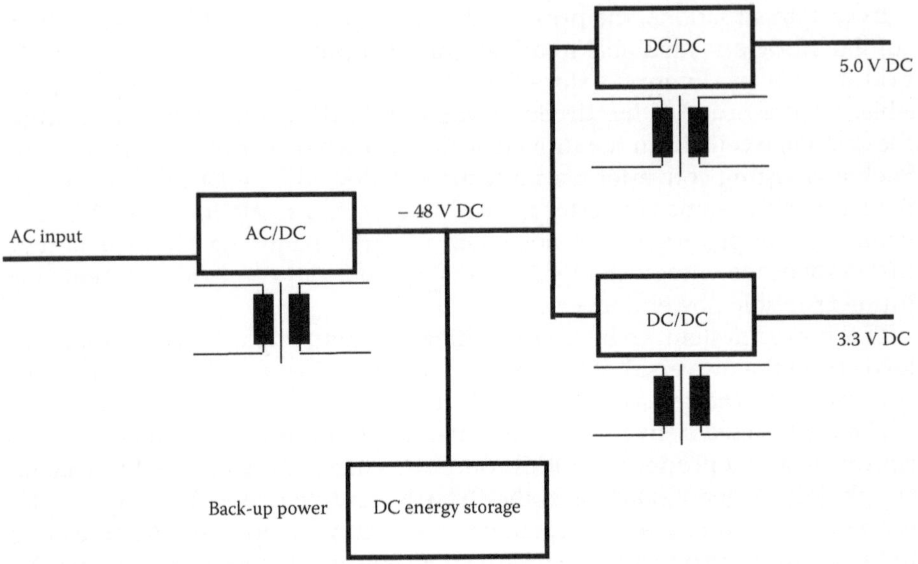

FIGURE 1.11
Sketch of a distributed power system.

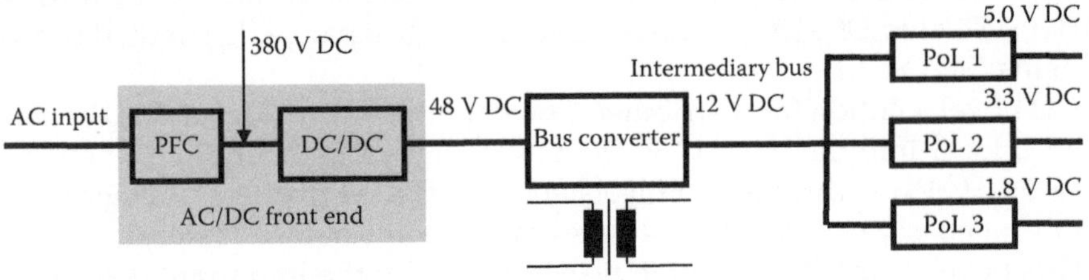

FIGURE 1.12
Novel solution for distributed power system.

circuits. Such a solution is obviously not optimal since it features multiple stages of energy conversion. However, it makes sense with the independent development of equipment modules for complex systems and physical separation of power loads. Other optimal architecture concepts are possible at lower power levels, including a direct 380 V DC→12 V DC conversion.

1.6 Rack Usage within Data and Telecom Centres

The impressive development of computing equipment for data or telecom centres has introduced a series of dedicated packaging techniques. This subsequently led to the deployment of standards and guides for good practices within the data or telecom centre.

The equipment dedicated to be packed inside a rack is called the rack mount, rack-mount instrument, rack-mounted system, rack mount chassis, subrack, rack-mountable system or simply shelf. In order to provide compatibility of equipment and rack systems, these are standardized.

Rack systems originated from railroad signalling relays from the 1900s. Interestingly enough, the size and standardization of racks stayed the same while the technology within changed continuously.

The most-used rack system is the 19-inch frame. A 19-inch rack is a standardized enclosure for mounting multiple equipment modules, and apparently was first introduced by AT&T in 1922. Each module or subsystem has a front panel that is 19 inches (480 mm) wide, including ears that protrude on each side for fastening to the rack frame with screws. The hole spacing for fasteners is 0.625 inches, or 15.9 mm. The height of each electronic module or subsystem is also standardized as multiples of 1.75 inches (44.45 mm); that is, one rack unit, often denoted with 'U'. Hence, there are frames with names like 1, 2, 3, 4, 5 or even 6 U. Over all systems, the most racks are sold in the 42-U size and are 36 inches deep. Other height sizes are rarely accepted. The depth of the rack is not the same as the depth of the electronic equipment; most equipment is shorter. Since the weight of the electronic equipment tends to lean towards the front, the common four-post racks feature a mirrored pair of rear mounting posts. The available rear space is used by vertical cable managers/ducts to run cables or, in rare cases, special cooling.

Computer servers with rack mounting also include some features like siding rails with possible locking and folding cables to allow operation during debugging of the equipment outside the rack, side handles, duplicate indicator lights on the front and rear of the rack and so on.

In some very rare cases, the telecom industry may use the so-called 23-inch rack. However, the 23-inch rack uses the same standardization for vertical dimensions: the 1–6-U size system.

While there is no dedicated standard for cooling of a rack, the most-used airflow has the intake in the front and the exhaust in the rear. For this reason, racks are grouped inside a data or telecom centre so that they are rear-to-rear placed, with an exhaust on the same side being easily drawn towards the outside of the building. Overall, cooling is a major reliability problem since fans usually fail before the electronics inside.

Rack organization within a data or telecom centre depends on the size of the centre.

1.7 Considerations for Power Systems within a Data or Telecom Centre

Considering the rack organization of equipment within a data or telecom centre, power distribution able to ensure uninterruptible power delivery is very important. To simplify any discussion on this topic, the Uptime Institute has introduced a tier classification for data and telecom centres. This classification is made based on performance, investment and redundancy. The I–IV Tier classification considers:

- Tier I = Basic Capacity data centre (99.671% availability). The centre has:
  - A single path of power supply, with a single uninterruptable power supply (UPS) to handle short outages
  - Dedicated cooling systems
  - Engine generators for extended outages
  - Its own dedicated space
  - Its own infrastructure for IT support outside of the office
  - No redundancy
- Tier II = Redundant Component data centre (99.749% availability). The centre has:
  - A single path for power and cooling distribution
  - Additional redundancy using multiple UPS modules, chillers or pumps and engine generators which protect it from interruptions in IT processes
- Tier III = Concurrently Maintainable data centre (99.98% availability). The centre has:
  - Redundant components and multiple distribution paths
  - The capability to operate without shut-down for maintenance, repair or replacement of equipment
  - Active power and cooling distribution paths
- Tier IV = Fault Tolerant data centre (99.995% availability). The centre has:
  - Multiple power and cooling distribution paths, with autonomous response to failure
  - Self-healing capability

Most power systems for data or telecom centres use the national AC grid voltage as the primary energy source. Since the power installed within a data or telecom centre is constantly increasing, new distribution systems based on DC voltage are currently being considered [15,16].

This success demonstrates that DC systems are less complex than AC systems, using fewer power conversions. Therefore, the DC system requires less real estate space and reduces costs of equipment, installation and maintenance.

Figure 1.13 shows the most-used national distribution systems for supplying the data centre. This incoming voltage is processed with UPS systems to secure the required availability percentage [17].

The design of the power system for data or telecom centres is not limited to distribution alone since the concerns of protection and fault management are extremely critical given the need for a high percentage of availability.

For instance, in 2012, ABB Corporation installed a 1-MW DC power distribution system for the 1,100-m2 Green Datacenter in Zurich West, Switzerland. DC power is achieved with thyristor rectification from a 16-kV local grid utility. The DC side switchgear is designed for an operating voltage of 400 V DC and can convey a maximum constant current of 3000 A. However, a more generic DC bus voltage for telecom and data centres is 380 V DC.

The most important concern about a fault of the power system is the potential arc fault, whose severity is increased by higher data centre capacities,

Location	Nominal medium voltage	Nominal low voltage
North America	4.16 kV 13.8 kV 34.5 kV	600 3-wire+ground (primarily Canada) 480 Y/277 4-wire+ground 480 3-wire+ground 208 Y/120 4-wire+ground 415/240 V 4-wire+ground (new data centres)
South America	6 kV 11 kV 13.8 kV 22 kV 23 kV	220 Y/127 60 Hz 4-wire+ground (ranges from 110 to 127 V) 380 Y/220 60 Hz 4-wire+ground (ranges from 220 to 240 V) 400 Y/230 50 Hz 4-wire+ground (ranges from 220 to 230 V)
Europe	10 kV 20 kV 35 kV	400 Y/230 50 Hz 4-wire+ground 480 Y/277 60 Hz 4-wire+ground
China	10 kV 35 kV	380 Y/220 4-wire+ground
Japan	6.6 kV 22 kV	200 3-wire+ground 1-phase 200/100 3-wire+ground 1-phase 100 2-wire+ground

FIGURE 1.13
Various AC distribution systems used to supply the entire data centre.

higher rack densities and higher-efficiency designs [18]. For instance, the installation of a 1-MW data centre introduces over 50 kA of fault current on the low-voltage side of the medium-voltage transformer. Therefore, the trends towards data centres of larger power capacities and increasing power densities beyond 4–5 kW/rack within a modern data centre both tend to increase the available fault current in the data centre room. Additionally, the trend towards higher-efficiency power distribution systems stimulates reduced resistance and inductance of the wires and transformer and also tends to create larger fault currents.

1.8 Lifetime and Reliability

Along with requirements for monitoring operation and recovery from failure modes, a series of recent efforts target the improvement of lifetime and reliability. For instance, monitoring of operation allows accurate calculation of the work time as well as storing the conditions of operation. This allows for a mathematical calculation of the lifetime and appropriate programming of the maintenance service.

Some examples:

- Laptop batteries have a function which can report on the remaining lifetime.
- Another improvement of the lifetime consists of using redundant power supplies.
- Many power supplies are composed of multiple converter branches which can replace each other under failure. The control algorithms can operate and deliver electrical energy in certain failure modes through the usage of a converter section instead of the entire system.
- The usage of digital control and PMBus interfaces allows improvements in each phase of the entire lifetime: in the design phase and during operation in dynamic mode for energy loss reduction, as well as in failure mode and service and testing for good operation.

1.9 Special Design Requirements and Software-Intensive Control Systems

The virtual explosion of telecom digital equipment, with more and more demand for standardized power supplies and optimized power system architectures, also produced new standards and design requirements [19].

Such requirements come in addition to the traditional efficiency and packaging size requirements for a power converter. Due to the convergence of solutions, the difference between IT and telecom loads is lost today [20], and both applications subscribe to the same power management principles.

Examples of such novel requirements include uninterrupted operation when the AC grid voltage fails, entering sleep mode during short periods of lack of operation, reduction of undesired DC components injected into the AC grid or interconnection of various power supplies during operation under live voltage (hot-swap).

Many such requirements can be implemented with a novel digital controller carried out within microcontrollers and a communication interface between the DC power supplies. Implementation of multiple tasks implies more software development. In the late 1980s and early 1990s, microcontrollers were successfully introduced to the control of power converters, and engineers were happy just to make things work. The simultaneous addition of novel tasks to the controller and the explosion of microcontroller technologies have all favoured the extensive development of software and the actual length of programs. More advanced programming skills are required today. This justifies the introduction of software development theory into power converter controllers. The most advanced trend now addresses the complex problems imposed by software-intensive control systems. This is an amazing new research direction in power electronics which will see more and more development over the forthcoming years. Some ideas pertaining to these new beginnings are presented in Chapters 4 and 5, where the controller is designed with the help of state space theory.

All this new software taking over research problems in power converters has also extended the development of novel peripherals for communication. Actually, this paradigm of seeing the connectedness of control and communication has recently been emphasized with numerous U.S. governmental and corporate programs related to C2 (control-communication) and C4 (command, control, communications and computer systems) concept systems. These started as military tasks and rolled into other applications like telecommunications systems [21].

In 2004, power supply manufacturers Artesyn Technologies and Astec Power Supplies, as well as a group of semiconductor suppliers, Texas Instruments, Volterra Semiconductors, Microchip Technology, Summit Microelectronics and Zilker Labs, formed a coalition to develop an open standard for communications with a set of commands dedicated to power supplies. This was the birth of what is today known as PMBus.

The most important peripheral extension of software-intensive control systems relates to recent advances of the low-cost standard that has been established around PMBus [22–24] in a hardware structure defined with circuits for serial communication of type I^2C. It allows fault management, nonlinear control, lifetime estimation, maintenance requirements estimation, optimal operation and so on. The PMBus is discussed in Chapter 8, with examples derived from a demonstration system with 14 independent power-supply channels controlled with LTPower software [25].

Summary

This chapter briefly discussed the role and requirements of power supplies used within telecom equipment. It provided a definition of the main power converter systems and a description of the book structure and scope. System architecture concepts and converse design requirements were also mentioned.

References

1. Gumhalter, H., 1986, *Power Supply in Telecommunications*, 2nd edition, Springer, Berlin, Heidelberg.
2. Soldano, M., 2007, "Power Monitoring in Server Power Supplies", in *IBM 2007 Power and Cooling Symposium*, Raleigh, NC, USA, 2007.
3. Anon, 2012, Digital Power Compendium, *Ericsson Corporation Document 1553-CXC*, 173 1142/2 Rev B, pp. 1–32.
4. Anon, 2013, Series NQB-N – Fully Regulated Advanced Bus Converters, www.cui.com, document date 11/07/2014, accessed on September 21, 2017, pp. 1–30.
5. Knauber, P., 2010, Package, Power, Board Space Are the Design Considerations, Murata Corporation, *EPN Power Management*. https://www.digikey.at/Web%20Export/Su://www.digikey.at/Web%20Export/Supplier%20Content/MurataPower_811/PDF/Murata_Selecting-a-Brick.pdf, accessed on September 21, 2017.
6. Anon, 2014, Design Guide and Applications Manual – For Maxi, Mini, MicroFamilies DC/DC converters, *Vicor Corporation White Paper.*
7. Anon, 2007, MPC7448 RISC Microprocessor Hardware Specifications, *Freescale Semiconductor Document Number MPC7448EC*, Rev. March 4, 2007.
8. Anon, 2006, Voltage Regulator-Down (VRD) 11.0 – Processor Power Delivery Design Guidelines – For Desktop LGA775 Socket, *Intel Corporation.*
9. Anon, 2014, Digital Power Comes of Age, *AMP Group documentation*, http://www.ampgroup.com/wp-content/uploads/2014/12/Digital-Power-Comes-of-Age.pdf. Accessed on September 21, 2017.
10. Jovanovic, M.M., 2011, *Power Conversion Technologies for Computer, Networking, and Telecom Power Systems – Past, Present, and Future, IEEE First International Power Conversion & Drive Conference*, St. Petersburg, Russia.
11. Lubrito, C., 2010, "Telecommunication Power System: Energy Saving, Renewable Sources and Environmental Monitoring", *Trends in Telecommunications Technologies*, C.J. Bouras (Ed.), *InTechOpen* (Internet book), https://www.intechopen.com/books/trends-in-telecommunications-technologies, accessed on September 21, 2017.

12. Raviprasad, V., Shanker, T., Ravindra, K.S., 2012, "Power Architectures for Telecommunications – A Review", in *Proceedings of International Conference on Advances in Electronics, Electrical and Computer Engineering*. July 7–9, 2012, Dehradun, Uttarakhand, India, pp. 70–74.
13. Salato, M., Zolj, A., Becker, D.J., Sonnenberg, B.J., 2012, "Power System Architectures for 380V DC Distribution in Telecom Datacenters", in *IEEE Conference*, December 6, 2012, Scottsdale, AZ, USA, pp. 1–7.
14. Anon, 2002, Power Supplies for Telecom Systems, *Maxim Integrated Application Note* no. 280, accessed on July 17, 2002.
15. Murrill, M., Sonnenberg, B.J., 2010, Evaluating the Opportunity for DC Power in the Data Center, Emerson *Network Power White Paper 124W-DCDATA/1010*.
16. Glinkowski, M., 2013, "Data Center Defined", *ABB Review*, 4, 8–10.
17. Hu, P., 2016, Electrical Distribution Equipment in Data Center Environments, *APC White Paper* #61, revision A, pp. 1–15, accessed on September 2, 2016.
18. Avelar, V., 2014, Arc Flash Considerations for Data Center IT Space, *APC White Paper #194*, version D, pp. 1–12. http://www.apc.com/salestools/VAVR-6KGRYW/VAVR-6KGRYW_R0_EN.pdf, accessed on September 21, 2017.
19. Mittal, S., 2014, "A Survey of Techniques for Improving Energy Efficiency in Embedded Computing Systems", *Int'l. Journal of Computer Aided Engineering and Technology*, 6(4), 440–459.
20. Grusz, T., 2017, Telecom and IT Power Trends and Issues, *Technical Documentation Emerson-Liebert Corporation*.
21. Anon, http://www.c4i.org/whatisc4i.html. Accessed on September 21, 2017.
22. White, R.V. 2016, PMBus™: Review and New Capabilities, Embedded Power Labs, *Seminar APEC*. March 20, 2016, Long Beach, California, pp. 1–154.
23. Anon, PMBus Specifications – Version 1.3.1, http://PMBus.org. Accessed on September 21, 2017.
24. Narveson, B.C., Harris, A., 2007, "Power-Management Solutions for Telecom Systems Improve Performance, Cost, and Size", *Texas Instruments Analogue Applications Journal*, 3Q. http://www.ti.com/lit/an/slyt276/slyt276.pdf, accessed on September 21, 2017.
25. Anon, 2013, LTpowerCAD TM II v2.0 – Design Tool User's Guide, *Linear Technology documentation*, version 2.0.

2

Power Semiconductor Devices

2.1 Context

Semiconductor devices [1,2] used within electronic power converters operate in switched mode. Switching a power semiconductor device means changing the conduction state from an open circuit to a closed circuit (*direct transition*) or from a closed circuit to an open circuit (*reverse transition*). The words *transition* and *commutation* are interchangeable in this context.

The power semiconductors used within power converters have an ideal characteristic in that they are able to define both a closed circuit, as a short circuit, and an open circuit [3,4]. Since high-energy signals are processed within power converters, energy loss is important to monitor and reduce [5,6]. The goal here is the reduction of energy loss in both direct and reverse commutation as well as during the conduction state.

A classification of power semiconductor devices in classes of devices yields

- Diodes
- Devices with control of turn-on processes only (*thyristors*)
- Devices derived from transistors able to control both the turn-on and turn-off processes (*insulated gate bipolar transistors* [IGBTs] or *metal–oxide–semiconductor field-effect transistors* [MOSFETs])

2.2 Diodes

A diode is a simple semiconductor device, and the operation of a diode is well known. A semiconductor diode is a piece of semiconductor material with a *PN* junction connected to two electrical terminals. The discovery of rectifying abilities of certain crystal materials was made by German physicist and Nobel prize awardee Ferdinand Braun (6 June 1850–20 April 1918) in 1874, and the first semiconductor selenium rectifiers were introduced to power systems in the 1930s.

When a positive voltage is applied across it, the diode enters the conduction state and allows the current circulation. If the direct current is decreasing under a certain value, the diode turns off through *natural reverse commutation*. If a negative voltage is applied across a diode while conducting current, the diode turns off through *forced reverse commutation*.

The operation of a diode through a commutation circuit is analyzed next. When applying a positive voltage, the diode conducts current with a very reduced voltage drop; thus, it can be considered a good approximation of an ideal switch.

A less obvious detail within small-signal diodes is reverse commutation, when a direct supply with a negative voltage across the device produces forced reverse commutation. When applying a negative voltage, the current through the diode suddenly decreases to zero and allows a reverse circulation of the current, produced through the discharge of the charge stored within the capacitance of the PN junction. Important parameters for characterization of this forced reverse commutation event are (Figure 2.1):

- Reverse recovery current, usually denoted by I_{RM}
- Reverse recovery time, usually denoted by t_{rr}

Both forward voltage drop and reverse recovery charge influence power loss in diodes. Design decisions produce conflicting effects on the two considerations: increased injection of the carriers can be used to reduce the forward voltage drop, but it produces more charge that needs to be removed at turn-off before the diode will be able to block voltage. Hence, the reverse recovery time is longer and negative effects are more important.

The negative effects of reverse recovery current consist of an increase in overall energy loss, higher EMI radiation and the appearance of a higher overvoltage in the circuit, with possible effects of damage to other power semiconductor devices from the converter. These effects will be also demonstrated later on when studying Figures 2.11 and 2.12 for commutation of a MOSFET transistor.

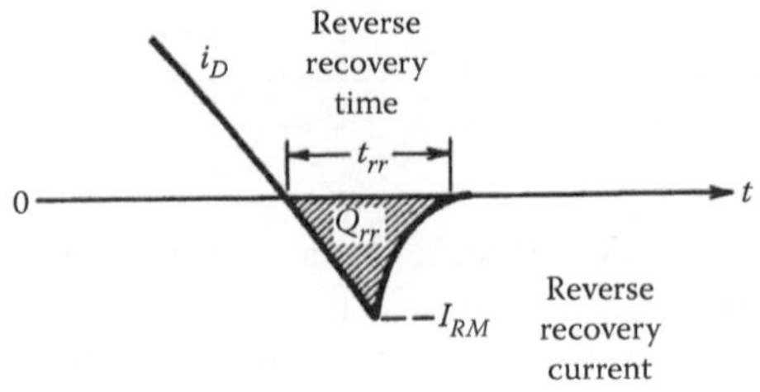

FIGURE 2.1
Turn-off characteristics of a diode.

Since this recombination phenomenon is the most important aspect linked to the use of diodes within the power converter, diodes can be classified as:

- Rectifier diodes – used at lower frequencies and operated with higher recovery current
- Schottky diodes – determine a decrease in conduction voltage and do not present the recovery phenomenon or display reduced energy loss
- Fast or ultrafast recovery diodes

Schottky diodes are based on a semiconductor–metal junction and were invented by German physicist Walter H. Schottky (23 July 1886–4 March 1976). The core invention relates to the Schottky barrier at the interface of a semiconductor and a metal which acts as a previously discovered emission of electrons from a metal into vacuum. Typical metals are molybdenum, platinum, chromium, tungsten or palladium silicide and platinum silicide, whereas the semiconductor would be the conventional n-type silicon.

While a conventional silicon diode has a typical forward voltage of 600–700 mV DC, the Schottky diode benefits from a lower forward voltage of 150–450 mV DC. Furthermore, Schottky diodes do not have a recovery time at reverse transition, since there is no charge carrier depletion region at the junction of the semiconductor and metal. The transition time is around 100 picoseconds for small-signal diodes and up to tens of nanoseconds for higher-voltage power diodes.

Despite all these apparent efficiency improvements, Schottky diodes are not used very often in practice:

- Schottky diodes have relatively low reverse voltage ratings, in the range of 50 V DC. Since a high electric field gradient occurs near the edges of the Schottky contact, the breakdown voltage threshold is limited. A solution of using guard rings to overlap the metallization and spread out the field gradient increases the area and compromises the voltage drop during conduction. At higher voltage ratings, the voltage drop is comparable to conventional diodes. For this reason, SiC Schottky diodes are more advantageous in high-voltage applications [7].
- In low voltage ranges, using a MOSFET as a synchronous rectifier is more advantageous than using a Schottky diode. More detail is provided in Sections 3.5 and 7.5.

Fast or ultrafast recovery diodes use some variation of a *PIN* structure in order to improve transition times. They are important devices for low-voltage switched-mode power supplies. While the conventional diode role is more and more often implemented with MOSFET synchronous rectifiers, ultrafast diodes are always used within RCD snubber construction.

2.3 Thyristors

A very important device in the history of power converters, the thyristor is not much used in telecom equipment. A thyristor, also called a *silicon-controlled rectifier* (SCR), is a semiconductor device with a PNPN structure, able to turn on on the basis of a control signal applied on the gate (Figure 2.2). Unlike a diode, a thyristor also has the ability to block direct voltage.

The secondary effects of this operation mode can be seen at higher voltages applied directly or in reverse, when the thyristor 'breaks' and allows uncontrolled current circulation. Hence, there is a mandatory requirement for protection at both forward and reverse overvoltages.

The oldest use of thyristors is within controlled rectifiers. A rectifier is a circuit able to convert energy from AC form into DC form. The DC component of the voltage on the load is calculated as an average of the load voltage (Figure 2.3). The turn-on moment can be delayed through a control so that the average value of the output voltage yielded is smaller or larger.

2.4 The Ideal Switch

The requirements for an ideal switch consist of:

- The capacity to block any voltage applied directly or in reverse.
- The ability to conduct relatively large currents whenever the circuit is turned on.

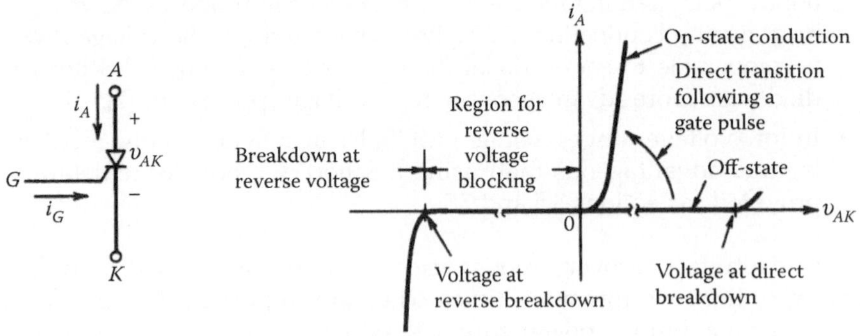

FIGURE 2.2
Characteristics of a thyristor device.

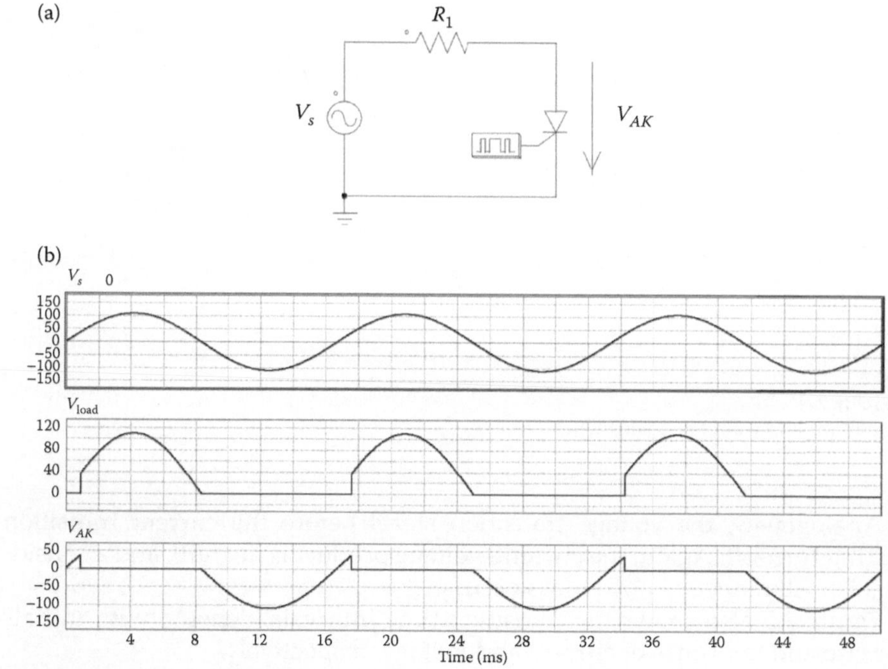

FIGURE 2.3
Operation of a thyristor within a controlled rectifier.

- The transition from conduction into the blocking state or the reverse transition from blocking into the conduction state should be done instantaneously, without energy loss.
- The transition from conduction into the blocking state or the reverse transition from the blocking state into the conduction state should be done through a command necessitating a low level of energy (zero gate control loss).

Figure 2.4 shows a simple circuit used for analysis of commutation processes.

All secondary effects due to recovery currents are first neglected, and changes of the power semiconductor conduction states are observed. Since power semiconductor circuits process high-level currents, it is important to understand the energy loss specific to any technology. Therefore, IGBT and MOSFET devices are analyzed as physical implementations of the ideal switch. Figure 2.5 illustrates a generic transition of current and voltage around a semiconductor switch, either IGBT or MOSFET. It can be seen that the current transition starts at turn-on of the semiconductor device. This produces a small superposition of current and voltage, yielding their nonzero product. This means a considerable instantaneous power loss.

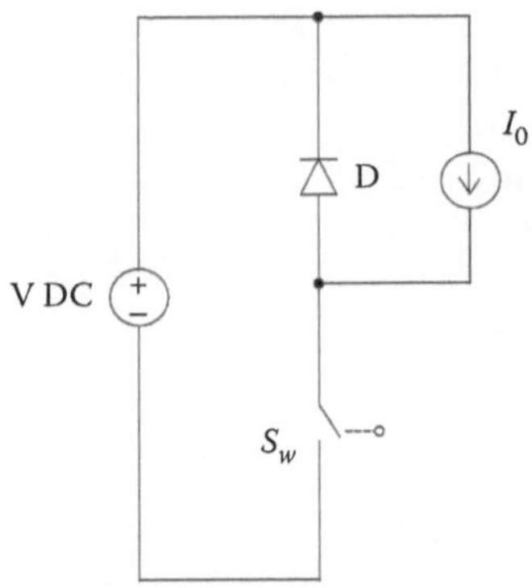

FIGURE 2.4
Circuit for analysis of an ideal switch.

Analogously, the voltage transition starts before the current transition during turn-off. Again, the current–voltage product is not null, and a considerable instantaneous power loss occurs.

Figure 2.5 also shows the transition time intervals, denoted with initials for rise and fall times of current and voltage, respectively.

2.5 Switched-Mode Power Metal–Oxide–Semiconductor Field-Effect Transistor

The switched-mode power MOSFET represents the principal option for power converters with low installed power and rated voltage under 200 V DC [8]. Some applications may reach a rating of 1000 V DC, while the range of currents is typically under 100 A DC. *N-channel MOSFET* transistors are the most used, and the main advantage is reduced energy loss.

A voltage needs to be applied to the gate circuit to maintain the device permanently in conduction. The main shortcoming consists of the relatively large voltage drop in the conduction state, which is even higher when transistors are dedicated to operation at higher voltages.

MOSFET symbol and static characteristics are shown in Figure 2.6.

Due to a large and well-established market, the MOSFET product lineup includes a wide variety of devices. For instance, the synchronous buck converter presented in Chapter 3 benefits from different MOSFET parts for the control switch and synchronous switch. Therein, the control switch is optimized to reduce power loss during transients, while the synchronous MOSFET is optimized to operate as a diode with reduced conduction loss. Other similar examples are possible.

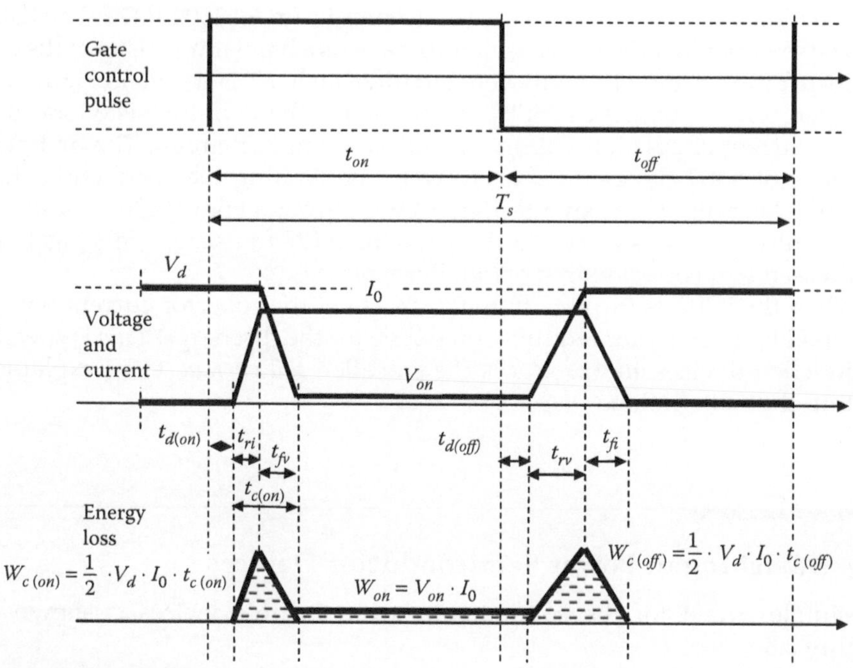

FIGURE 2.5
Switching characteristics of an ideal switch.

2.6 Insulated Gate Bipolar Transistor

The IGBT device combines the advantages of MOSFETs and bipolar transistors. The similarities with MOSFETs include high impedance within the gate circuit, useful for working with a voltage control at reduced current. The similarities to bipolar transistors consist of a reduced voltage drop in

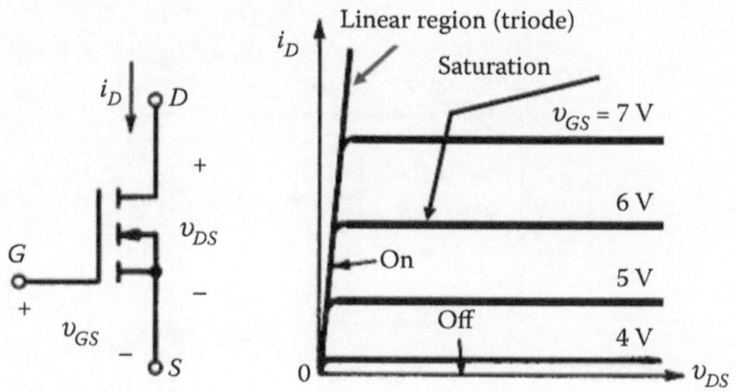

FIGURE 2.6
Symbol and example for the static characteristics of a MOSFET – channel N transistor.

conduction, even for devices dedicated to circuits under 1000 V DC. The IGBT can be seen as a MOSFET in gate circuitry and as a bipolar transistor at the collector (drain) circuit. The equivalent circuit of an IGBT is shown in Figure 2.7.

When turning on the MOSFET in the model, the PNP transistor starts to allow current circulation within the collector–emitter circuit. The first generations of IGBT devices had a problem maintaining the conduction state through latch-up due to an internal reaction through body-region resistance (also called *drift resistance*) and the parasitic *NPN* transistor. This has been addressed and corrected in more modern devices.

When the IGBT is turned off, the decrease of the collector current within the final PNP transistor requires some time for the discharge. This is specific to junction devices and produces the so-called *tail current*, to be explained within the next section.

2.7 Switching of Power Semiconductor Devices

A simple circuit for the commutation of switching devices is shown in Figure 2.8.

The previous discussion of the ideal switch is next expanded with a study of switching within a practical device's circuitry. The effects of gate circuits are similar for MOSFET and IGBT devices and they are studied along with the diode recovery phenomenon associated with switching the power transistor. The dynamic regime is usually characterized by the equivalent model for the transistor (Figure 2.9). This is mainly composed of the gate-collector and gate-emitter capacitances. These capacitances do not have constant values during operation, and their variations with the circuit parameters are illustrated in Figure 2.10. Based on this dynamic model, details of the direct and reversed transitions for the power transistors MOSFET and IGBT are next discussed.

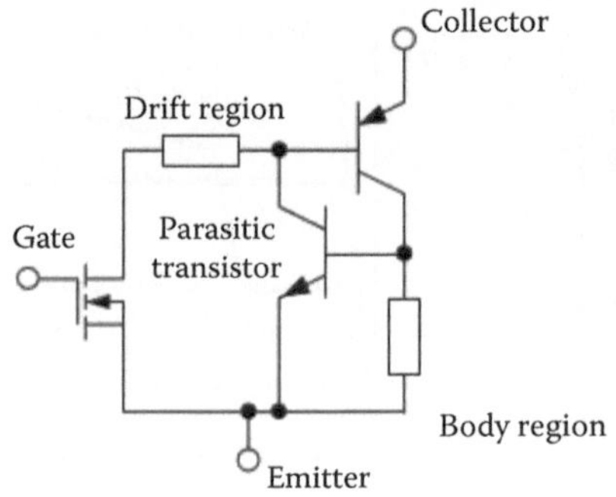

FIGURE 2.7
Equivalent circuit for an IGBT device.

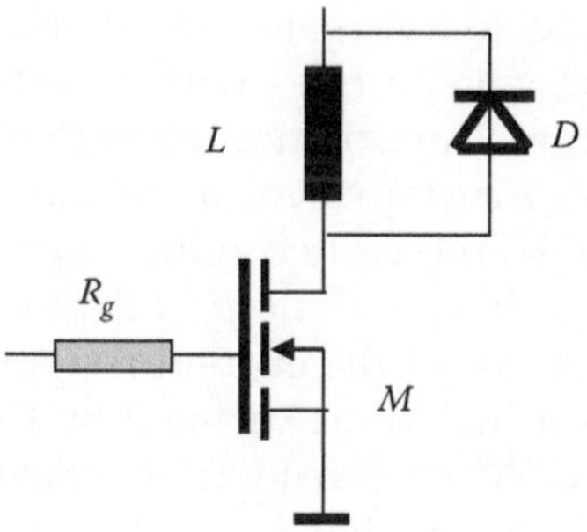

FIGURE 2.8
MOSFET switching circuit.

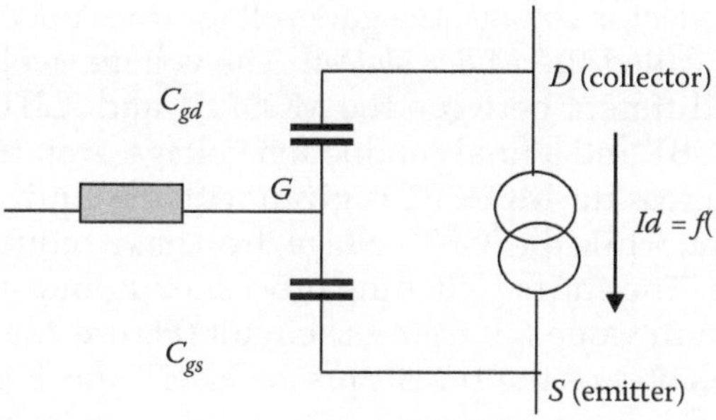

FIGURE 2.9
Dynamic model of a MOSFET device.

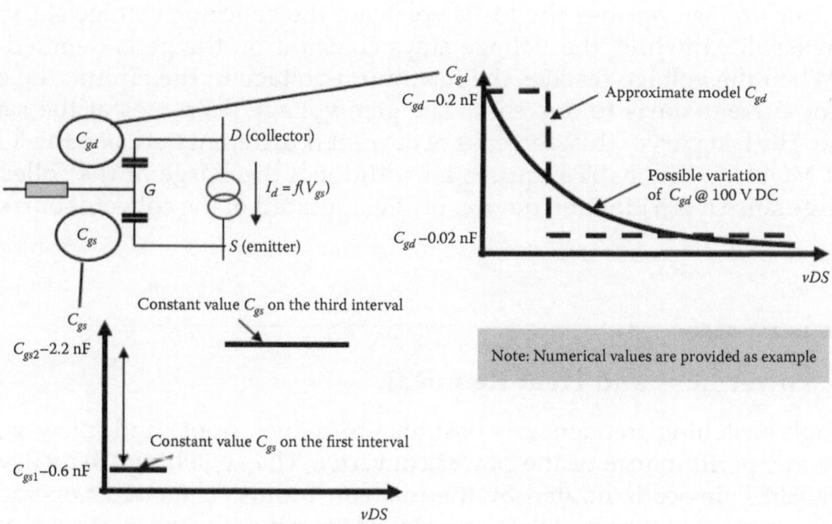

FIGURE 2.10
Actual variation of the model capacitances.

When applying voltage to the gate circuit, the gate current can rise suddenly, while the gate voltage rises with a certain slower slope due to charging the equivalent gate capacitance through the series gate resistance. When this voltage reaches a certain level of voltage, called the *threshold*, V_{GSth}, the device (MOSFET or IGBT) starts to conduct current. The rise of the collector current is made with a characteristic *di/dt* defined by the external circuit. For an inductive load, the rise of the current is approximately linear. When this current reaches a nominal value dictated by the external circuitry, the voltage starts to decrease. When considering a diode present in the circuit, the load current yields are directed through the diode towards the load before the transistor is asked to turn on. After the turn-on command, the diode would turn off, possibly with a recovery current, depending on the diode choice. The diode's recovery current can be seen as an increase over the collector current, shown with a dotted line. During the interval featuring a decrease of the collector voltage, the gate voltage does not change, and it is clamped to a level called the *Miller plateau*. The voltage evolution after this moment is a little different between the MOSFET and IGBT, with a slower transition for the IGBT and a final conduction voltage drop for both devices. The voltage drop across the MOSFET is given with the equivalent resistance and the load current, while the IGBT voltage drop has a reduced dependence on the load current. The entire switching process ends long after the current reaches the maximum value I_0 within the circuit (Figure 2.11).

The reverse transition of the transistors MOSFET and IGBT is shown in Figure 2.12.

At a sudden decrease in control voltage, the voltage in the device's gate decreases slowly and the gate current discharges the gate charge. When the gate voltage reaches the Miller plateau, the collector voltage starts to increase. Meanwhile, the voltage stays constant on the gate, denoted by G. When the voltage reaches the maximum voltage in the circuit, the collector current starts to decrease. The gate voltage decreases at the same time. The last part of this decrease of current is different between the IGBT and MOSFET. The IGBT requires an additional discharge of the collector charge since it is a junction device, producing a tail of the collector current.

2.8 Power Loss and Heat Removal

A high switching frequency is desirable to reduce ripple and allow good dynamic performance of the power converter. The switching frequency of a MOSFET device is limited by thermal constraints. The circuit operation determines the power loss, the power loss determines the thermal behaviour of the system and the thermal profile determines the reliability, failure rate and lifetime of the converter.

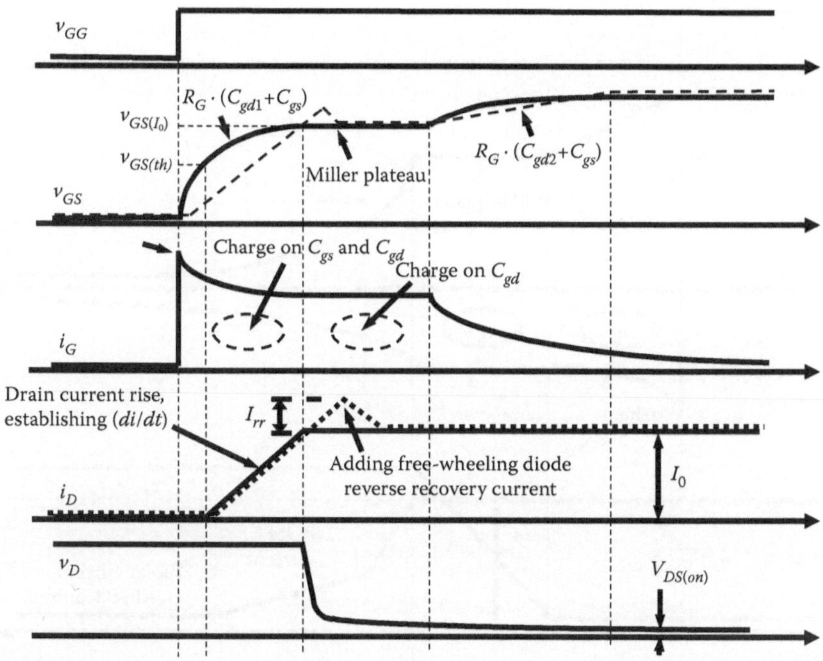

FIGURE 2.11
Direct commutation of a MOSFET/IGBT device.

The total power loss within a power MOSFET depends on two components: conduction and switching loss.

The conduction loss for a single device can be calculated with:

$$P_{con} = \frac{1}{T} \cdot \int_{0}^{T} u_{on}(t) \cdot i(t) dt \tag{2.1}$$

where T represents the pulse period (for DC converters) or fundamental period (for AC converters), $u_{on}(t)$ is the actual on-state voltage drop, and $i(t)$ represents the instantaneous value of the current. The conduction loss is calculated for intervals of conduction, also known as the on state.

For both the MOSFET and IGBT, the on-state voltage drop is further approximated with a fixed voltage value and a series resistive component. Obviously, the MOSFET uses R_{dson} for the series resistive loss and a minimal voltage drop at zero current. Hence, the conduction loss is easier to estimate. For a DC/DC converter, it yields:

$$P_{con} = D \cdot R_{dson} \cdot I_D^2 \tag{2.2}$$

where D is the nominal duty cycle.

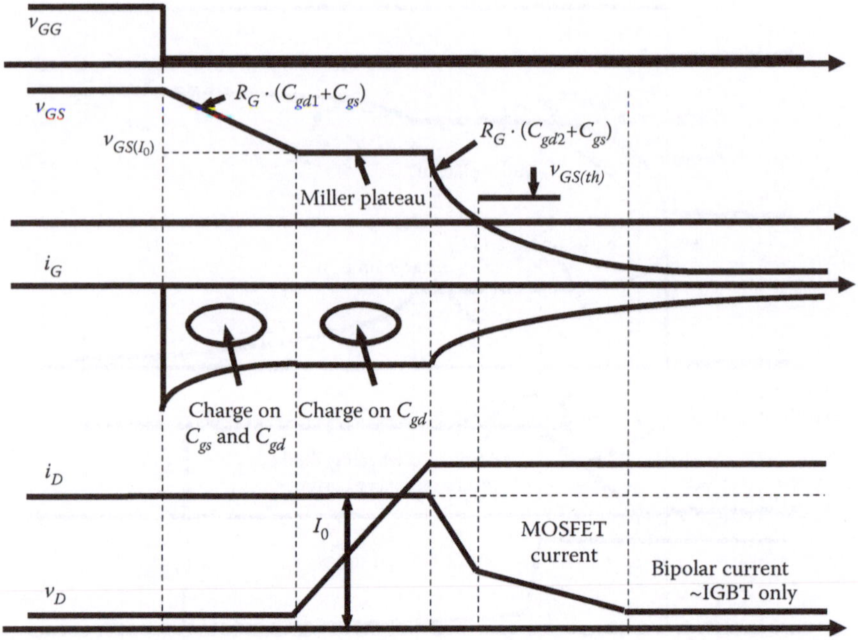

FIGURE 2.12
Reverse commutation of a MOSFET/IGBT device.

For most applications, the switching loss is the most important component. Chapter 7 introduces a series of converter topologies used to reduce switching loss. The switching loss calculation follows the calculation for conduction loss, equating power with an instantaneous voltage–current product. The loss within DC/DC converters can more easily be estimated with oscilloscopes as per-pulse power loss, which is an average of the instantaneous voltage–current product. AC/DC or DC/AC converters deal with variable pulse width and make the calculation more difficult.

Figure 2.11 can be used to demonstrate turn-on loss energy as:

$$E_{on} = \int v(t)\cdot i(t)dt = \frac{1}{2}\cdot V_D \cdot (I_D + I_{RRM})\cdot t_{ri} + \frac{1}{2}\cdot V_D \cdot \left(I_D + \frac{2}{3}\cdot I_{RRM}\right)\cdot t_{fv} \quad (2.3)$$

where V_D is the maximum supply voltage.

Figure 2.12 can be used to demonstrate turn-on loss as:

$$E_{off} = \int v(t)\cdot i(t)dt = \frac{1}{2}\cdot V_D \cdot I_D \cdot t_{rv} + \frac{1}{2}\cdot (V_D + V_{pk})\cdot I_D \cdot t_{fi} \quad (2.4)$$

The total switching loss yields:

$$P_{sw} = f \cdot (E_{on} + E_{off}) \tag{2.5}$$

In some cases, E_{on} and E_{off} are calculated in datasheets and reported as graphs with respect to the current.

The entire power loss needs to be withdrawn as heat loss through the device's package. Hence, package selection determines the maximum allowable losses and the switching frequency. For instance, a high-voltage 30-kW converter has power losses in the range of 130 W [1] that have to be removed from an area of 1 cm^2; that is, a heat density of 130 W/cm^2. This is more than the heat density of a stovetop or light bulb. This is the actual problem: not the absolute value of power loss (merely 130 W), but the high power density (130 W/cm^2).

This requirement encourages work on packaging of power semiconductor devices. Packages for power semiconductor devices can be classified as discrete or modular. Discrete packages are common in low power ranges, and they mostly use TO packages for through-hole and surface-mount applications. TO is a packaging family of standardized packages where the silicon chip is soldered directly to a solid copper base. The package is created without electrical insulation. Furthermore, the three terminals of the device are connected through a transfer mould process which links a lead directly to the copper base, and the other two leads are soldered to the silicon chip with aluminium wires. This solution is not very reliable since the difference in the thermal expansion coefficient of the silicon chip, copper base and aluminium wires may produce material breakage. Some improvement occurs when the copper base is replaced with a ceramic substrate.

Due to the cost constraints on building power supplies, a solution with all components soldered as surface-mount devices (SMDs) is more and more desirable. This is easier on the automatic soldering machines and supports an increased use of multilayer printed-circuit boards (PCBs). A PCB is smaller, allowing better packaging, better power density and easier interconnection of subsystems. Sensitive to this trend, most power MOSFET transistors today are packaged in a surface-mount outfit, which also benefits from a reduction of parasitic inductance by more than 33% when compared to the through-hole solution. The most used SMD packages are DPAK (TO-252), D^{2}Pak or DDPAK (TO-263) and Super-D^{2}Pak [9].

As an example, a DPAK package used for a low-voltage MOSFET transistor can allow removal of around 6 W of loss power if a printed-circuit board is used as a heat sink, with a more than 80 cm^2 copper plane and a maximum junction temperature of 120°C from a 25°C ambient temperature; that is, only a 15-K/Watt change in temperature [9].

Whatever package technology is used to extract heat from the power semiconductor device, the thermodynamics law is a differential equation of the first order. Since the first-order differential equation looks similar to the Kirchhoff laws for electrical circuits, an analogy is often used. The thermal resistance of a circuit is defined as:

$$R_{th(a-b)} = \frac{T_a - T_b}{P} = \frac{\Delta T}{P} \tag{2.6}$$

where T_a and T_b represent the temperatures of two materials, *a* and *b*, when a power loss *P* is applied to the two materials. This way, thermal resistances between the junction and case, between the case and heat sink or between the heat sink and ambient temperature can be defined. The larger the thermal resistance, the harder it is for heat to propagate to the next material. The thermal resistance of the silicon itself is only 2%–5% of the total thermal resistance. Hence, the packaging is important.

Obviously, this is a very simplified view of the temperature of a component since it neglects the temperature gradient within the materials, each component having its own internal temperature distribution.

Similarly to an electrical circuit, the temperature variation is characterized by an equivalent capacitance which can store thermal energy. This can show the dynamics and the evolution of temperature. Hence, a thermal capacity is defined for each material, while the power applied to the material is electrically represented with a current source. The heat capacity represents the rate of change of the heat energy *Q* with respect to the material's temperature *T*. The heat capacity per volume yields:

$$\frac{dQ}{dT} = C_v \tag{2.7}$$

A physically proper model for the power semiconductor device, able to show geometrical definition of the temperature within the system, is shown in Figure 2.13. This model is also called the *Cauer model*. The ground of the electrical circuit used as the model represents the ambient temperature, and each node represents the temperature of a material: junction semiconductor, capsule or heat sink. Between nodes, thermal resistances can model the thermal transfer.

The resistances and capacitances from the Cauer model can be calculated from the material properties:

$$\begin{cases} R_{th} = \dfrac{1}{\lambda} \cdot \dfrac{d}{A} \\ C_{th} = c \cdot \rho \cdot d \cdot A \end{cases} \tag{2.8}$$

where *d* is the material thickness, *A* the section, *c* the heat capacity and ρ the material's specific density. After these parameters are calculated

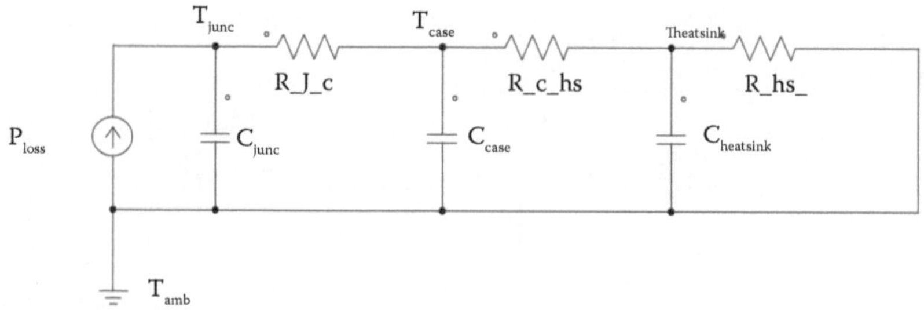

FIGURE 2.13
Cauer model for a power semiconductor device.

from the data, the thermal behaviour of the system can be determined by simulation.

The thermal impedance Z_{th} obtained as a result of this dynamic model can help in understanding the thermal dynamics of a device. It is also called *transient thermal resistance,* and it is defined in datasheets mostly based on repetitive pulses of loss power at a fixed 50% duty cycle and variable time duration. This typically features a small 2:1 variation: the longer the pulse duration, the higher the thermal impedance.

When the premises are changed and the transient thermal impedance is reported for a fixed frequency and a variable duty cycle, a change of duty cycle from 0.50 to 0.05 produces a 10:1 decrease in transient thermal impedance since there is more time to cool the device off between pulses.

Modern simulation tools provide dual electrical and thermal modelling of power semiconductor devices when used within power converters.

2.9 Gate Driver Circuits

A gate driver used for either MOSFET or IGBT devices can be reduced to a voltage source, a pulse generator and gate resistance (Figure 2.14) [10].

Both the control information and power to the gate circuitry have to be transmitted into the gate circuit. This transmission is made with or without galvanic isolation. Since most power converters are built with devices on different power groundings, the gate driver of the high-side device needs to have grounding separated from the low-side circuitry (Figure 2.15).

Multiple configurations are possible for the delivery of gate control information and power supply to the gate driver. Examples of architecture for the galvanically isolated gate driver are:

- Gate driver circuits with inductive transfer of energy (for instance, 1-MHz switched-mode power converter) and a direct transfer of control information

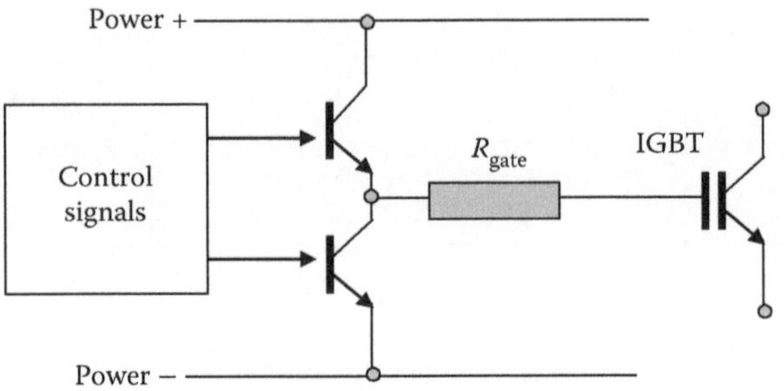

FIGURE 2.14
Schematic of the gate driver circuitry.

- Gate driver circuits with inductive transfer of energy (for instance, 20-kHz switched-mode power converter) and an opto-coupler for transmission of control information

Competition for low-cost equipment has pushed forward the development of gate drivers without ground isolation. Champions in this category are the *bootstrap-type circuits* (Figure 2.16) for power supply of high-side devices, with a floating ground and level shifting for information transmission.

The bootstrap power supply (Figure 2.16) has two states of operation:

- When the low-side IGBT turns on, the bootstrap capacitor charges through the diode towards the gate supply voltage, usually at 15 V DC.
- When the low-side IGBT turns off, the voltage across the bootstrap capacitor can supply the gate driver to maintain the conduction state of the high-side IGBT.

Gate drivers can be classified based on the number of channels which need to be controlled:

- Simple gate driver with input and output at the same ground level

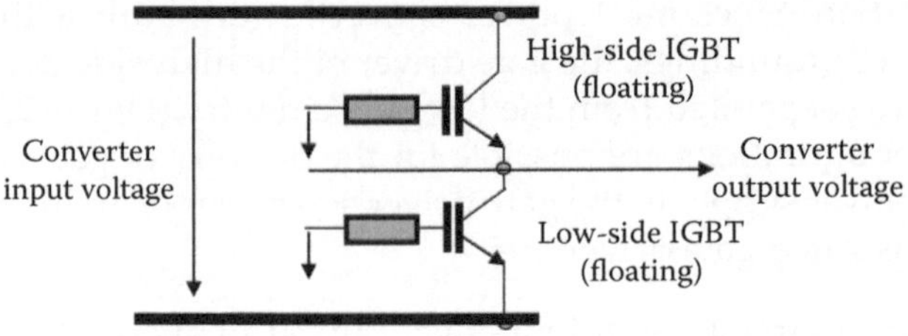

FIGURE 2.15
Grounding circuitry along gate drivers.

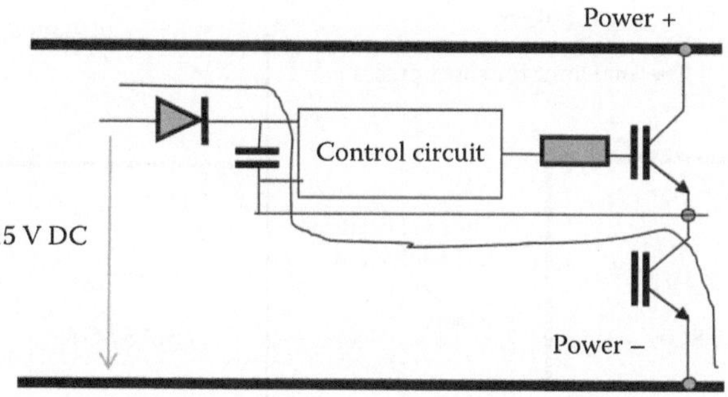

FIGURE 2.16
Bootstrap power supply.

- Gate driver for high-side circuits, with input and output at different voltage levels
- Dual gate driver, with the same circuit for both devices (high side and low side), which is also called a half-bridge
- Gate driver for three-phase converters featuring six channels; that is, three dual gate drivers

2.10 Protection of Power Semiconductor Devices

Protection of power semiconductor devices can be secured by avoiding operation near the operational limits:

- The maximum collector current is determined from a constraint of avoiding the latch-up phenomena.
- The maximum gate-emitter voltage is determined by avoiding gate oxide breakdown.
- The maximum collector current which can circulate during a short-circuit within an IGBT device when operated under maximum gate voltage is 4–10 times higher than the nominal or rated current through the circuit. In such conditions, the IGBT device will work in the active region, with collector-emitter voltage corresponding to the blocking state.
- The maximum collector-emitter voltage comes from the breakdown constraint of the internal PNP transistor.
- The maximum junction temperature should be under 150°C unless a special device with a higher rating is considered.

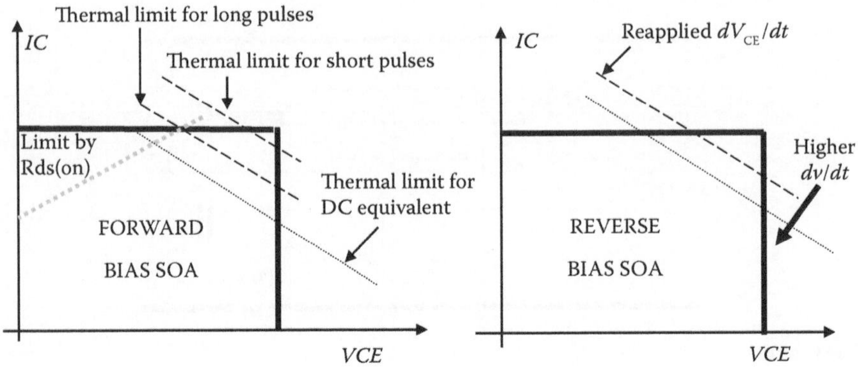

FIGURE 2.17
Safe operation area.

It is important to understand the *safe operation area* (SOA). This is defined for various typical operation conditions and reported in the device's datasheet. Both IGBT and MOSFET devices have a rectangular SOA (Figure 2.17). The MOSFET's SOA also illustrates a limit by $R_{ds(on)}$ at a lower voltage drop and high current.

Within a real circuit, the operation points are along the trajectory shown in Figure 2.18 due to a switching inductance, the series inductance for the controlled circuit. A proper design necessitates that these points be inside the rectangular SOA shown within the datasheet (Figure 2.17). The trajectories in Figure 2.18 depend on parasitic inductances and parasitic capacitances, as well as the dynamic performance of the power semiconductor device itself such as *dv/dt* and *di/dt*. For instance, the package of an IGBT device has a parasitic inductance of approximately 15 nH. Finally, let us mention that *di/dt* and *dv/dt* are not fixed or constant datasheet information, and they can be somewhat adjusted through the gate driver [10,11]. For instance, a decrease

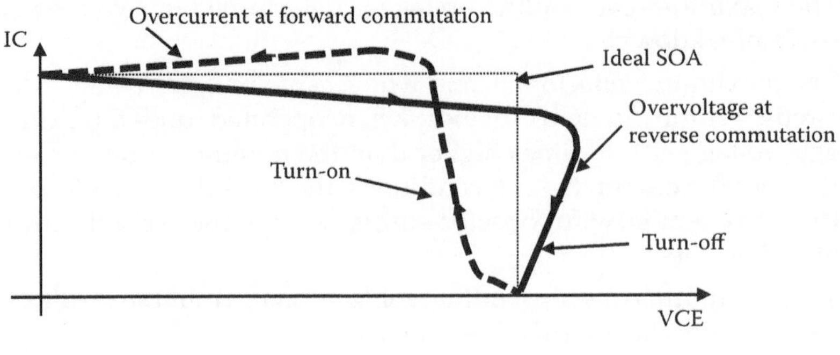

FIGURE 2.18
Actual switching trajectory within the safe operation area.

of the gate resistance can determine a decrease of the current's slope through the device, producing a faster transition.

Most IGBT devices can sustain operation at short circuit for 5 or 10 μsec depending on the semiconductor design. This property comes from demands from standards for various application fields:

- UPS, 0 … 5 μs
- Appliances and white goods, 5 μs
- Motor drives, mostly 10 μs, rarely 5 μs
- Renewable energy applications, 10 μs
- Generic power factor converters, 10 μs

In order to accommodate survivability across a longer short-circuit time, the device needs a larger die area. This means a lower transconductance in the internal model MOSFET, which means a higher $R_{ds(on)}$, which directly increases the $V_{CE(sat)}$ of the IGBT. Longer short-circuit times are traded for conduction loss. Tougher industrial applications, like motor drives, requiring a robust IGBT device may allow more conduction loss to save the equipment.

2.11 Gate Turn-Off Devices

These devices (Figure 2.19) are used at high currents and may disappear in time, replaced by IGBT devices. Their use within telecom equipment is extremely rare. Gate turn-off (GTO) devices overcome thyristor shortcomings with a possible control of the turn-off event, hence the name *gate-turn-off thyristors*.

Similarly to conventional thyristors, the GTO thyristor device can be brought into conduction with a gate control impulse. After turn-on, the

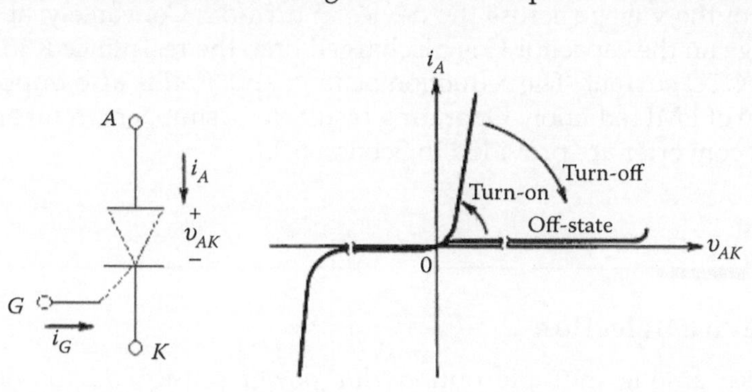

FIGURE 2.19
Symbol and characteristics of a GTO device.

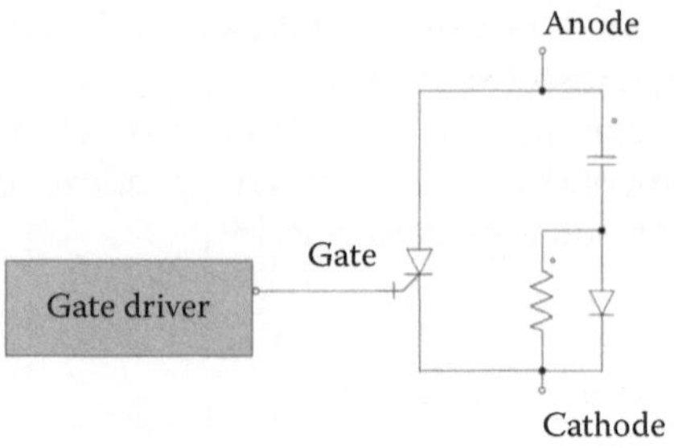

FIGURE 2.20
Snubber circuit.

GTO device can maintain the conduction state even when a current/voltage gate signal is missing as long as current circulation is maintained through the device anode. Contrary to the conventional thyristor, the GTO device can be turned off with a negative gate impulse. Similar to the turn-on process, the gate current (now negative) should circulate for a reduced interval only until the power device is turned off. After the device is turned off, the negative gate current can also be annulated. Unfortunately, this negative gate current should be relatively large, up to a third of the total anode current.

Another shortcoming for GTO devices possibly occurs when they are operated with inductive load. When a turn-off signal is applied on the GTO's gate and the load is inductive, the inductance reacts with a counter-electromotor voltage when current decreases. A fast-varying voltage appears across the GTO device (dv/dt), aiming at maintaining the current circulation. GTO devices do not tolerate such variation, and they will produce a return to the conduction state due to the internal currents produced by the sudden change in voltage (as in $C \cdot dv/dt$).

In order to prevent this from happening, a protection circuit called a *snubber* is employed (Figure 2.20). The snubber capacitor limits the speed of the increase of the voltage across the device at turn-off. Conversely, at turn-on, the voltage on the capacitor C is discharged onto the resistance R in order to limit the GTO current. The reduction of dv/dt and di/dt is also important for limitation of EMI radiation. Operating results for a snubber circuit applied to a flyback converter are provided in Section 6.3.4.

2.12 Device Selection

The choice among multiple options for power semiconductor devices is mainly dictated by the application, voltage and current ratings. This means that grid-side converters, also called telecom rectifiers, will work at higher voltages of several hundred volts and will be implemented with IGBT devices and diodes. Isolated or local point-of-load converters will work at voltages of tens of volts and can be implemented with MOSFET devices.

Given the tremendous competition within the low-voltage power supply market, efficiency is the main performance target. Hence, selection of power-switching devices is mainly made based on power loss. Dedicated devices are proposed for each application class.

A novel trend is the replacement of conventional silicon-based MOSFET devices with GaN transistors [12]. This is a very dynamic field of research and development. The advantage of GaN transistors is reduced loss. For instance, a conversion from $V_{in} = 12$ V DC to $V_{out} = 1.2$ V DC at $f_{sw} = 1$ MHz can achieve over 90% efficiency for input currents from 8 to 34 Amp when using a single pair of switching GaN transistors.

Another trend in semiconductor device technology is that the size of the semiconductor die is decreasing by 33% at each generation upgrade (2–3 years) for the same current processed.

Semiconductor die shrinking is typical of processor manufacturers and consists of creating an identical circuit using a more advanced fabrication process, usually involving an advance in the lithographic node. The die-shrinking decision reduces overall costs for a chip company (fab). For instance, in January 2010, Intel released Clarkdale Core i5 and Core i7 processors fabricated with a 32-nm process, down from a previous 45-nm process used in older iterations of the Nehalem processor microarchitecture.

As an example from the power semiconductor market, Cree has introduced third-generation SiC technology of power semiconductor devices as a die shrink able to address previous channel mobility issues.

While it reduces the cost per amp, the die-shrink approach applied to high-power applications can create a design dilemma as smaller dies typically compromise die temperature and dramatically increase energy capability, which can jeopardize overall system performance and reliability.

Summary

The most important power semiconductor devices were analyzed. All switch-mode power semiconductors have similar top-level characteristics. These are characterized by important loss during reversed transition from the conduction state to the off state or during direct transition from the blocked state to the conduction state. Additionally, conduction loss is seen during current circulation when the device is in conduction.

Gate driver circuits and protection circuits make operation of power semiconductor devices possible within power converters. Principles for these circuits were explained, and a variety of application circuits are possible.

Modern trends in power semiconductor devices were outlined, and several trade-offs were enumerated to incite interest for following the future of the technology within this market.

References

1. Lutz, J., Schlangenotto, H., Scheuermann, U., De Doncker, R., 2011, *Semiconductor Power Devices – Physics, Characteristics, Reliability*, Springer, Berlin.
2. Neacsu, D., 2004, "Power Semiconductors and Control for Automotive Applications", in *Tutorial Presentation at IEEE APEC2004*, Anaheim, CA, February 2004, pp. 1–130.
3. Anon, 2017, Annual power semiconductor market share report 2017 – The power semiconductor market recovered in 2016", IMS Research, August 30, 2017, https://technology.ihs.com/595076/annual-power-semiconductor-market-share-report-2017-the-power-semiconductor-market-recovered-in-2016, accessed on September 21, 2017.
4. Anon, 2015, Global power semiconductors market: Key research findings 2015, Yano Research, https://www.yanoresearch.com/press/pwww.yanoresearch.com/press/pdf/1499.pdf, accessed on September 21, 2017.
5. Seki, Y., Hosen, T., Yamazoe, M., 2010, "The Current Status and Future Outlook for Power Semiconductors", *Fuji Electric Review*, 56(2), 47–50.
6. Fujihira, T., Kaneda, H., Kuneta, S., 2006, "'Fuji' Electric Semiconductor: Current Status and Future Outlook", *Fuji Electric Review*, 52(2), 42–47.
7. Adamowicz, M., Giziewski, S., Pietryka, J., Krzeminski, Z., 2011, "Performance Comparison of SiC Schottky Diodes and Silicon Ultra Fast Recovery Diodes", in *2011 7th International Conference-Workshop on Compatibility and Power Electronics*, Tallinn, Estonia, June 1–3, 2011, pp. 144–149.
8. Brown, J., Moxey, G., 2003, "Power MOSFET Basics: Understanding MOSFET Characteristics Associated with the Figure of Merit", *Vishay-Siliconix Application Note AN-605*, September 2003.
9. Ejury, J., 2003, The SMD-Package Selection for MOSFETs Considering Thermal Issues, Infineon Application Note, pp. 1–4.
10. Zverev, I., Konrad, S., Voelker, H., Petzoldt, J., Klotz, F. 1997, "Influence of the Gate Drive Technique on the Conducted EMI Behaviors of a Power Converter", *IEEE PESC*, 22, 1522–1528.
11. Gerster, Ch., Hofer-Noser, P. 1996 ",Gate Controlled dv/dt and di/dt – Limitation in High Power IGBT Converters", *EPE Journal*, 5(3/4), 11–16.
12. Lidow, A., 2015, GaN technology will transform the future, EDN Network, Pubished January 20, 2015, http://www.edn.com/electronics-blogs/from-the-edge-/4438419/GaN-technology-will-transform-the-future, accessed on September 21, 2017.

3

Buck and Boost Converters

3.1 The Role of Direct Current/Direct Current Power Converters

Many electronic circuits need to be supplied with direct current with a precisely controlled voltage and protection with limitation of the maximum current [1–3]. DC converters take energy from a DC source, which can be a battery or rectifier, and change the level of this voltage to the required level of the load current (Figure 3.1).

DC/DC converters can be classified using several criteria:

- After the grounding connection
  - Converters with direct transfer of energy
  - Converters with galvanic isolation, where the input and output circuits do not share the same grounding
- After the operation mode
  - Switched-mode converters
  - Analogue (linear)-mode voltage regulators

3.2 Direct Conversion (without Galvanic Isolation)

Consider the ideal switching circuit in Figure 3.2.

When transistor is turned on (*Sw* = *on*), the input voltage is applied towards the output circuitry, whereas when the transistor is turned off (*Sw* = *off*), there is no energy transfer from input to output. This sequence of the on and off states of the transistor produces a train of pulses, wherefrom one can extract a DC component with a low-pass filter. The DC level of the output voltage can thus be adjusted through the conduction interval for the transistor. A linear dependency can be established theoretically between the pulse

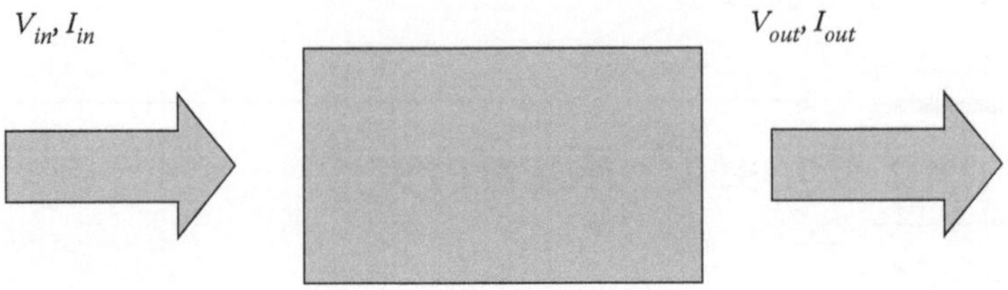

FIGURE 3.1
The idea of DC/DC converters.

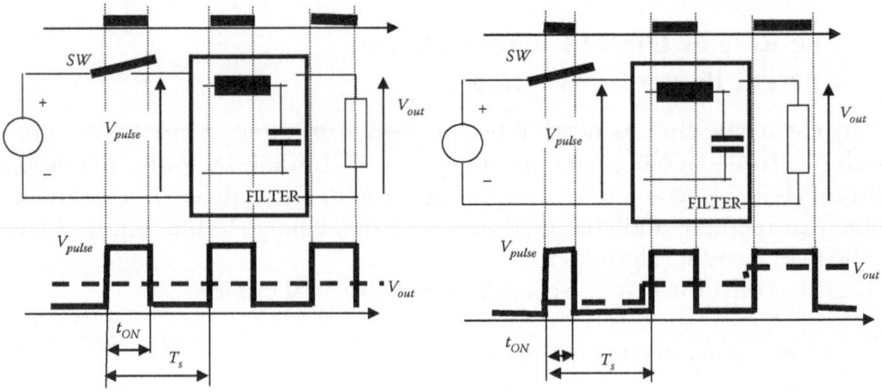

FIGURE 3.2
An ideal switching converter.

duration and the output voltage. Successive control of the on and off states of the transistor is called *pulse-width modulation* (PWM).

3.3 Buck Converter

A power converter able to decrease the level of DC voltage from the input source to the output-connected load is considered herein. Such a step-down converter is called *buck converter* (Figure 3.3). The circuit looks similar to the previous theoretical solution. A diode is added to the practical implementation to offer the current a circulation path during the time the transistor is in the off state.

Immediately after the transistor is turned off, the inductive component within the filter tends to maintain the current circulation through the generation of a voltage ($v = L \cdot di/dt$) at the first tendency of current decrease. This induced voltage would have the polarity to favour or maintain the previous

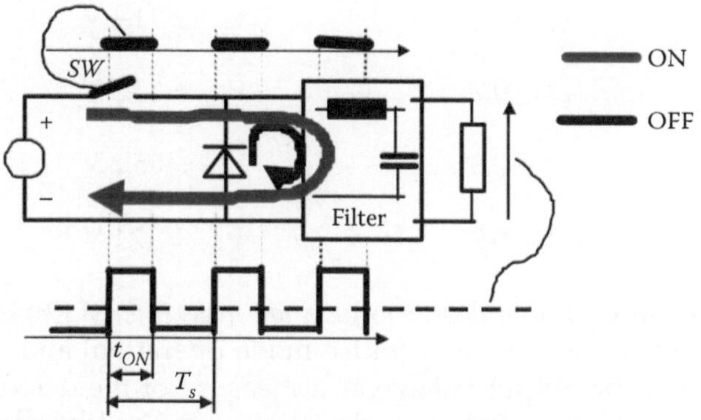

FIGURE 3.3
The Buck converter.

current circulation. The level of such a voltage could be as high as necessary to break a path for the current. Eventually, it can produce transistor breakdown due to overvoltage. However, in the case of the buck converter, the introduced diode turns on and offers a path for the inductive current. It thus protects the transistor as well.

The voltage before the low-pass filter is shown in Figure 3.3 as a train of pulses, with the active pulse having a pulse width equal to the on time of the transistor. The average value of the output voltage is:

$$V_o = \frac{1}{T_s} \cdot \int_0^{T_s} v_o(t) = \frac{1}{T_s} \cdot \left[\int_0^{t_{ON}} V_{in} dt + \int_{t_{ON}}^{T} 0 dt \right] = \frac{t_{ON}}{T_s} \cdot V_{in} = D \cdot V_{in} \quad (3.1)$$

where $D = t_{on}/T$ is called the *duty cycle*. Control of the output DC voltage can be achieved by variation of the duty cycle *D*.

3.3.1 Example

$V_{in} = 19$ V DC	$D = 0.1$	$\rightarrow$	$V_{out} = 1.9$ V
	$D = 0.2$	$\rightarrow$	$V_{out} = 3.8$ V
	$D = 0.3$	$\rightarrow$	$V_{out} = 5.7$ V
	$D = 0.4$	$\rightarrow$	$V_{out} = 7.6$ V
	$D = 0.5$	$\rightarrow$	$V_{out} = 9.5$ V

$$D = 0.6 \quad \rightarrow \quad V_{out} = 11.4\ \text{V}$$

$$D = 0.7 \quad \rightarrow \quad V_{out} = 13.3\ \text{V}$$

$$D = 0.8 \quad \rightarrow \quad V_{out} = 15.2\ \text{V}$$

$$D = 0.9 \quad \rightarrow \quad V_{out} = 17.1\ \text{V}$$

To benefit from this operation mode, a *low-pass filter* (LPF) is required to eliminate the effect of switching (pulse-mode operation) and to extract the average value of the output voltage as a measure of the desired DC value. What remains after the filtration of the voltage ripple depends on the components within the filter structure. Thus, a finite value of capacitance would determine an important ripple. Since the filter has a typical L-C second-order structure, the magnitude characteristics will drop by 40 dB/decade (Figure 3.4). The frequency characteristics of this filter should be selected such that the harmonic components multiple of the PWM frequency is attenuated with more than 60–80 dB; that is, 1 V DC goes into a 1-mV DC ripple.

A practical buck converter is shown in Figure 3.5. All the circuit currents depend on the load connected at the output. The load current comes from the load voltage and load resistance. The average value of the inductance current equals the DC current within the load, and the AC component of the inductance current ideally goes into the capacitor and not into the load. The integral of the inductance voltage needs to be zero during operation in the steady state when considering the same initial values. The formation of the inductive current from the pulsed voltage is shown in Figure 3.6. The details of operation are easier to follow through a numerical example for a conversion from $V_d = 5.0$ V DC to $V_0 = 3.3$ V DC. The characteristic waveforms are shown in Figure 3.7, superimposed for load resistances of 10 and 5 Ω, respectively.

The waveforms for the output voltage as well as the voltage drop on the inductance do not depend on the value of the load resistance (in this case 10 or 5 Ω). The previous analytical relationship between input/output voltages and the duty cycle holds true for both values of load resistance.

The inductance current is composed of a DC component equal to the average value of the load current, which also equals the ratio between the load voltage and the load resistance, and an AC component, called ripple, is approximately the same for any load resistance. The two waveforms for the inductor current look as if they slide up or down with the amount of load DC current.

The operation mode shown in Figures 3.6 and 3.7 is referred to as *continuous conduction mode* (CCM) due to a continuous circulation of current through the inductance. While within this operation mode, the converter behaves like a DC voltage transformer, where the duty cycle D can be seen as the transfer ratio.

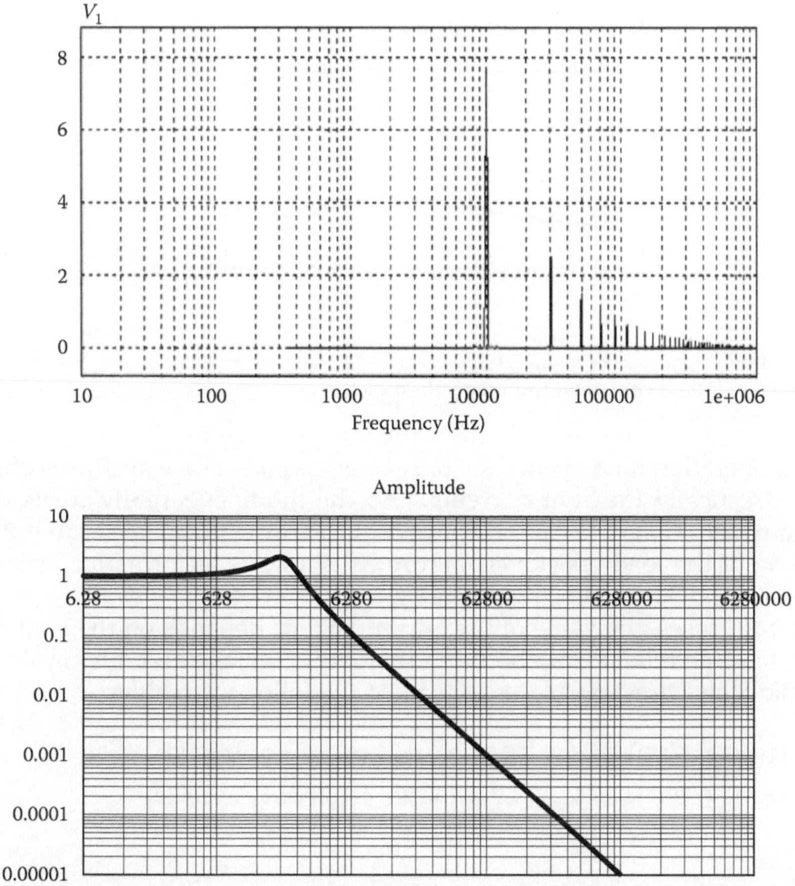

FIGURE 3.4
PWM harmonics and filter characteristics. Example for a low-pass filter with resonant frequency at 300 Hz, providing an excessive 60 dB (1:0.001) to the PWM frequency at 10 kHz.

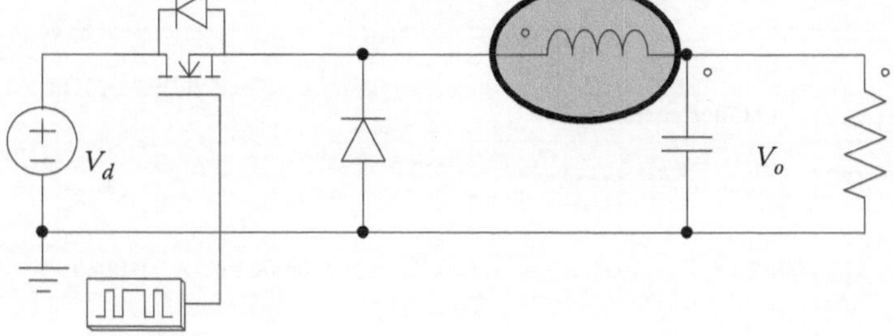

FIGURE 3.5
A practical buck converter circuit.

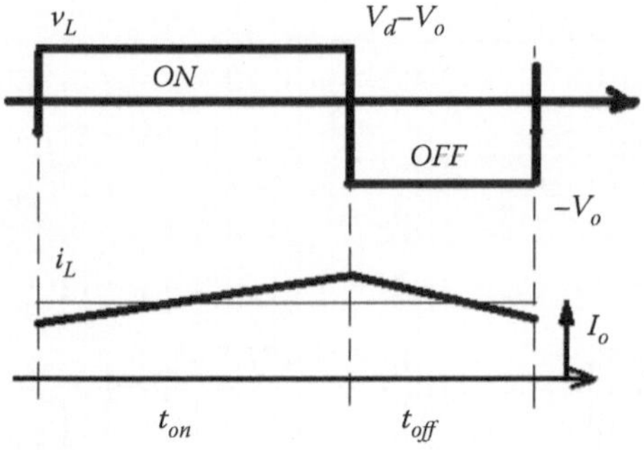

FIGURE 3.6
Formation of inductive current from the pulsed voltage.

This operation mode is not the only one possible. The waveforms change more drastically for light currents, and the diode eventually turns off at zero current. This case can be studied in the same practical setup with an increase of the load resistance. Figure 3.8 shows waveforms for operation with 100- and 50-Ω loads, respectively.

For both cases, the value of the output voltage depends on the load resistance, and the relationship between the output voltage and duty cycle does *not* hold true. Direct measurements yield the following results:

- For $R = 50\ \Omega \Rightarrow V_0 = 3.8$ V DC (at same duty cycle D)
- For $R = 100\ \Omega \Rightarrow V_0 = 4.2$ V DC (at same duty cycle D)

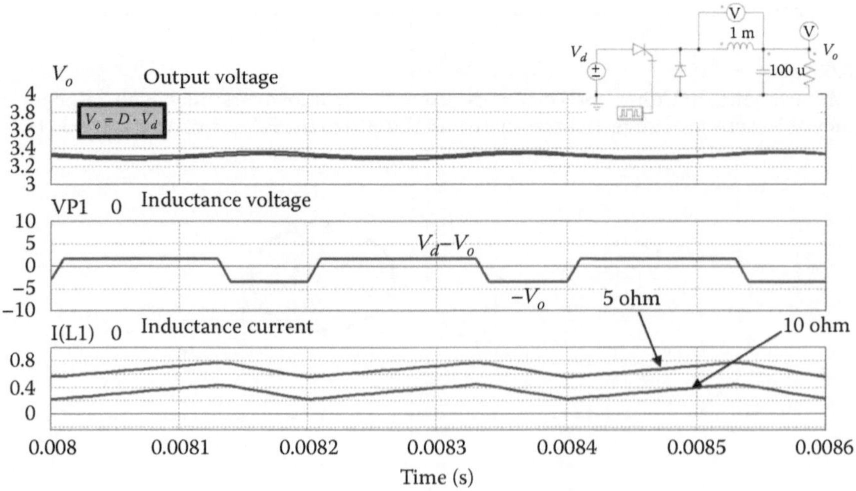

FIGURE 3.7
Waveforms for operation with a load of 10 and 5 Ω, respectively.

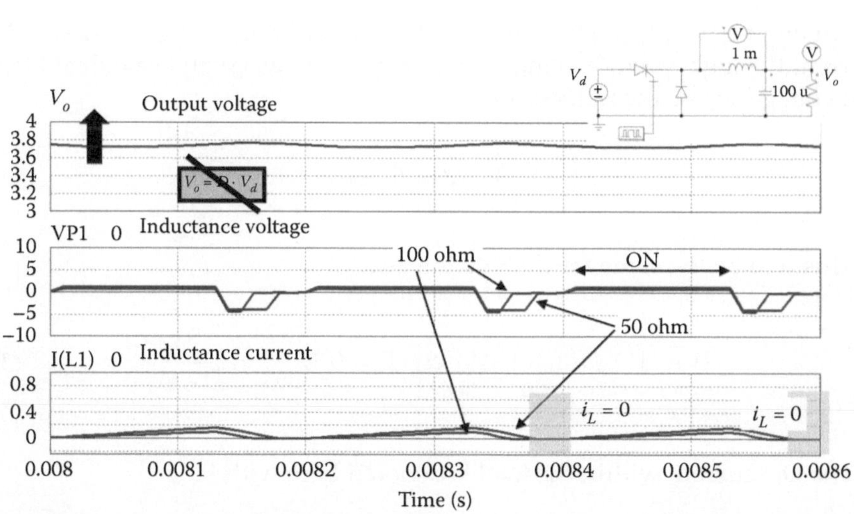

FIGURE 3.8
Waveforms for 100 and 50 Ω.

The average value of the output voltage can be calculated based on both the load resistance and duty cycle, and not directly from the duty cycle. The inductance current has time intervals when it is zero. The output voltage grows higher as the load resistance is higher and the inductor's zero current interval becomes larger. When the inductance current is zero, the inductance voltage is zero, which also means there is no variation of the current. This operation mode is referred to as *discontinuous conduction mode* based on the discontinuous circulation of the current through inductance, also explained with Figure 3.9.

Figure 3.9 shows the voltage and current through the inductor and characteristics of the operation of the buck converter in discontinuous

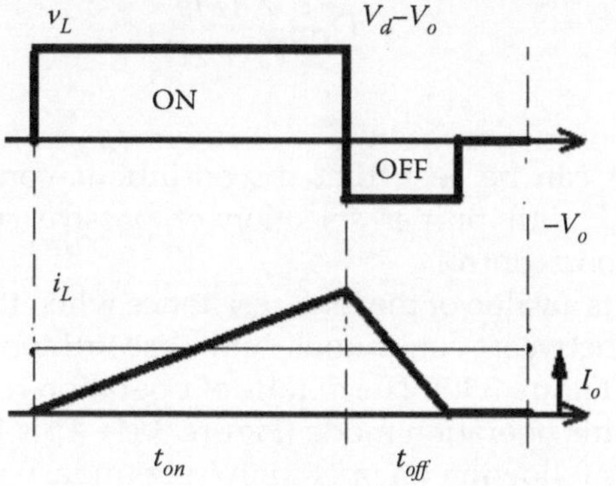

FIGURE 3.9
Waveforms characteristic for a load with discontinuous conduction mode.

conduction mode. It can be used to calculate the transfer characteristic. The current through the inductance can be expressed as Δi and calculated from the voltage across the inductance:

$$V \cdot \Delta t = L \cdot \Delta i \tag{3.2}$$

This will be the same for the two intervals:

$$(V_d - V_o) \cdot D \cdot T_s + (-V_o) \cdot \Delta_1 \cdot T_s = 0 \Rightarrow \frac{V_o}{V_d} = \frac{D}{D + \Delta_1} \tag{3.3}$$

The DC current within the load is denoted here with I_0:

$$\begin{cases} L \cdot \dfrac{0 - i_{L,peak}}{\Delta_1 \cdot T_s} = -V_0 \\ I_0 = \dfrac{1}{T_s} \displaystyle\int_0^{T_s} i(t)dt \end{cases} \Rightarrow \begin{cases} i_{L,peak} = \dfrac{V_0}{L} \cdot \Delta_1 \cdot T_s \\ I_o = i_{L,peak} \cdot \dfrac{D + \Delta_1}{2} \end{cases} \tag{3.4}$$

$$\Rightarrow I_o = \frac{V_0}{L} \cdot \Delta_1 \cdot T_s \cdot \frac{D + \Delta_1}{2} = \frac{V_d \cdot T_s}{2 \cdot L} \cdot D \cdot \Delta_1 \Rightarrow \Delta_1 = \frac{2 \cdot L \cdot I_0}{V_d \cdot T_s \cdot D}$$

The transfer characteristic yields:

$$\frac{V_o}{V_d} = \frac{D}{D + \dfrac{2 \cdot L \cdot I_0}{V_d \cdot T_s \cdot D}} \tag{3.5}$$

In conclusion, it can be seen that discontinuous conduction mode can mostly be avoided with proper selection of passive components and/or through a closed-loop control.

Moreover, there is a value of the load resistance when the operation stands at the boundary between continuous and discontinuous conduction, for example, $R \sim 30\ \Omega$ (Figure 3.10). The details of operation reveal a possible loss reduction within this operation mode (Figure 3.11). This is the basis of modern control circuits following an operation in boundary mode for improved efficiency of the power converter [4].

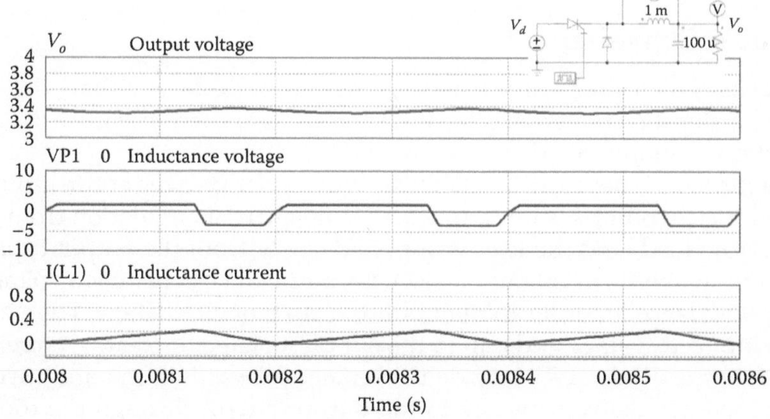

FIGURE 3.10
Waveforms for operation at boundary between discontinuous and continuous conduction mode.

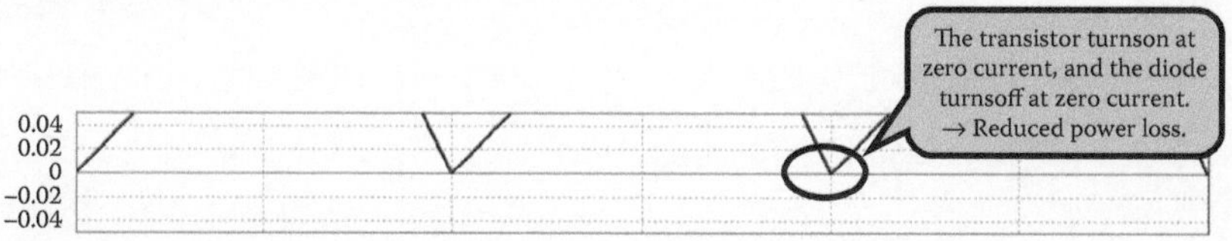

FIGURE 3.11
Details of operation at boundary mode.

The average value of the critical current in boundary mode is:

$$I_{LB} = \frac{1}{2} \cdot i_{L,peak} = \frac{t_{on}}{2 \cdot L} \cdot [V_d - V_o] = \frac{D \cdot T_s}{2 \cdot L} \cdot [V_d - V_o] = \frac{D \cdot T_s \cdot V_d}{2 \cdot L} \cdot [1 - D] = I_{OB} \quad (3.6)$$

This is further represented in Figure 3.12. There is discontinuous conduction at currents lower than the boundary current, and there is continuous conduction at currents larger than the boundary current.

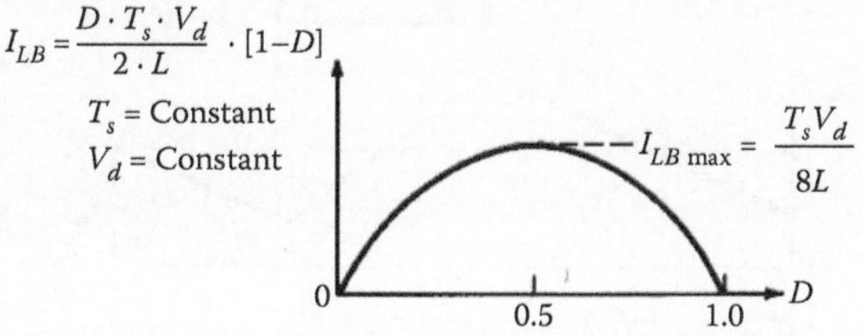

FIGURE 3.12
Current at boundary between discontinuous and continuous conduction modes for various duty cycles (D).

3.4 Boost Converter

Whenever the desired output voltage needs to be larger than the input voltage, a step-up converter is built and used as shown in Figure 3.13. This converter is usually called a boost converter.

For a turned-on device ($Sw = ON$), the inductance sees the entire input voltage and the current rises linearly. There is no current circulation from input towards the load, and the load is supplied solely from the output capacitor. When the switch turns off ($Sw = OFF$), the inductance generates a voltage, no matter how large, in order to maintain the current circulation. The polarity of this induced voltage adds up to the supply voltage to create a larger voltage before the diode *D*. The diode *D* becomes forward biased and turns on. The inductance delivers energy to the output circuit through the diode *D*. The output capacitor is thus charged from the input voltage plus the voltage induced in the inductance. The generic waveforms for inductance voltage and current are shown in Figure 3.14, and the analytical relationship yields:

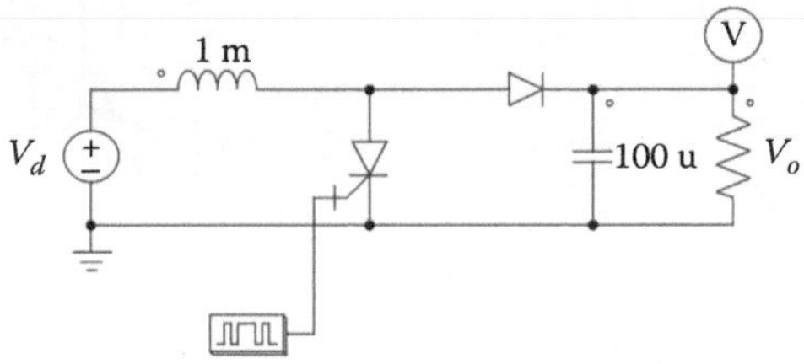

FIGURE 3.13
Circuit for a boost converter.

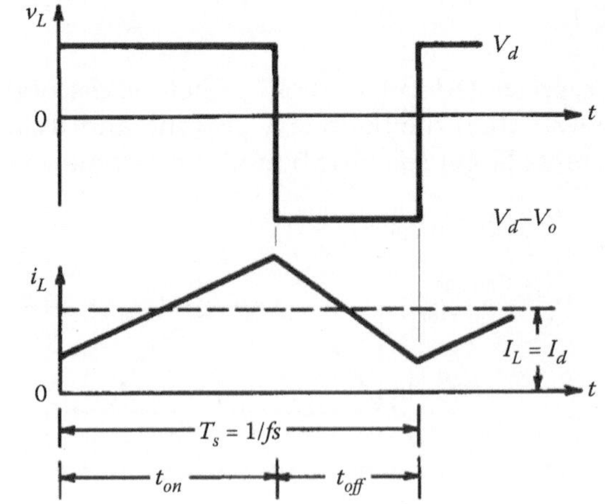

FIGURE 3.14
Waveforms for the boost converter in continuous conduction mode.

$$\frac{V_o}{V_d} = \frac{T_s}{t_{off}} = \frac{1}{1-D} \tag{3.7}$$

$$V_d \cdot I_d = V_o \cdot I_o \Rightarrow \frac{I_o}{I_d} = 1-D \tag{3.8}$$

Figure 3.15 shows details of operation at continuous conduction mode of a boost converter for a load resistance of 10 and 50 Ω, respectively. A conversion from 5 V input to 12 V output is considered herein. The definition of conduction mode refers to the inductance current. The voltage drop on the inductance is approximately identical for the same duty cycle for either a 10- or 50-Ω load. This produces the same ripple of the inductance current with a different average value, and the current waveform seems to slide up and down with the average value. The input power can be calculated as a product of the input voltage (5 V DC) and the average input current.

The ripple of the output voltage depends on the load resistance: the larger the resistance, the smaller the ripple. During continuous conduction mode, the converter behaves as a DC transformer, and the output voltage can be calculated with Equation 3.7. The inductance current is identical to the input current and not the output current, as it was in the case of the buck converter. Further on, the relationship between the input and output currents also follows the duty cycle, as in Equation 3.8.

Figure 3.16 shows the case of a load resistance of 150 Ω for the same conversion from a voltage source of 5 V DC to a load at 12 V DC. Note the appearance

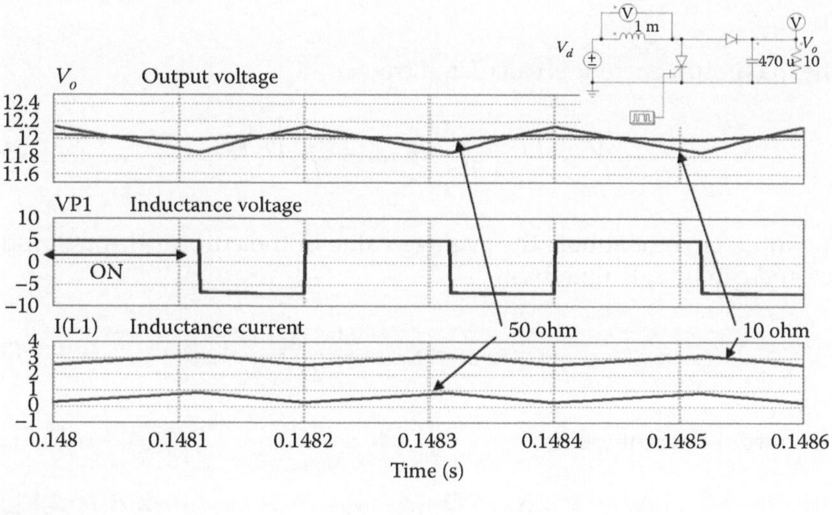

FIGURE 3.15
Waveforms for the continuous conduction mode of a boost converter.

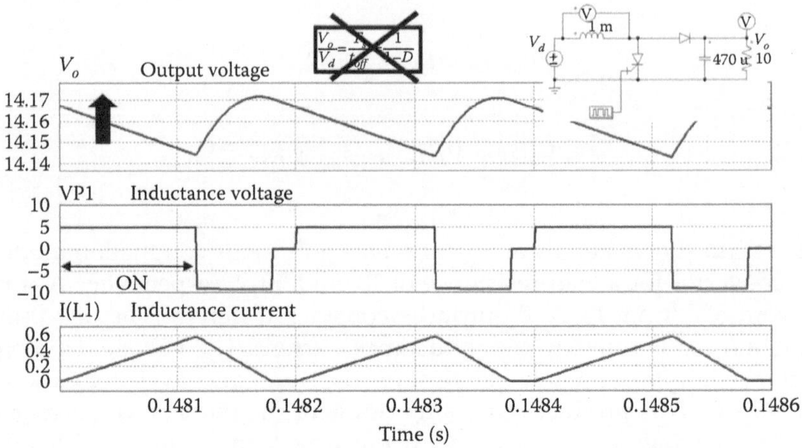

FIGURE 3.16
Waveforms for the case of a resistance of 150 Ω and discontinuous conduction mode.

of zero-current intervals within the inductance current, which coincide with intervals of zero voltage across the inductance.

This operation mode with intervals of zero current through inductance is called discontinuous conduction mode. Due to the zero-current intervals, the output voltage is higher than expected for the same duty cycle. A different calculation of the output voltage becomes mandatory. From Figure 3.16, the average value for the inductance voltage should be zero. It yields:

$$(V_d)\cdot D\cdot T_s+(V_d-V_o)\cdot \Delta_1\cdot T_s=0\Rightarrow \frac{V_o}{V_d}=\frac{D+\Delta_1}{\Delta_1}\Rightarrow \frac{I_o}{I_d}=\frac{\Delta_1}{D+\Delta_1} \tag{3.9}$$

The maximum current is calculated from:

$$V_d=L\cdot\frac{i_{L,peak}}{D\cdot T_s}\Rightarrow i_{L,peak}=\frac{V_d}{L}\cdot D\cdot T_s \tag{3.10}$$

The input current equals the average value of the current through inductance and can be calculated as:

$$I_d=I_{L,av}=i_{L,peak}\cdot\frac{D+\Delta_1}{2} \tag{3.11}$$

The output current yields:

$$I_o=I_d\cdot\frac{\Delta_1}{D+\Delta_1}=\frac{\Delta_1}{D+\Delta_1}\cdot\frac{D+\Delta_1}{2}\cdot\frac{V_d}{L}\cdot D\cdot T_s\Rightarrow \Delta_1=\frac{2\cdot L\cdot I_0}{V_d\cdot D\cdot T_s} \tag{3.12}$$

The transfer characteristic is expressed as:

$$\frac{V_o}{V_d} = \frac{D+\Delta_1}{\Delta_1} = \frac{D+\dfrac{2\cdot L\cdot I_0}{V_d\cdot D\cdot T_s}}{\dfrac{2\cdot L\cdot I_0}{V_d\cdot D\cdot T_s}} \tag{3.13}$$

Discontinuous conduction mode can be avoided through a proper selection of passive components, or a (closed-loop) feedback control can be used in this respect.

Analogous to the buck converter, there is a value of the load resistance which produces an operation at the boundary between continuous and discontinuous conduction. For the converter previously analysed, a value of the load resistance of $R{\sim}100\ \Omega$, leads to the critical or boundary conduction mode (Figures 3.17 and 3.18).

The average value of the inductance current within this boundary conduction mode (Figure 3.19) yields:

$$\begin{cases} I_{LB}\dfrac{1}{2}\cdot i_{L,peak} = \dfrac{t_{on}}{2\cdot L}\cdot[V_d] = \dfrac{D\cdot T_s}{2\cdot L}[V_o]\cdot[1-D] \\ I_{0B} = I_d\cdot[1-D] = I_{LB}\cdot[1-D] \end{cases} \Rightarrow I_{0B}$$

$$= I_{LB}\cdot[1-D] = \frac{D\cdot T_s}{2\cdot L}\cdot[V_o]\cdot[1-D]^2 \tag{3.14}$$

Both the output and inductance currents can be represented graphically with dependency on the duty cycle D when the boundary constraints are met.

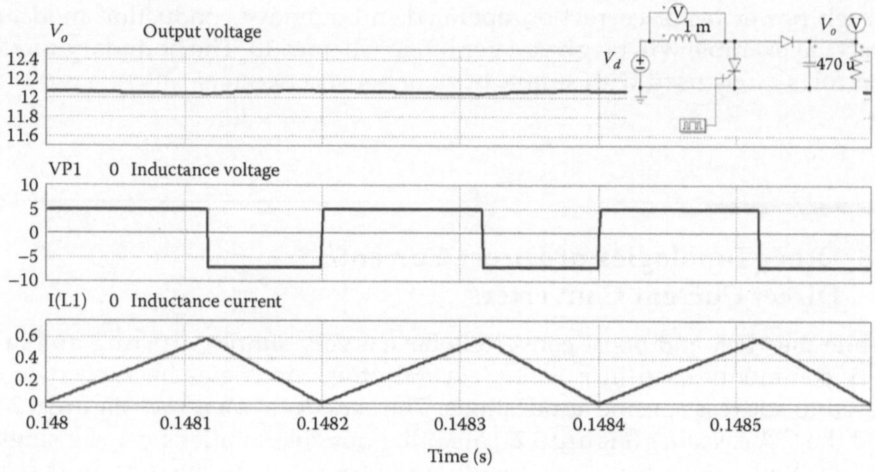

FIGURE 3.17
Buck converter operated at boundary mode.

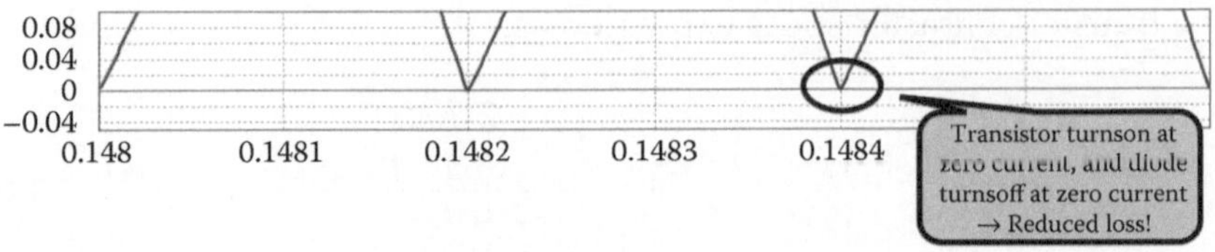

FIGURE 3.18
Detail of operation at boundary conduction mode.

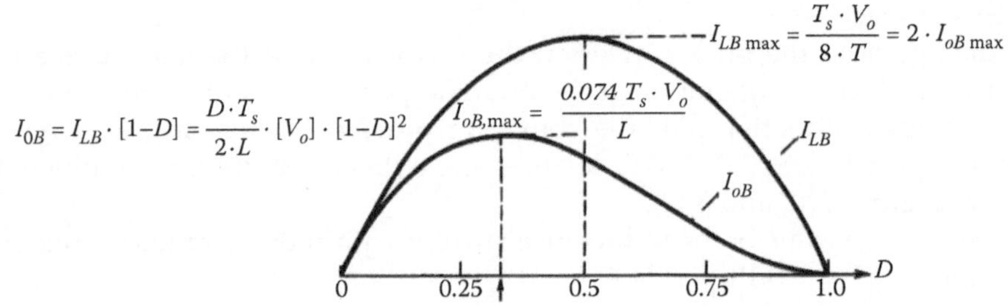

FIGURE 3.19
Boundary current in dependency with duty cycle for the boost converter.

The operation in boundary conduction mode features substantial power loss reduction and efficiency improvement, as suggested in Figure 3.18. This property is used within certain integrated circuits which modify the switching frequency (pulse frequency) to maintain the boundary constraint when voltage or current varies. This property has been explained before with the buck converter. Since the case when it is most used corresponds to the single-switch power factor correction operated in boundary conduction mode, a practical example will be given later on, in Chapter 10. The boundary mode control is rarely used with simple buck or boost converters.

3.5 Other Topologies of Direct Current/ Direct Current Converters

While the buck and boost converters have a very simple structure and are very commonly used in practice, several other topologies have been proposed to address specific applications. The *buck–boost converter* (Figure 3.20) and the *Cuk converter* (Figure 3.21) are other possible solutions using a single switch. Both converters can provide an output voltage magnitude that is either greater than or less than the input voltage magnitude. Both converters invert the polarity of the voltage. The Cuk converter is essentially a boost

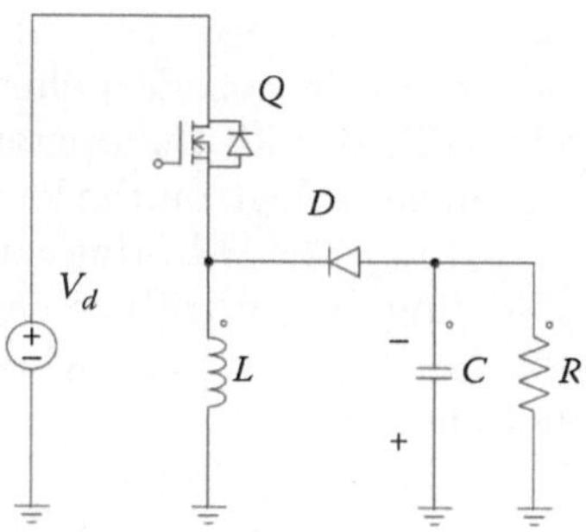

FIGURE 3.20
Buck–boost converter.

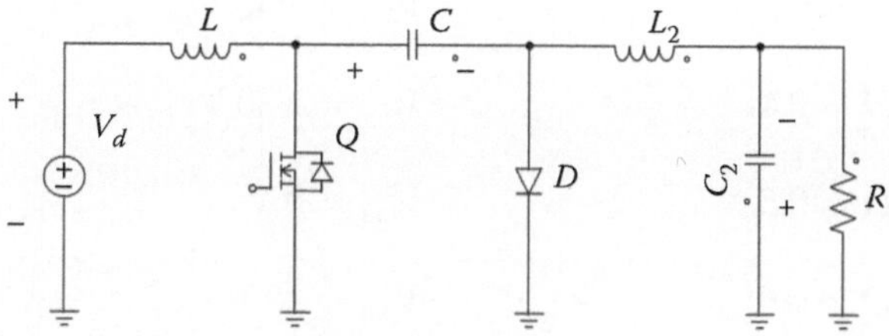

FIGURE 3.21
Cuk converter.

converter followed by a buck converter with a coupling capacitor in between. Details of its operation are beyond the scope of this book.

The *H-bridge converter* (Figure 3.22) has the advantage of continuous control of the output voltage since the output voltage always passes through a controlled device. It features two devices in series, leading to an increase of the voltage drop and loss. Depending on the application, it has the shortcoming or advantage of a bipolar voltage on the load. A diagonal of the H-bridge

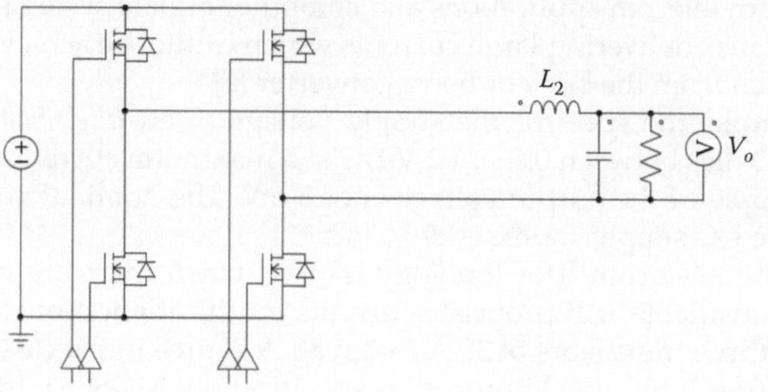

FIGURE 3.22
H-bridge converter.

determines a positive voltage on the load. The other diagonal determines a negative voltage that constitutes the second polarity of the voltage across inductance. The duty cycle defined with the operation of one or the other diagonals determines the average voltage on the load, which can be positive or negative, varying from $-V_d$ to V_d. The H-bridge converter is generally used at higher input voltages, allowing design with conventional transistors.

Examples of variation of the output voltage when considering the input voltage $V_{in} = 100$ V DC are as follows:

$$D = 0.1 \quad \rightarrow \quad V_{out} = -80\ \text{V}$$

$$D = 0.2 \quad \rightarrow \quad V_{out} = -60\ \text{V}$$

$$D = 0.3 \quad \rightarrow \quad V_{out} = -40\ \text{V}$$

$$D = 0.4 \quad \rightarrow \quad V_{out} = -20\ \text{V}$$

$$D = 0.5 \quad \rightarrow \quad V_{out} = 0\ \text{V}$$

$$D = 0.6 \quad \rightarrow \quad V_{out} = 20\ \text{V}$$

$$D = 0.7 \quad \rightarrow \quad V_{out} = 40\ \text{V}$$

$$D = 0.8 \quad \rightarrow \quad V_{out} = 60\ \text{V}$$

$$D = 0.9 \quad \rightarrow \quad V_{out} = 80\ \text{V}$$

3.6 Multiphase Converters

Most modern telecom applications and computer or data systems necessitate power supplies delivering large currents which cannot be achieved with the single switch from the buck or boost converter [5].

For example, the spec for the supply voltage of an *Intel mobile processor* (Figure 3.23) lies between 0.6 … 1.2 V DC at a maximum current of 80 A. The allowed ripple of the output voltage is of 5 mV. The nominal voltage at the input of the DC supply source is 12 V DC.

The multiphase converter topology is used when there are no MOSFET transistors available at appropriate currents for the application, for instance, the use of three transistors of 30 A for an 80-A source in the case of a power supply for the Intel mobile processor specification. MOSFET transistors in surface mount technology, working under 20 V DC, can currently carry up to 30 A without special and expensive heat-sink mounting. Hence, it makes

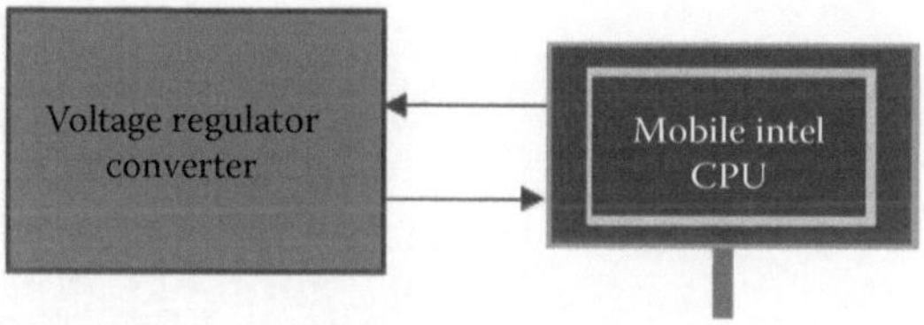

FIGURE 3.23
Supplying the Intel mobile processor while this processor may adjust its supply voltage.

sense to use multiple transistors rated at 25–30 A to accomplish an operation at higher currents. MOSFET transistors can easily be connected in parallel. However, as long as three transistors are connected in parallel anyway, control of the new topology is expanded further and the control pulses can be phase-shifted from each other.

Another example relates to a graphics card, the high-end Radeon HD 5870 Lightning with a 15-phase VRM: 12 phases for the graphical processing unit and 3 for memory.

Multiphase converters are used to control several identical buck converter sections (Figure 3.24) of the power converter with a constant phase-shift so that the load current ripple decreases. Control is achieved with a phase shift between control pulses for each buck converter, a procedure called *interleaved*

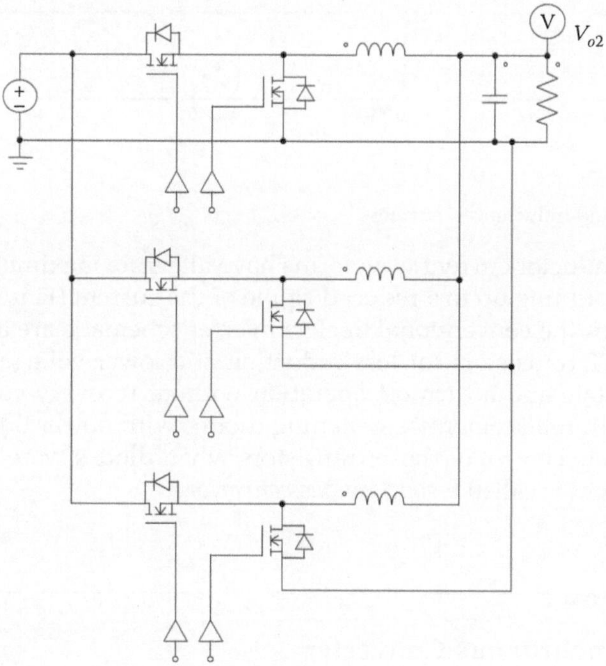

FIGURE 3.24
Multiphase buck converter.

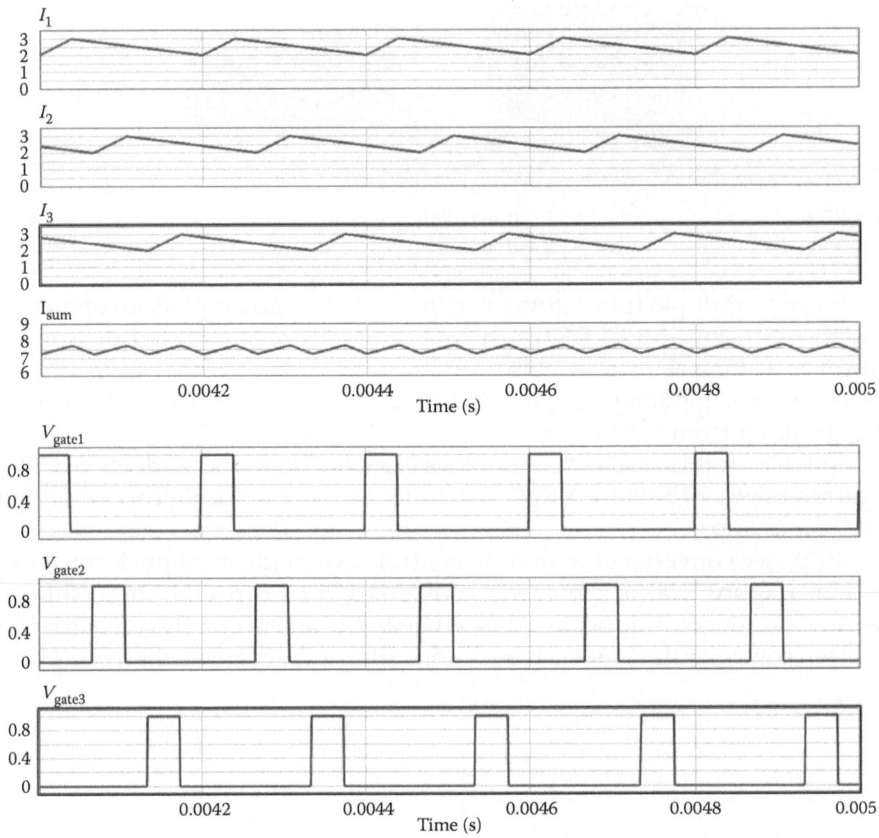

FIGURE 3.25
Control pulses and inductances' currents.

control. The inductor current waveforms have alternate maximum and minimum values, adding up to a reduced ripple of the current (Figure 3.25).

Diodes from the conventional buck converter schematic are also replaced with MOSFET transistors for loss reduction at a lower voltage drop in the conduction state and a turn-off operation without recovery currents. Such operation with replacement of switching diodes with power transistors and the appropriate control of those transistors when diodes were supposed to conduct current is called a *synchronous convertor.*

3.7 The Synchronous Converter

A MOSFET transistor is used to reduce conduction loss when replacing a switching diode within a power converter [6]. In a conventional buck converter, when the main buck transistor turns off, the diode turns on and the MOSFET transistor sees negative voltage. The command of this transistor determines conduction in the third quadrant (at $V < 0, I < 0$). The voltage drop across the transistor (also measured as drain-source voltage) is given by the product between the drain-source resistance ($R_{ds} \sim 5 \ldots 20\ \Omega$) and the converter current. It is lower than the voltage drop across a diode. Furthermore, the MOSFET replacing the diode turns off faster, without recovery phenomena. A dual effective reduction in the power loss results.

Since the power MOSFET needs to be controlled when the diode conducts current, the converter is called a synchronous converter. However, the diode is still used in the circuit to address possible inadequate control of the power transistor when trying to match the right turn-on moment. Thus, the diode will turn on quickly as an effect of the voltage induced by the buck inductance when the main transistor turns off. The synchronous command comes shortly, and the synchronous MOSFET may see the diode already in conduction. The synchronous MOSFET turns on, and its drain-source voltage drop becomes lower than the voltage drop across the diode. This means the diode will turn off and the transistor stays turned on.

Further loss reduction is achieved with usage of dedicated MOSFET transistors, highly optimized for either a buck transistor or synchronous transistor operation. The conduction loss depends on the $R_{ds(on)}$. However, a transistor with a low resistance will impose a construction with a large switching loss. Design curves are generally drawn for finding the best compromise between the $R_{ds(on)}$ and the switching loss. Additionally, the efficiency will depend on the load current. A good design finds an equilibrium able to keep a high efficiency throughout the operation range [7]. For instance, one can use a pair of MOSFET transistors composed of IRF6617 for the control transistor and IRF8891 for synchronous rectification when a buck converter under 12 V is built.

Different transistor structures are useable within the same buck converter.

3.8 Selection of Passive Components

While buck and boost converters can be used within multiple applications, with voltages from 0.6 V DC up to 1000 V DC and currents from miliAmperes (mA) to hundreds of A, the focus here relates to the usage of these power stages in telecom, computer or data processing systems.

Telecom applications, especially, feature digital electronic loads which require supply voltages in the range 0.6–3.3 V DC and currents from 10 to hundreds of A. Since the domain is very well defined, the construction of these buck converters has specific design practices with the goal of delivering supply to computation electronics with low voltages.

Contemporary solutions converge towards the usage of multiphase converters with synchronous rectification equipped with dedicated MOSFET devices for high-side (principal buck transistor) or low-side (diode) synchronous rectification, manufactured with surface-mount technology, with ratings in the range of 20–30 A.

These premises impose technological restrictions on the selection of passive components, buck inductors and output capacitor banks. For instance, the *MPC7448 RISC Microprocessor Hardware Specifications* [8] defined a fixed ± 50-mV ripple allowance for all supply voltages in the range 1.0 V DC … 1.2 V DC and ±5% for voltages in the range 1.5 V DC … 2.5 V DC.

3.8.1 Selection of the Buck Inductor

Buck inductors used within point-of-load power supplies in telecom equipment are mostly selected using surface-mount technology, with a rated current up to 25–30 A DC and a saturation current of 35–50 A DC. The actual design of the buck inductor considers a current ripple under 20% of nominal; that is, up to 10 A peak to peak. An example for a quick estimation at a conversion from 5 to 0.9 V is similar to:

$$L = \frac{V_{in} - V_{out}}{\frac{\Delta i}{D \cdot T}} = \frac{5 - 0.9}{\frac{10}{0.18 \cdot (3.33 \ldots 2.00) \cdot 10^{-6}}} \approx 150 \ldots 246\,\text{nH} \tag{3.15}$$

which can provide the basis for selecting a 200-nH SMD inductor. Adopting an inductor with larger values reduces ripple and supports continuous conduction mode for the buck converter at the risk of compromising the dynamic performance of the converter. The ripple of the inductor current directly influences loss. Selection of an inductor towards the lower end of values helps with the dynamic response of the converter.

Inductors operated at switching frequencies in the range of 300–500 kHz are made with materials like ferrite, SUPERFLUX and WE-PERM.

After the inductor is selected, the output capacitor bank can be assembled. The main requirement here is the voltage ripple, and traditionally this is restricted to the selection of a capacitor with a small ESR. With technology improvements, the ESR value has decreased, as has its priority in calculation. The holdup of the output voltage during a perturbation because of, for example, either a glitch in input supply voltage or a sudden change in the load current affects the choice of capacitor. The output capacitor bank needs to keep the output voltage within the spec.

A numerical example is:

$$C = \frac{i}{\frac{\Delta v}{\Delta T}} = \frac{25}{(50 \cdot 10^{-3}) \cdot (500 \cdot 10^{3})} \approx 1mF \tag{3.16}$$

This complex set of requirements today defines a combination of capacitors within the output capacitor bank. Both ceramic capacitors and aluminium or tantalum electrolytic capacitors are involved up to the required energy storage. The capacitor bank is finally completed with a high-frequency capacitor. This choice is heavily dependent on capacitor technology evolution, and the capacitor bank adopted in 2005 does not look anything like the solution adopted in 2017 for the same design specifications (Figure 3.26) [9].

A complete design example at the technological level of 2017 follows, and it has been adapted from [11].

A Murata GRM32ER60J107ME20 ceramic capacitor rated as 100 μF with operation up to 6.3 V DC is considered representative of mainstream surface-mount ceramic capacitor technology. The actual capacitance is first derated when the capacitor is used at 1.2 V DC instead of the nominal 6.3 V DC. If the operating temperature stays within 60°C, the capacitance remains within 1%; hence, this is not a factor.

The frequency variation of the impedance of a group of ceramic capacitors connected in parallel is shown in Figure 3.27a. It can be seen that the self-resonance is in the mid-100s of kHz.

Ride-through capability is ensured with a bank of electrolytic capacitors able to provide more capacitance in a small surface-mount footprint. The Panasonic 2R5TPE470M9 can be considered a competitive polymer tantalum electrolytic, able to add on sets of 470 μF per package.

Finally, a high-frequency ceramic capacitor is used to provide a lower impedance at higher frequencies. This has to ride the inductive character of the other capacitors within the capacitor bank. Such a low-value high-frequency capacitance does not influence the analysis of stability and control-loop design.

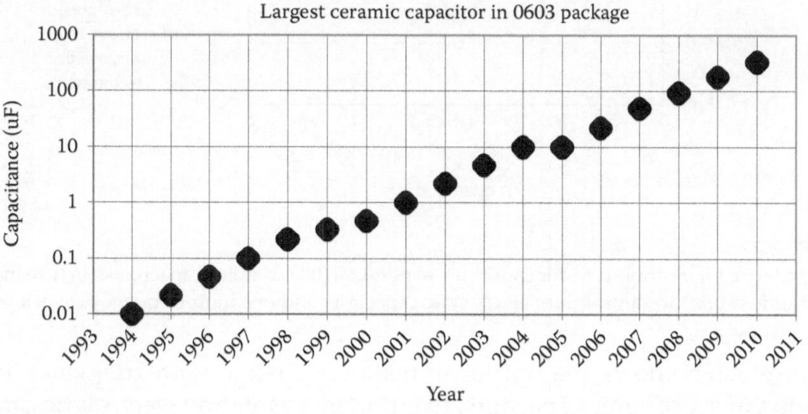

FIGURE 3.26
Historical performance of ceramic capacitors packaged within a 0603 package, provided as example for the dynamics of this field, with influence on design of output capacitor bank.

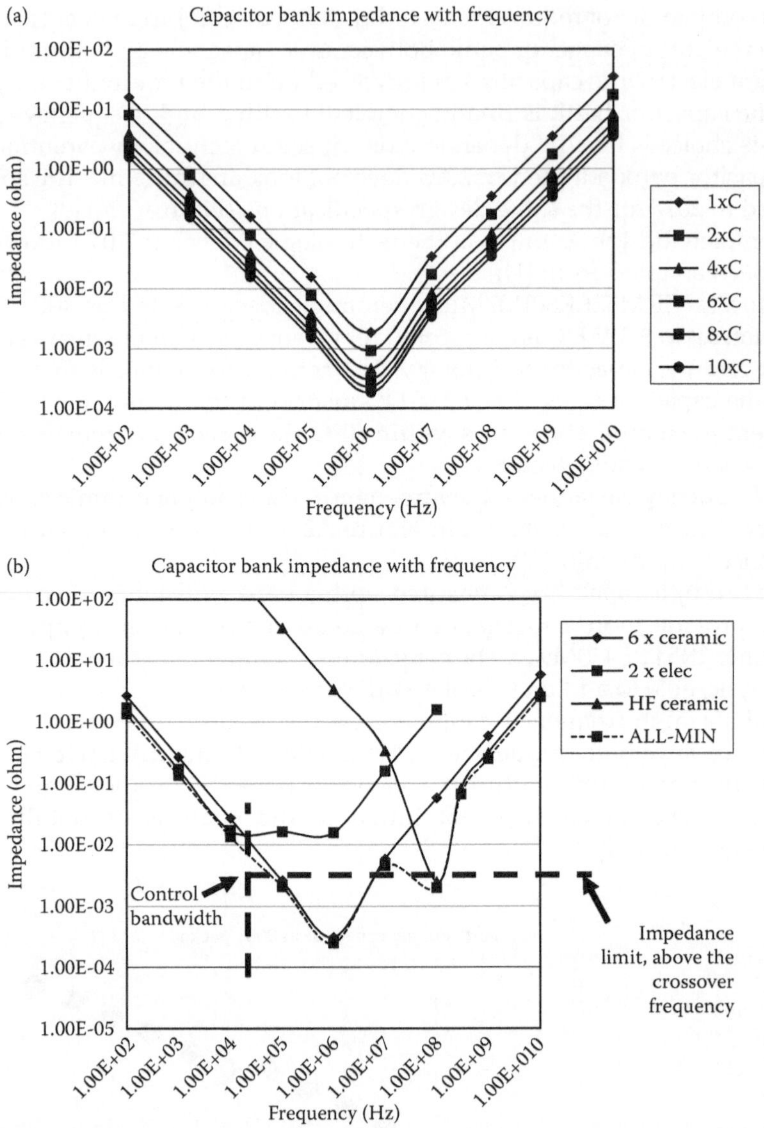

FIGURE 3.27
(a) Impedance for multiple ceramic capacitors in parallel; (b) a complex capacitor structure including ceramic capacitors, aluminium electrolytic capacitors and one high-frequency capacitor.

Figure 3.27b shows the output impedance change with frequency for a selection of six ceramic capacitors, two polymer tantalum electrolytic capacitors and an HF ceramic capacitor [11]. The envelope dotted curve follows the impedance of the parallel combination of these capacitors. The group of μF ceramic capacitors is the overall dominant curve, while the electrolytic dominates at low frequencies and the HF ceramic dominates at very high frequencies. Further improvements in both ceramic and electrolytic capacitors may simplify this solution and further reduce the number of individual components. However, such evolution is in competition with the maximum current demand from the point-of-load converters, which is also increasing with newer generations of digital equipment.

The equivalent impedance of the bank of output capacitors needs to be below a certain value for any frequency above the control-loop bandwidth. At frequencies under the control-loop bandwidth, the control loop should be able to address ripple or voltage variation. Above this frequency, the filtering provided by the capacitor bank should help with the ripple. In this respect, the impedance should stay below a value equal to the ratio between the allowed voltage ripple and the maximum current step change in the load. This can be approximated with the maximum load current.

If the design does not satisfy the requirements and the ripple becomes larger than expected, more complex structures are considered. First, one can try a cascaded double LC filter, or even a trap or notch filter structure [10,11]. In more advanced research approaches, active filtering is considered [12–14].

3.9 Control Circuits for Buck/Boost Converters

Due to a very large application field for buck/boost converters, the control functions, including feedback control and PWM generation, have been implemented as integrated circuits since early on. The PWM control chip was invented by Bob Mammano in 1975 and introduced to the market in 1976 by the Silicon General Company as SG1524. By the 1970s, PWM control ICs had already been developed by multiple corporations, with products like Motorola MC3420, Texas Instruments TL454, Signetics NE5560 and Ferranti ZN1066 [15]. All such products have been developed as mixed-mode IC technology with a simple and well-known current structure. The technology development coincided with the advent of switching power supplies in the late 1970s. It satisfied a clear market need and was able to add value for customers.

The launch of these integrated circuits represented the creation of a new market with the paradigm 'vendors competing for customers', able to add a new set of customers. This advent in integrated circuit technology enabled other new power supply technologies and subsequently supported more incremental development. Offspring technologies like current-mode ICs, cycle-by-cycle current-limiting protection, single-ended push-pull supplies, LDOs, hot-swap, soft-switching and others were released later based on the original success of Mammano's chip. Today, the power control IC industry has a market of over $5 B and is growing at a very high annual rate of almost 10%.

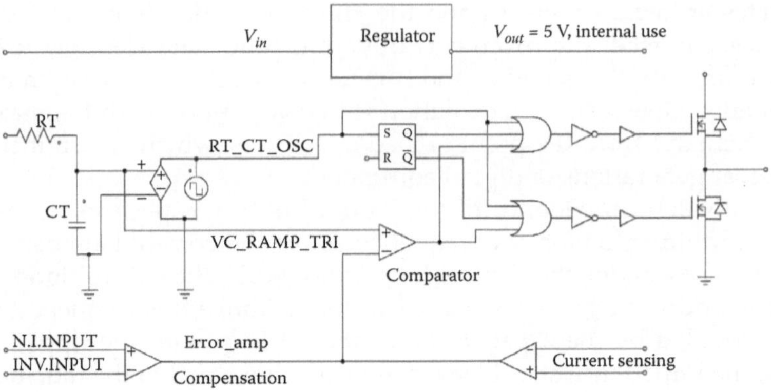

FIGURE 3.28
Principles of the PWM integrated circuit for control of power converters.

The original structure of the PWM control circuit, which is also used today, is shown in Figure 3.28. An oscillator is set up to create a triangular waveform at a fixed frequency equal to the desired PWM frequency. This frequency can be set with an external capacitor which determines the slope of the triangular frequency. This triangular signal is compared with a signal derived from voltage or current error, depending on whether a voltage control loop or a current control loop is involved. This comparison adjusts and synchronises the pulse width for the control of power devices.

Additionally, several protection circuits are included to block the gate control signals in case of a fault.

The generic application circuit is shown in Figure 3.29.

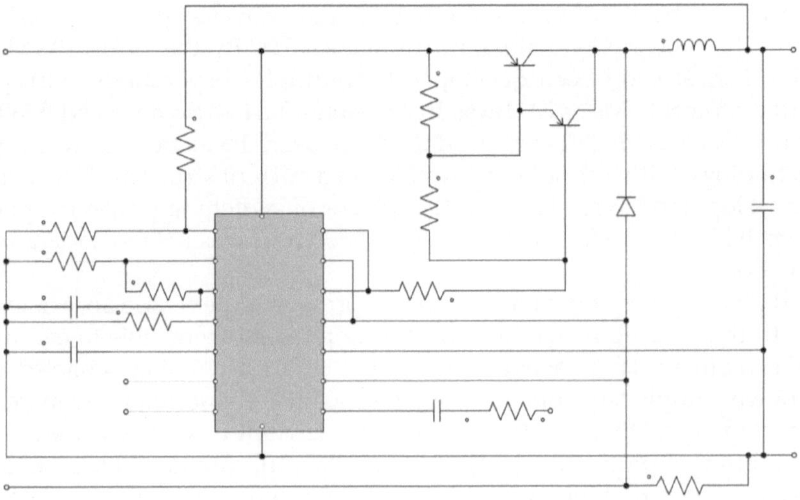

FIGURE 3.29
Application circuit for a generic PWM IC controller.

Year	Process (%CMOS)	Size	Complexity (gates)	Prod cost (Cent/gate)
1980s	Bipolar	3.0 ... 7.0 μm	100	2
1990s	Bi-CMOS	3.0 ... 7.0 μm	400	0.5
2000s	Bi-CMOS	1.0 ... 3.0 μm	3,000	0.03
2000s	CMOS	~0.25 μm	100,000	0.001
Future	CMOS	~0.13 μm	200,000	0.0001

FIGURE 3.30
History of PWM chip, with technology evolution.

The history and technology evolution for this class of integrated circuits are explained in Figure 3.30. This is the most-used class of ICs for both non-isolated and isolated power supplies, featuring the entire controller in an 8-pin package.

Even though the core circuitry has been the same since 1976, the addition of more control and fault-management features, as well as the improvement of the circuit structure for basic functions, has evolved the concept:

- Technology of the 1980s – was introduced as UC3842 with a simple structure, 144 transistors in 7.5-mm technology, sold for $1.75 (at the time)
- Technology of the 1990s – features improvements, marketed as UCC3802, 478 transistors in 3.0-mm technology, sold for $0.85 (at the time)
- Technology of the 2000s – has more features, marketed as UCC38600, 1158 transistors in 0.5-mm technology, sold for $0.45

Modern implementation solutions for the control of DC/DC converters tend to transition towards digital solutions, either implemented in software with microcontrollers or in digital hardware like FPGA or ASIC.

Summary

Direct DC/DC converters without isolation are used to convert electrical energy from a DC source to a DC load with a different DC level requirement. Most DC converters have a topology based on a single switch. Buck or boost converters can operate in discontinuous or continuous conduction mode depending on the continuity of the current through the inductance. These converters and operation limits were analysed.

Other topologies include the buck/boost converter and the Cuk converter for converters with a single switch, or the H-bridge converter with bipolar operation. Synchronous converter operation is also used for efficiency improvement.

Modern telecom applications feature high currents and impose the use of multiphase converters. Multiphase converters are buck or boost converters with multiple similar branches (parallel hardware) and an interleaved control achieved with dedicated integrated circuits. A special case was given with synchronous converters, where the diode is replaced with another MOSFET transistor.

References

1. Erickson, R., Maksimovic, D., 2001, *"Fundamentals of Power Electronics First and Second Editions"*, Springer, New York.
2. Mohan, N., Undeland, T.M., Robbins, T.M., 2002, *"Power Electronics"*, Wiley, New York.
3. Luo, F.L., Ye, H., 2016, *"Advanced DC/DC Converters, Second Edition"*, Taylor and Francis, Boca Raton, FL, USA.
4. Chiang, C.Y., Chen, C.L., 2005, "Integrated Circuit Approach for Soft Switching in Boundary-Mode Buck Converter", in *31st Annual Conference of IEEE Industrial Electronics Society*, 6–10 November 2005, Raleigh, NC, USA, pp. 1–5.
5. Gumhalter, H., 1995, *"Power Supply in Telecommunications"*, Springer, Berlin.
6. Nowakowski, R., Tan, N., 2009, "Efficiency of Synchronous versus Non-Synchronous Buck Converters", *Texas Instruments Analog Applications Journal*, 4Q, 15–20.
7. Mößlacher, C., Guillemant, O., 2012, "Optimum MOSFET Selection for Synchronous Rectification", *Infineon Application Note AN 2012-05*, version 2.4, May 2012.
8. Anon, 2007, "MPC7448 RISC Microprocessor Hardware Specifications", *Freescale Semiconductor Document Number MPC7448EC*, Revision 4, March 2007.
9. Randall, M., Skamser, D., Kinard, T., Qazi, J., Tajuddin, A., Troller-McKinsky, S., Randall, C., Ko, S.W., Dechakupt, T., 2007, Thin Film MLCC, 2007 Electronic Components, Assemblies & Materials, Arlington, VA, USA, March 2007, vol. 1, pp. 1–12. Also at http://www.kemet.com/Lists/TechnicalArticles/Attachments/63/2007%20CARTS%20-%20Thin%20Film%20MLCC.pdf, accessed on 3 October 2017.
10. Ridley, R., 2000, "Second-Stage LC Filter Design", Switching Power Magazine, July 2000, 8–10. Also at http://ridleyengineering.com/images/phocadownload/1%20second%20stage%20filter%20design.pdf, accessed on 3 October 2017.
11. Neacsu, D., Butnicu, D., 2017, "A Review and Ultimate Solution for Output Filters for High-Power Low-Voltage DC/DC Converters", in *IEEE International Symposium on SCS 2017*, Iasi, Romania, pp. 1–5.
12. Shan, Z.Y., Tan, S.C., Tse, C.K., 2013, "Transient Mitigation of DC–DC Converters for High Output Current Slew Rate Applications", *IEEE Transactions on Power Electronics*, 28(5), 2377–2388.

13. Shan, Z., Tan, S.C., Tse, C.K., 2011, "Transient Mitigation of DC–DC Converters using an Auxiliary Switching Circuit", in *Proceedings of IEEE ECCE Conference*, 17–22 September 2011, Phoenix, AZ, USA, pp. 1259–1264.
14. Neacşu, D.O., 2010, "Towards an All-Semiconductor Power Converter Solution for the Appliance Market", in *IEEE IECON Conference 2010*, 7–10 November 2010, Glendale, AZ, USA, pp. 1677–1682.
15. Mammano, B., 2007, "A Historical Perspective of the Power Electronics Industry", Plenary session, *IEEE APEC 2007*.

4

Point-of-Load Converters and Their Feedback Control Systems

4.1 Context

Chapter 1 presented various solutions for the architecture of a power supply system for telecommunications equipment. The current chapter focuses on PoL converters, which are DC/DC converters without isolation used to directly and locally supply electronic loads. Figure 4.1 shows the placement of a point-of-load converter within a telecom power system architecture.

The main function of a point-of-load converter is to regulate the voltage supplied to a certain electronic circuit (load) in a nearby location. This way, the input voltage and the load current can vary over a wide range without impeding performance. The output voltage control, according to design requirements, is made with a feedback control system, and output voltage adjustment is called voltage regulation.

Design requirements are usually defined for steady-state operation around a biased operation point. Output voltage variation with important operation parameters is characterised by the following performance indices:

- Regulation coefficient (*dVout/dVin*)

$$\frac{1}{S_0} = \left(\frac{\partial V_2}{\partial V_1}\right)\Bigg|_{\substack{I_2 = const \\ \theta = const}} \cong \left(\frac{\Delta V_2}{\Delta V_1}\right)\Bigg|_{\substack{I_2 = const \\ \theta = const}} \tag{4.1}$$

- Internal (output) resistance (*dVout/dIout*)

$$R_i = -\left(\frac{\partial V_2}{\partial I_2}\right)\Bigg|_{\substack{V_1 = const \\ \theta = const}} \cong -\left(\frac{\Delta V_2}{\Delta I_2}\right)\Bigg|_{\substack{I_2 = const \\ \theta = const}} \tag{4.2}$$

- Temperature coefficient *(dVout/dT)*

$$K_\theta = \left(\frac{\partial V_2}{\partial \theta}\right)\Bigg|_{\substack{V_1 = const \\ I_2 = const}} \cong \left(\frac{\Delta V_2}{\Delta \theta}\right)\Bigg|_{\substack{V_1 = const \\ I_2 = const}} \tag{4.3}$$

Additionally, an important steady-state characteristic is the steady-state error. Dynamic characteristics, like transient response and regulation, are also defined for a power supply with measures like system stability, start-up time, bandwidth or several other performance indices derived from step response, like overshoot, rising time and stabilisation time. The latter are just an abstract way of defining performance used with frequency models, yet they are widely accepted.

Voltage regulator hardware is composed of either analogue control circuits or a combination of analogue and digital circuits. Their design is based on requirements expressed with the previous performance indices.

4.2 Implementation within Analogue-Mode Power Supply Circuits

Operation relies on the series connection of a variable resistance between the input and output (Figure 4.2). This resistance is able to take up the

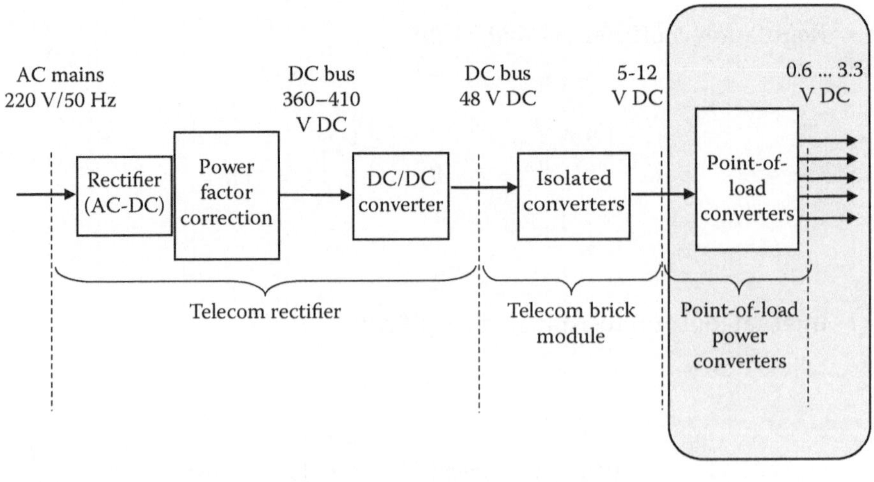

FIGURE 4.1
Place of point-of-load converters within the power system architecture.

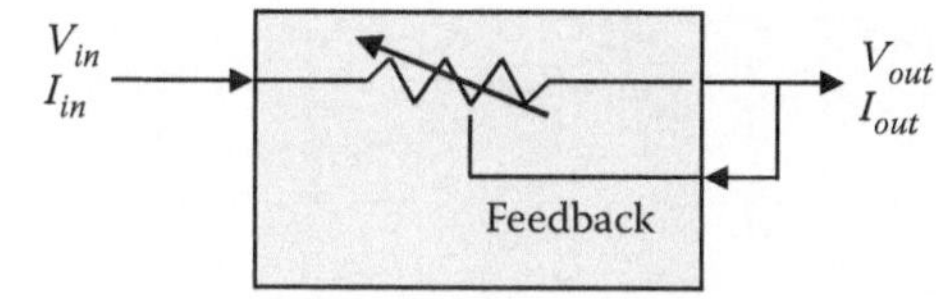

FIGURE 4.2
Principle of a point-of-load converter with analogue implementation.

instantaneous difference between the input voltage and the constant (fixed) output voltage. Obviously, this operation mode leads to a very large power loss.

The actual circuit emulates the series resistance with a power transistor, often called a *series regulator element* (SRE) or *pass device.* The collector-emitter voltage for this SRE takes over the large voltage drop (Figure 4.3). The entire voltage source converter has just three terminals for drop-in circuit connection. The most-used principal circuitry is shown in Figure 4.4.

The output voltage V_{out} is measured with a resistive divider with R_1 and R_2. The internal voltage reference can be carried out with

- a Zener diode biased with a current source when using a discrete implementation or
- through a band-gap reference (usually ~1.22 V) when using an integrated circuit.

The error amplifier compares a measure of the output voltage with the reference voltage and determines the operation point for the SRE. If the output voltage is too large, the SRE operates with a larger collector-emitter voltage drop, which in turn will reduce the output voltage. If the output voltage is too small, the SRE needs to operate with a reduced collector-emitter voltage drop so that the output voltage increases.

An example of an actual circuit is provided in Figure 4.5.

The output voltage is sensed with resistive divider R_1-R_2, while the voltage reference is provided by Zener diode Z_1. The diode Z_1 is biased through the resistance R_z, and this could be tied to the input terminal as well. The sense and reference voltages are compared on the base-emitter junction of the control transistor Q_{cntrl}. If the error voltage increases on the base-emitter junction, the collector current of Q_{cntrl} increases and the collector-emitter voltage

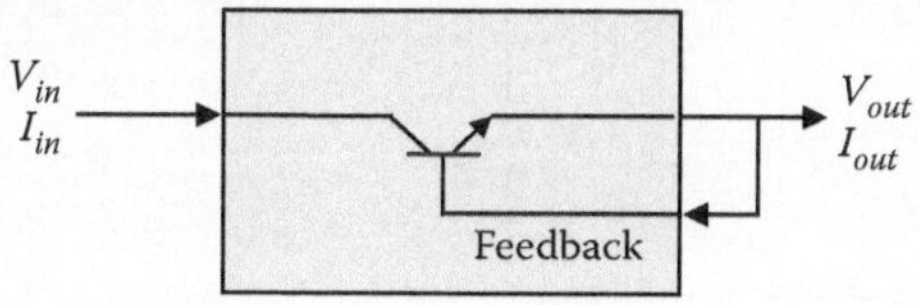

FIGURE 4.3
Implementation with a power transistor.

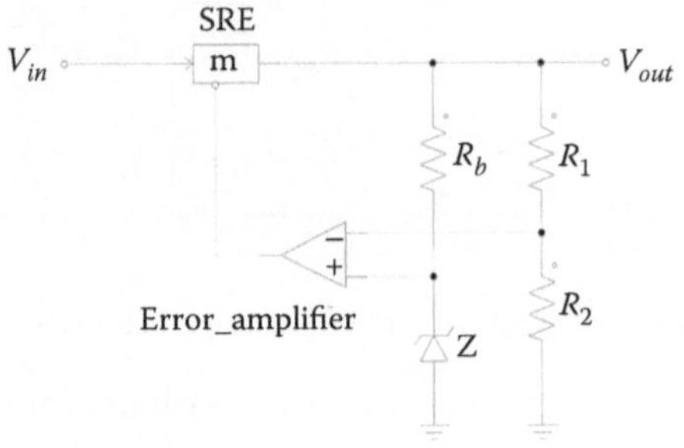

FIGURE 4.4
Typical circuitry for an analogue-mode power supply.

for the series transistor decreases. The base current for the series transistor Q_{ers} increases, as does the base-emitter voltage of Q_{ers}. The collector–emitter voltage across Q_{ers} decreases; thus, the load voltage also decreases. The R_p resistance has a role in circuit startup through initial polarisation of the transistor Q_{ers}.

The series regulator element is the most important component with the largest energy loss. Its main parameter is the dropout voltage. It is very economical to bring the input voltage as close as possible to the output voltage in order to reduce the voltage drop across the pass device. The source will reject large ripples and occasional variation in input voltage so that the output is fairly regulated. Such operation is referred to as a low-dropout voltage circuit.

Possible circuit solutions for the series regulator element are shown in Figure 4.6, and a comparison of performance is provided in Table 4.1. This comparison outlines certain conflicting constraints: operation with low-dropout voltage can be achieved by inverting topologies, which, however, can pose stability problems. This will be explained in detail later in this chapter.

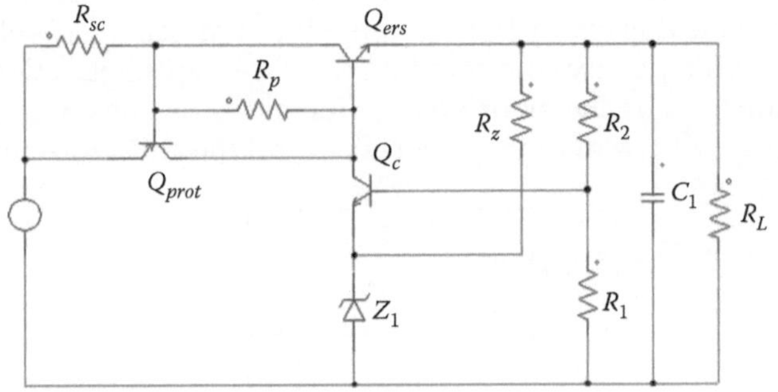

FIGURE 4.5
Actual complete circuit for an analogue power supply.

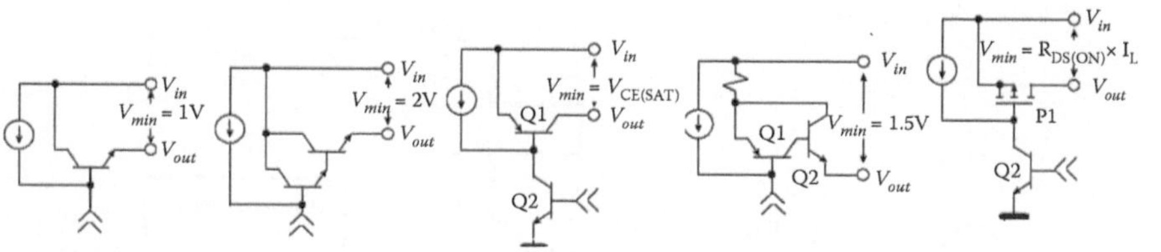

FIGURE 4.6
Options for the SRE: Q1 and Q2 are bipolar transistors, and P1 is a MOS transistor.

TABLE 4.1
Performance Comparison (CC = Common Collector, and CE = Common Emitter)

	(a) NPN	(b) Darlington NPN	(c) PNP	(d) PNP/ NPN	(e) PMOS (External)
Min. dropout voltage	1V	2V	0.1V	1.5V	Rdson* I_L
Current	<1A	>1A	<1A	>1A	>1A
Topology	follower, CC	follower, CC	inverting, CE	inverting, CE	inverting, CE
Output impedance	small	small	large	large	large
Bandwidth	large	large	small	small	small
Output capacitance for stability?	no	no	yes	yes	yes

4.3 Design of Feedback Control Systems

4.3.1 Definitions

Previous chapters have mentioned the need for operation with feedback control (closed-loop systems) of the output voltage without presenting details concerning the design and implementation of the control systems. A circuit able to adjust and stabilise the output voltage is called an automatic control system [1,2]. The generic structure of a control system is shown in Figure 4.7. For instance, the feedback amplifiers are a specific case.

The electronic circuit which implements the frequency compensation law *D*(*s*) is called the controller. The implementation with linear electronic circuits (amplifiers) defines a linear control. There are multiple ways to design a feedback control system. Among them, design with the frequency analysis method (also called Bode diagrams) is considered in this chapter, while the next chapter uses state-space–based design [3,4]. The design reduces to the definition of the control law *D*(*s*) able to satisfy a set of dynamic requirements along with securing the power supply stability.

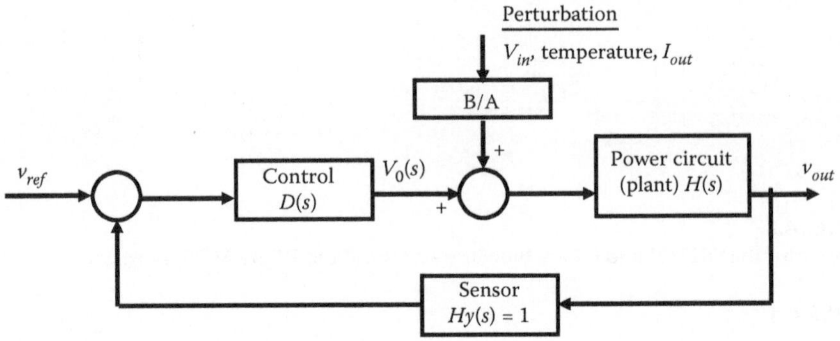

FIGURE 4.7
Principles of an automatic control system.

In order to establish the compensation law $D(s)$, a small-signal model of the converter is first required. Small-signal modelling [5] is a very common technique used in electronics to approximate the dynamic behaviour of a circuit around a known bias operation point. This is possible if the dynamic (AC) component of the signals has a magnitude considerably smaller than the bias (DC) component of the signal. A small-signal model is inherently linear, and it can approximate nonlinear systems with linear systems for small signals around a bias operation point. When applied to power converters, the small-signal model allows the description of system dynamics around an operation point defined with expected quasiconstant input voltage, output voltage, duty cycle and load current.

A system is stable if the linear conditions converge to zero (the transient response decreases) and unstable if the output becomes divergent in time. Furthermore, a linear and time-invariant system is stable if all the denominator roots (also called poles) have a negative real component. Another form of the same definition says that all systems' poles are placed within the left side of the complex plane.

If the dynamics of the power converter ('plant' in control systems theory terms) is described with a Laplace transfer function $G(s)$ and the compensation law is described with $D(s)$, the closed loop transfer function, after a negative feedback, is:

$$H_{cl}(s) = \frac{D(s) \cdot G(s)}{1 + D(s) \cdot G(s)} \tag{4.4}$$

This becomes unstable if the denominator equals zero. Hence, the stability is secured when:

$$1 + D(s) \cdot G(s) \neq 0 \Leftrightarrow D(s) \cdot G(s) \neq -1 \tag{4.5}$$

Since the *s*-domain (also called Laplace) is based on a two-dimensional complex plane, Equation (4.5) is false when the magnitude equals unity and the phase equals −180°; the latter is for the negative sign.

Since the frequency small-signal model for the power supply can be defined somewhat approximately, considering a possible variation of the parameters, the measurement (definition) of the systems' margins towards instability becomes very important. The Bode diagrams can thus define a *gain margin* and a *phase margin* as the distances to the instability constraint from Equation 4.5 (see Figure 4.8).

The gain margin is the distance from unity gain to the gain measured at the frequency where the phase is −180°, while the phase margin is measured as the phase distance to 180° when the gain is unity; that is, 0 dB. The gain margin shows how much the gain can be increased until the instability condition. The phase margin shows how much the phase can be decreased before the instability condition. Basically, one condition from Equation 4.5 is fixed and the other is estimated as the distance to instability.

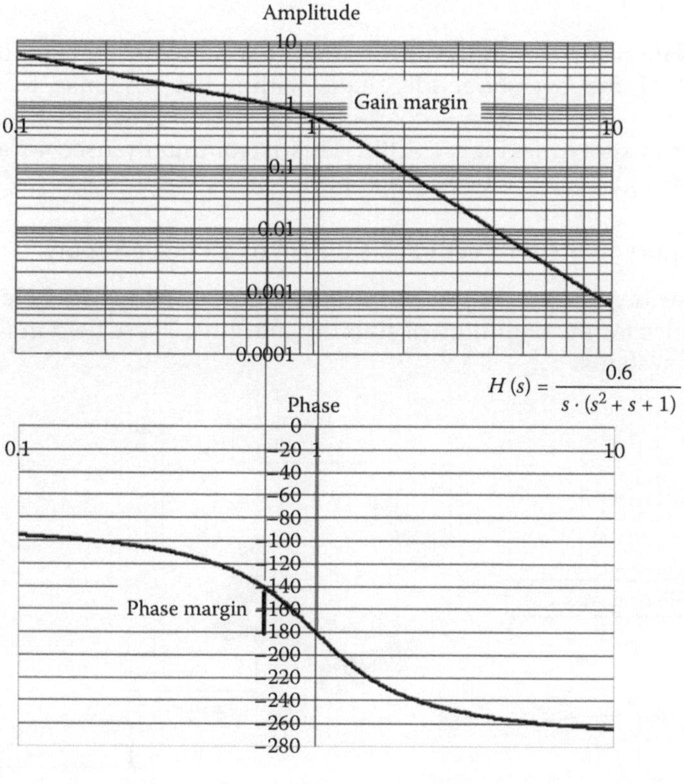

FIGURE 4.8
Definition of phase and gain margins.

Aside from establishing a safety margin to stability, the phase margin also corresponds to details of the system's transient response. In most cases, the overall system may be approximated with a second-order dominant system, and design rules are established for simplicity on this basis for second-order systems.

For a second-order system, a direct relationship between phase margin, attenuation and overshoot exists, as is explained in Figure 4.9. A system with unity feedback yields:

$$\begin{cases} open\_loop & G(s)\cdot D(s) = \dfrac{\omega_n^2}{s\cdot(s+2\cdot\xi\cdot\omega_n)} \\ closed\_loop & T(s) = \dfrac{\omega_n^2}{s^2+2\cdot\xi\cdot\omega_n\cdot s+\omega_n^2} \end{cases} \tag{4.6}$$

This determines a phase margin of (given herein without demonstration):

$$PM = \tan^{-1}\left(\frac{2\cdot\xi}{\sqrt{\sqrt{1+4\cdot\xi^4}-2\cdot\xi^2}}\right) \tag{4.7}$$

This relationship is plotted in Figure 4.10a and can be combined with the result of the overshoot identified within the response to an input step variation applied to a second-order system, plotted in Figure 4.10b. The result is shown in Figure 4.10c. This is commonly used as a starting point for design of the compensation law.

4.3.2 Requirements for Feedback Control of a Power Supply

The similarities encountered with various DC power supplies helped set up general rules for the definition of stability margins. These help in the design of control systems.

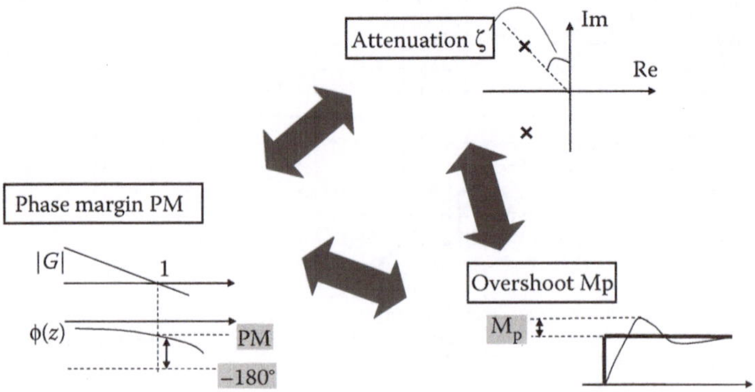

FIGURE 4.9
Relationship between phase margin, attenuation and overshoot for a second-order system.

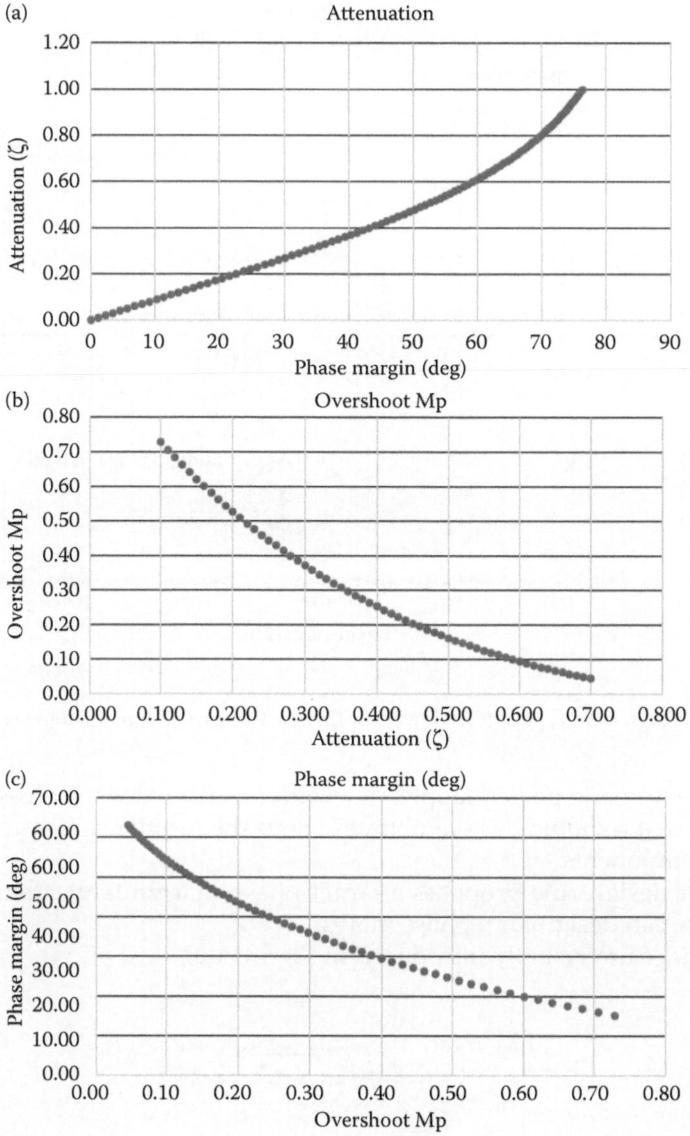

FIGURE 4.10
Dependency within a second-order system.

There are three rules usually employed at designing the compensation (Figure 4.11). A large DC gain at zero frequency is desired in order to reduce stationary (also called steady-state) errors. This can be achieved with an integral element (like a capacitor) placed within the compensation law.

A large crossover frequency is required in order to achieve a fast response. This is, however, sought for at least under 1/5 of the switching frequency

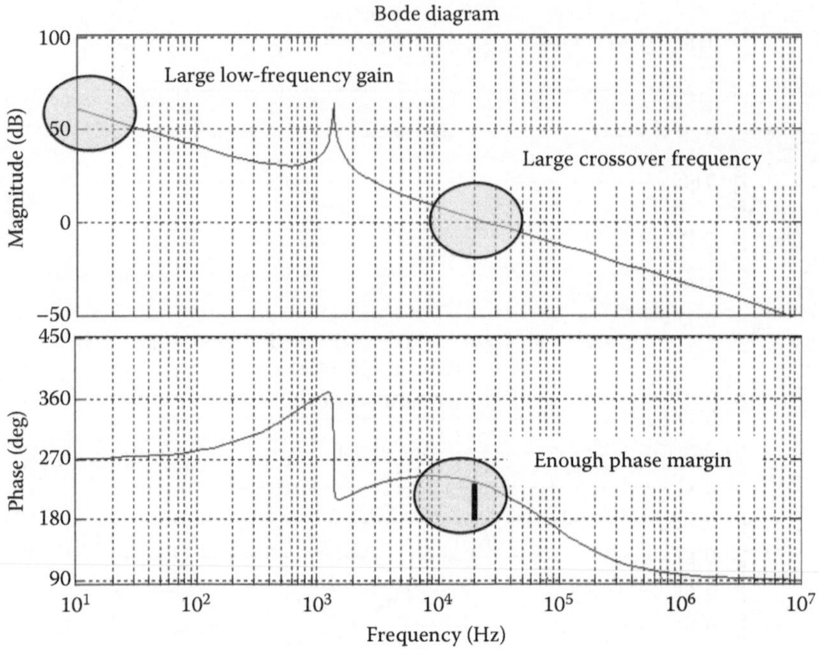

FIGURE 4.11
Example of design rules typical for a power supply; the three rules explained in text are herein encircled.

of the switch-mode power supply. A second condition for the crossover frequency is that it should be several times above the resonance frequency from passive components.

Another design rule proposes a target phase margin larger than at least 45°, but we can design for a phase margin of 60°.

These rules are visually introduced in Figure 4.10.

$$PM = \tan^{-1}\left(\frac{2\cdot\xi}{\sqrt{\sqrt{1+4\cdot\xi^4}-2\cdot\xi^2}}\right)$$

$$M_p = e^{-\frac{\pi\cdot\zeta}{\sqrt{1-\zeta^2}}}$$

4.3.3 Using Lead-Lag Compensators

When the gain of the converter's small-signal model at zero frequency is not enough for steady-state reduction, a proportional-integrator (PI) control:

$$D(s) = \frac{K_g}{s}\cdot\left(s+\frac{1}{T_I}\right) \tag{4.8}$$

or a lag compensator:

$$D(s) = K \cdot \frac{T \cdot s + 1}{K \cdot T \cdot s + 1} \quad K > 1 \tag{4.9}$$

is used. While the PI control introduces a pole at the origin and reduces the steady-state error to zero, the lag compensator works with a zero-pole pair, where the pole is at low frequency and the zero at high frequency (Figure 4.12). Steady-state error reduction now depends on the zero-frequency gain of the compensator.

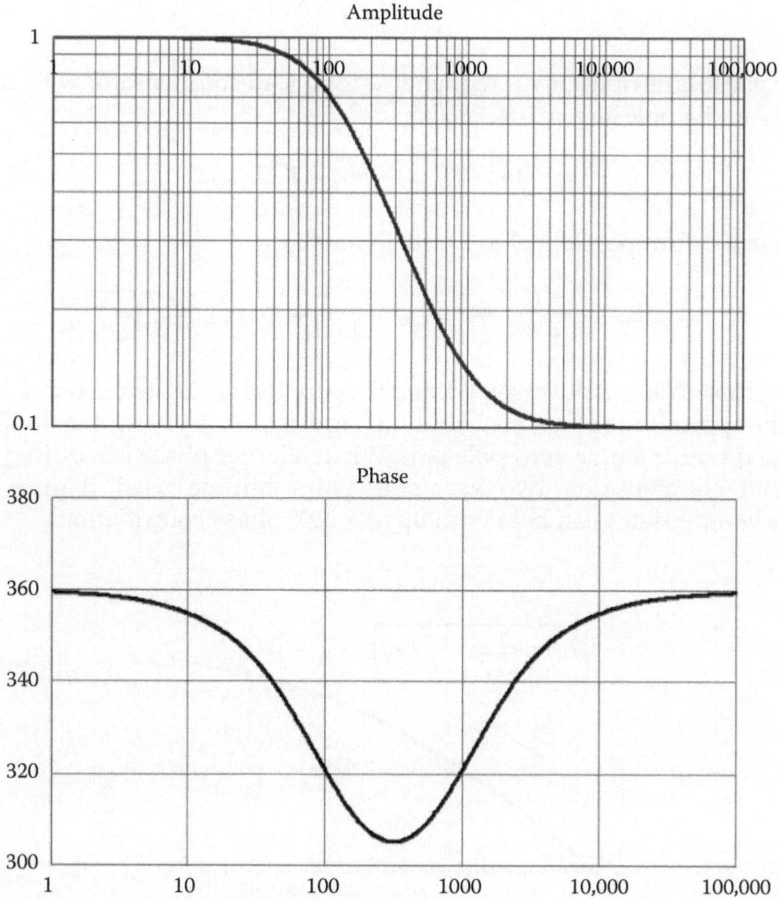

FIGURE 4.12
Example of lag compensation with a zero placed at 1,000 rad/sec and a pole located at 100 rad/sec.

The amount of gain introduced at low frequency equals K_g in Equation 4.9. The design objective herein is the addition of gain at low frequencies for improvement of the steady-state response without compromising dynamic performance. Between the frequencies of zero and pole, an important phase lag can be seen (Figure 4.12), hence the name.

If the phase margin derived from the design requirements is not what we already have for the converter model, we need to add the necessary phase difference with a Proportional Derivator (PD) control:

$$D(s) = (T_D \cdot s + 1) \tag{4.10}$$

or a lead compensator:

$$D(s) = \frac{T \cdot s + 1}{K \cdot T \cdot s + 1} \quad K < 1 \tag{4.11}$$

The phase contribution at frequency ω for a lead compensator with zero at $z = 1/T$ and a pole at $p = 1/(K \cdot T)$ is:

$$\varphi = \tan^{-1}(T \cdot \omega) - \tan^{-1}(K \cdot T \cdot \omega) \tag{4.12}$$

The maximum possible phase contribution is:

$$\sin \varphi_{max} = \frac{1-K}{1+K} \quad @ \quad \omega_{max} = \frac{1}{T \cdot \sqrt{K}} \Leftrightarrow \omega_{max} = \sqrt{|z| \cdot |p|} \tag{4.13}$$

The problem is usually reversed, and the gain K is calculated in order to provide the appropriate phase contribution (Figure 4.13). A phase of up to 60° can be added with a single zero-pole pair. When a larger phase is required from the lead compensator, two zero-pole pairs can be used (Figure 4.14). Such a compensator can help with up to a 120° phase contribution.

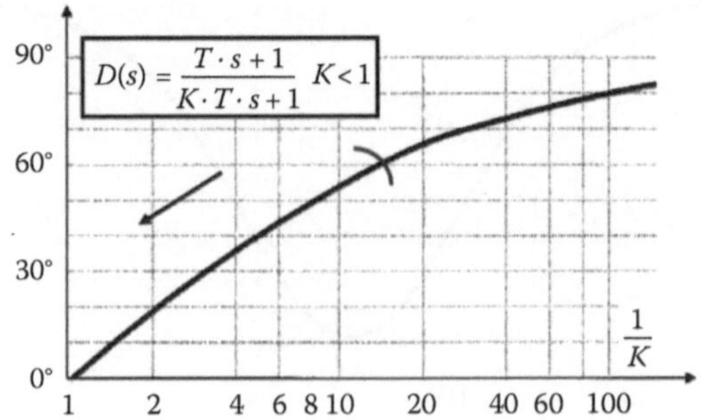

FIGURE 4.13
Design curve for selection of K to provide certain phase contribution.

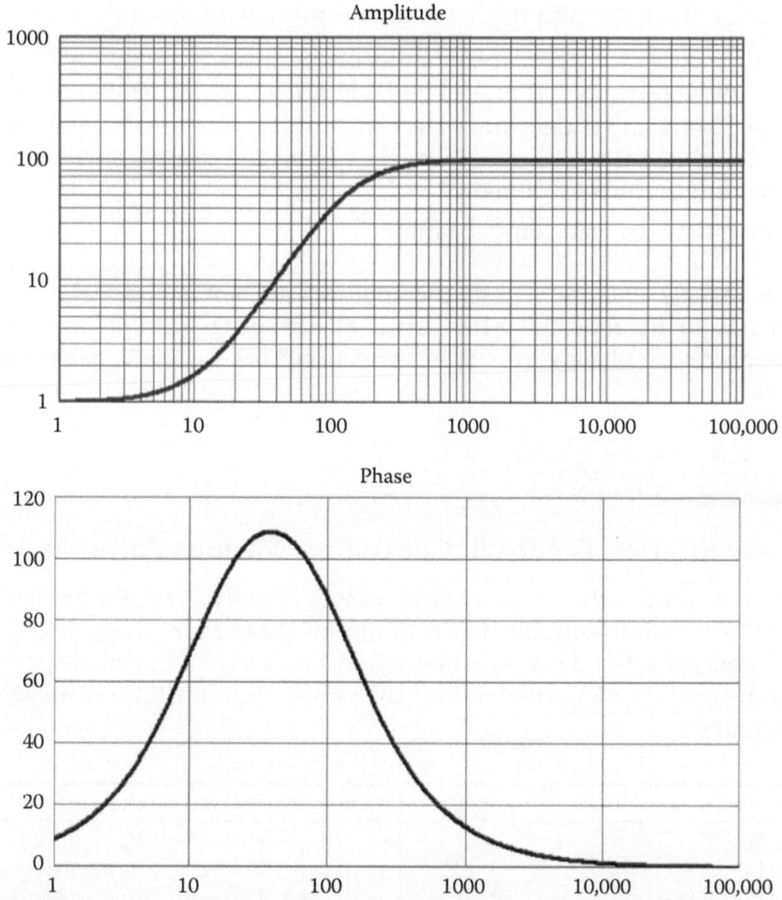

FIGURE 4.14
Double lead compensator with zeros at 10 and 15 Hz and poles at 100 and 150 Hz. A maximal phase contribution of 110° can be observed at a frequency of 38.72 Hz.

The design steps are:

- Define the desired low-frequency gain and phase margin from the design requirements for the power supply.
- Calculate the low-frequency gain and estimate the phase margin for the converter model.
- Define K_g as a ratio between the desired gain and existing gain in the converter model.
- Define the frequency of zero ($1/T$) from the lag compensator law to be at least a decade below the crossover frequency.

- Define the pole from the lag compensator law to be equal to $\omega = 1/(K_g \cdot T)$.
- Formulate the lag compensation from zero, pole and gain.
- Use Figure 4.13 to determine the constant K from the lead compensator, and use the crossover frequency to sit the lead compensator in the right frequency domain.
- Formulate the lead compensation.

Since various choices within the method are somewhat arbitrary, the design has to be repeated often until all design requirements are met. A more precise analysis is provided in Section 4.6.2. for the Venable K-factor method.

4.4 Case Studies: Feedback Control for Various Power Supplies

Figure 4.15 shows the most typical power supply configurations. Most point-of-load power supplies fall into one of these four categories. Design of their compensation laws is explained in this section. In this respect, first the small-signal model is defined for the power circuitry (in control systems terms, 'plant').

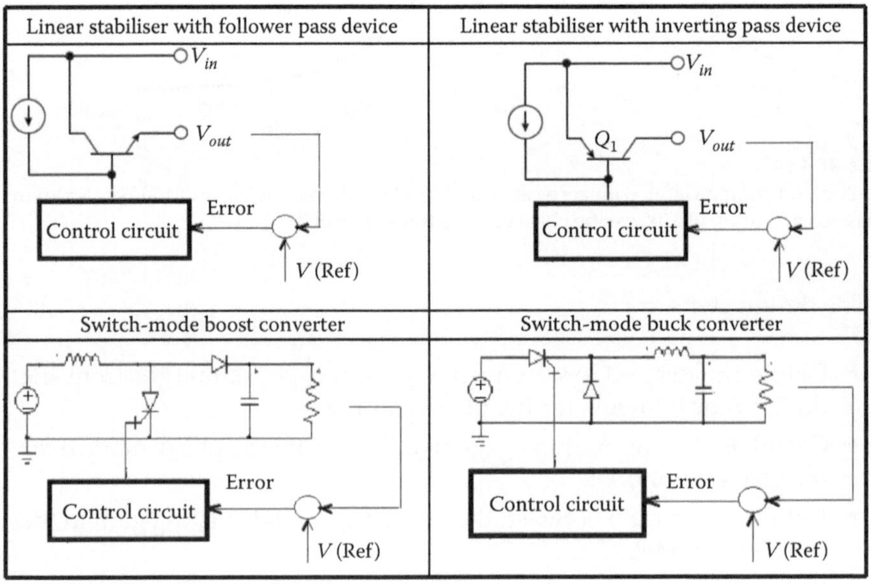

FIGURE 4.15
Typical power supply circuits considered as design examples.

4.4.1 Example 1 = Linear Circuitry, Follower (NPN Power Transistor with Common Collector)

Stability and dynamic performance analyses require the calculation of the power circuit's transfer function. The electrical circuit shown in Figure 4.16 can be translated into the small-signal circuit of Figure 4.17, represented further with the equivalent circuit from Figure 4.18. The latter can be described with the equation for the transfer function.

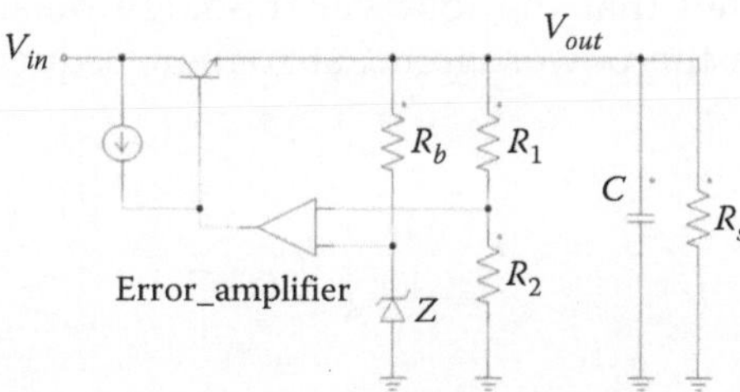

FIGURE 4.16
Circuit for the linear circuitry and follower (NPN power transistor with common collector).

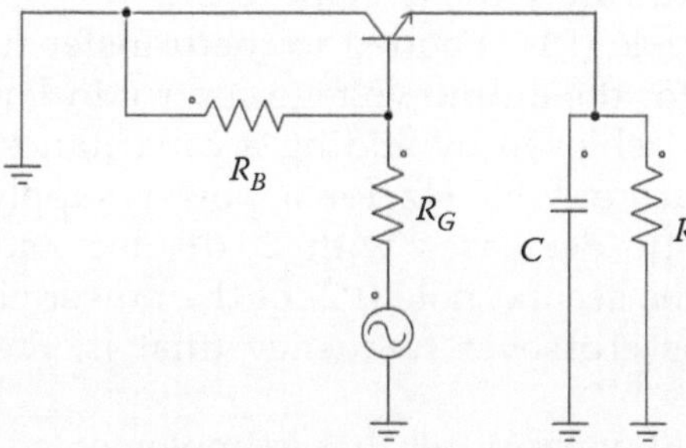

FIGURE 4.17
Small-signal circuit.

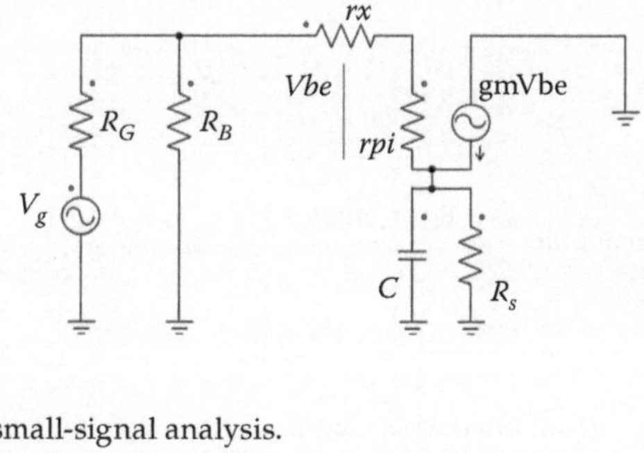

FIGURE 4.18
Equivalent circuit for small-signal analysis.

$$\begin{cases} A_{vi}(s) = \dfrac{V_o}{V_i} = \dfrac{(\beta+1)\cdot(C_s\,||\,R_s)}{r_\pi + r_x + (\beta+1)\cdot(C_s\,||\,R_s)} = \dfrac{(\beta+1)\cdot R_s}{(r_\pi + r_x)\cdot(1+s\cdot R_s\cdot C_s) + (\beta+1)\cdot R_s} \\ (C_s\,||\,R_s) = \dfrac{R_s}{1+s\cdot R_s\cdot C_s} \end{cases}$$

$$A_{vi}(s) = \frac{(\beta+1)\cdot R_s}{s\cdot(r_\pi + r_x)\cdot(R_s\cdot C_s) + [r_\pi + r_x + (\beta+1)\cdot R_s]} \tag{4.14}$$

It can be concluded that the follower topology determines a displacement of the pole in the power circuit at higher frequencies; that is, from [1/RsCs] to

$$\left[1+\frac{(\beta+1)\cdot R_s}{(r_\pi + r_x)}\right]\cdot\frac{1}{R_s\cdot C_s} \tag{4.15}$$

The generally accepted solution for compensation implies a capacitor within the feedback path of the error amplifier. Due to the location of this capacitor, it can also be implemented internally in the integrated circuit supporting the power supply. The circuit is as shown in Figure 4.19, while the Bode characteristics are shown in Figure 4.20.

A low-frequency pole (P1) is added to the transfer function which was previously derived for the output voltage control in Equation 4.8. Adding a pole is practically achieved by adding a capacitance. Usually, the pole (P1) is considered around 100 Hz for a power supply. The closed-loop system's characteristic decreases with 20 dB/dec and a phase around −90°. The effect of the second pole (P2) of the power circuitry is at larger frequencies than the crossover frequency (that is, the frequency where Gain = 1 or 0 dB).

This is a very well-known result that is implemented within many integrated circuits. An example of such a classical integrated circuit (the *first*

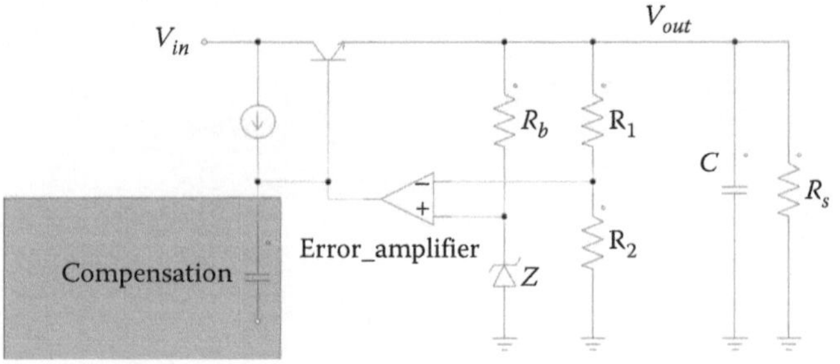

FIGURE 4.19
Compensation circuitry with an internal capacitor.

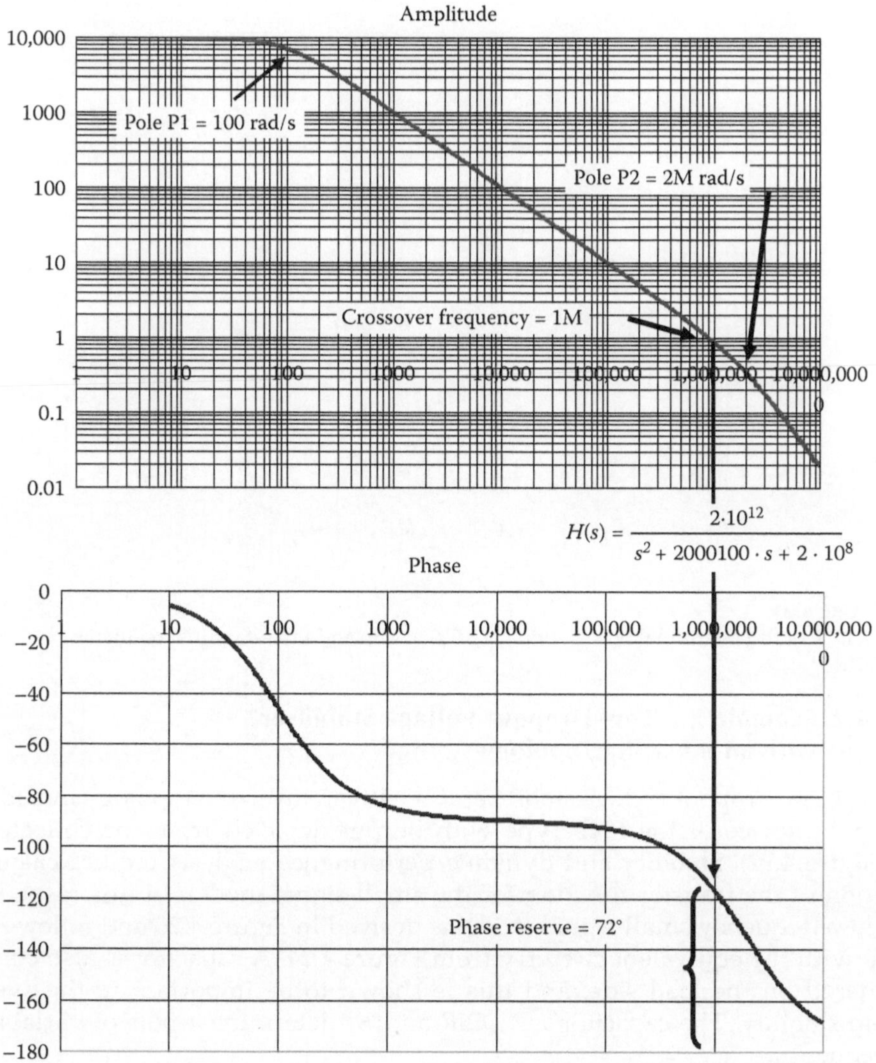

FIGURE 4.20
Bode characteristics after capacitor compensation.

integrated solution) is provided in Figure 4.21, and it is based on a pass device in a follower circuit. Note the internal compensation with C_1. For a practical example, the integrated circuit LM309 is a voltage stabiliser with three terminals for output at 5V/1A. This is not a high-performance circuit, as it requires a large operating current I_{ground} of around 5mA.

Other similar circuits belong to the series 780× (x = output voltage 7,803, 7,805, 7,812, …).

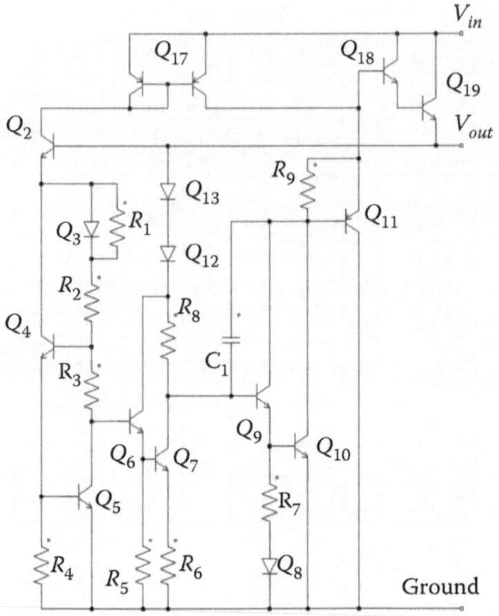

FIGURE 4.21
Example of three-terminal power supply IC, 5V/1 Amp and three-terminal regulator.

4.4.2 Example 2 = Low-Dropout Voltage Stabiliser with an Inverting Topology

Most low-dropout voltage stabiliser ICs with an inverting topology include a pass device of the PNP type with output depicted from the collector (Figure 4.22). Stability and dynamic performance analysis require calculation of the transfer function for the small-signal model. In this respect, a low-frequency small-signal model is derived in Figure 4.23 and followed up with the equivalent circuitry from Figure 4.24. A capacitor is also considered on the load side, and this is shown to be important to the system stability. The capacitor's C_L, ESR and R_L determine a pole of variable frequency.

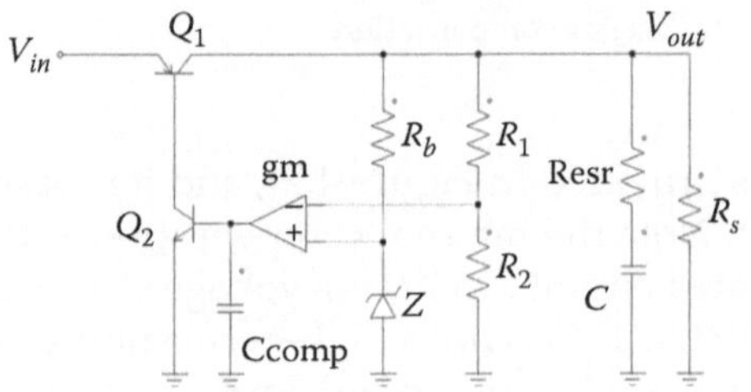

FIGURE 4.22
LDO circuit with an inverting topology.

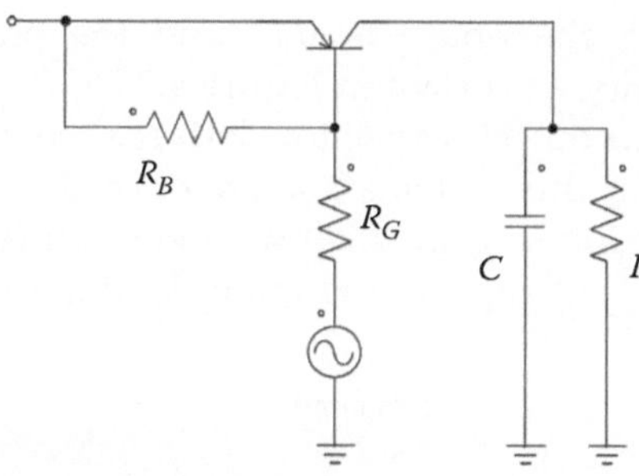

FIGURE 4.23
Small-signal model of the LDO stabiliser with an inverting topology.

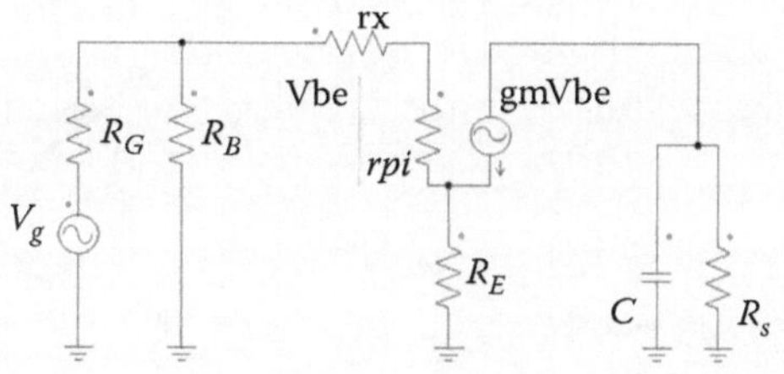

FIGURE 4.24
Equivalent circuit for the LDO stabiliser with an inverting topology.

After neglecting r_x and R_E from the model, the transfer function yields:

$$\begin{cases} |A_{vi}(s)| = \dfrac{V_o}{V_i} = g_m \cdot (C_s \parallel R_S) = \dfrac{g_m \cdot R_s}{1 + s \cdot R_s \cdot C_s} \\ (C_s \parallel R_S) = \dfrac{R_s}{1 + s \cdot R_s \cdot C_s} \end{cases} \quad (4.16)$$

Inverting the transistor connection determines a pole with variable frequency in the power circuit at relatively low frequencies. Fortunately, the capacitors are not ideal and we have a zero produced with the series connection of the capacitor's ESR resistance.

$$|A_{vi}(s)| = \frac{V_o}{V_i} = g_m \cdot (X_s \parallel R_s) = g_m \cdot \frac{\left(R_{esr} + \dfrac{1}{s \cdot C_s}\right) \cdot R_s}{R_{esr} + \dfrac{1}{s \cdot C_s} + R_s} = \frac{g_m \cdot R_s \cdot (1 + s \cdot R_{esr} \cdot C_s)}{1 + s \cdot C_s \cdot (R_{esr} + R_s)}$$

(4.17)

If a simple compensation with a single pole P1 (capacitor) is used along this power supply already having a pole PL at low frequencies due to the ideal output capacitor, the phase is 180° and the phase margin is zero. This produces instability, as shown in Figure 4.25.

When considering an output capacitor with real characteristics with loss, this produces a zero in the system. The effect of this zero can be seen on the loop transfer function (Figure 4.26) with enough phase margin. Hence, the selection of the output capacitor is made for stability of the closed-loop

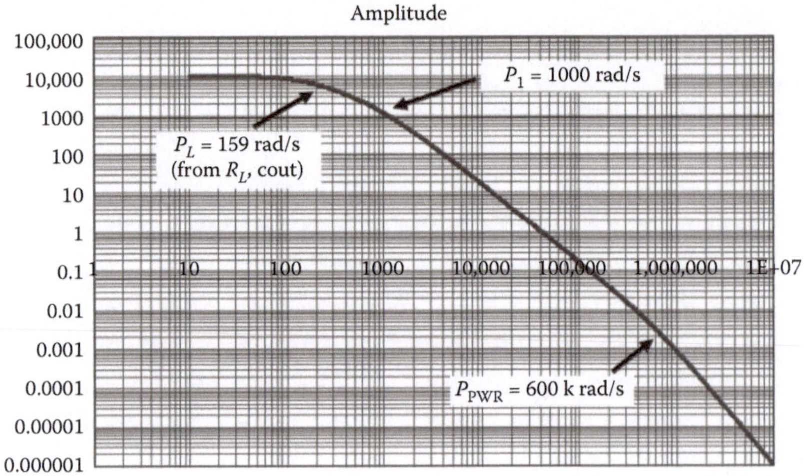

$$H(s) = \frac{10^{15}}{s^3 + 601159 \cdot s^2 + 695559000 \cdot s + 95400000000}$$

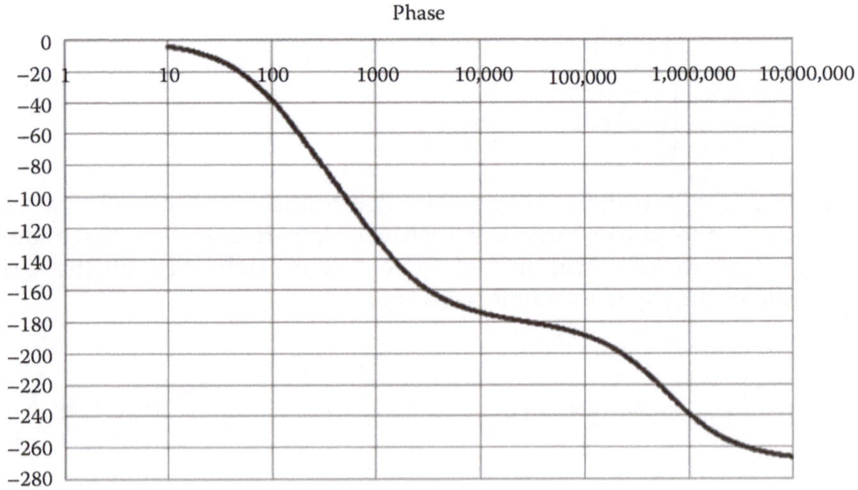

FIGURE 4.25
Instability with a conventional control system applied to an LDO stabiliser, with load $R_L = 100\ \Omega$, $C_{out} = 10\ \mu F$.

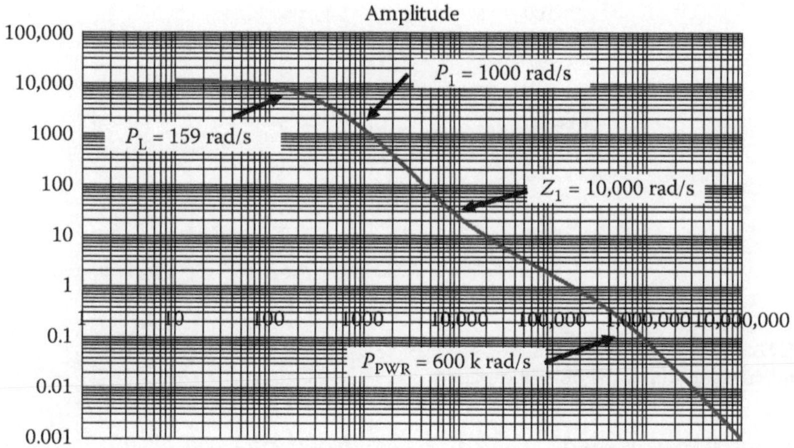

$$H(s) = \frac{10^{11} \cdot (s + 10^4)}{s^3 + 601159 \cdot s^2 + 695559000 \cdot s + 95400000000}$$

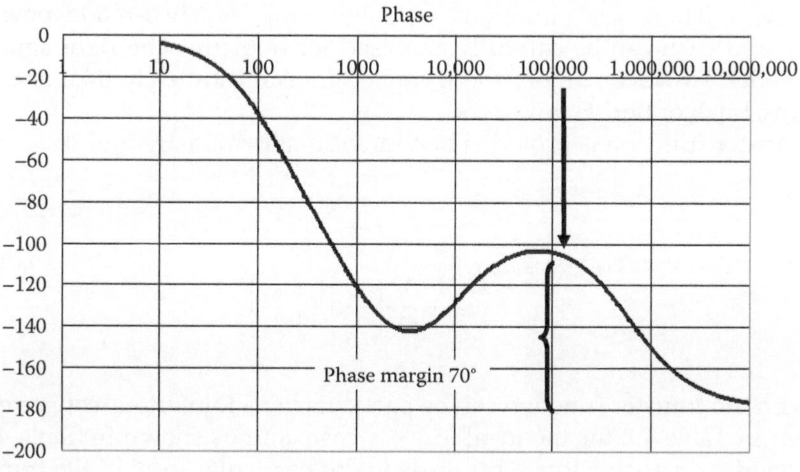

FIGURE 4.26
Stability achieved with a proper output capacitor.

system, and a finite ESR is welcome since it adds phase to the overall characteristics.

The range of design constraints is shown in Figure 4.27. It can be seen that a suitable ESR can secure stability for any load current.

Higher-order transfer functions of the compensation circuits can be used with the goal of adding more phase at the crossover frequency and to improve stability. Such a circuit for implementation of the compensation law needs to have one or more zeros within the transfer function in order to add phase into the feedback loop.

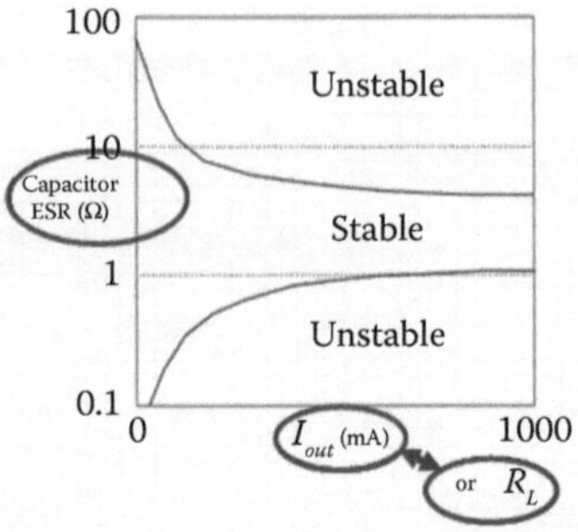

FIGURE 4.27
Constraints in selection of the output capacitor.

4.4.3 Example 3 = Switched-Mode Boost/Buck Converters

For a switched-mode converter, the system's transfer function is calculated from a variation of the control duty cycle (D) to a variation of the output voltage (V_{out}). There are numerous topologies for switched-mode power supplies, and each can be carefully analysed for depicting the right small-signal model. However, any of these converters can ultimately take one of two forms: buck or boost converters.

The transfer function is provided herein without demonstration:

$$G_{d-v}(s) = G_{d0} \cdot \frac{\left(1 - \frac{s}{\omega_0}\right)}{1 + \frac{1}{Q} \cdot \left(\frac{s}{\omega_0}\right) + \left(\frac{s}{\omega_0}\right)^2} \tag{4.18}$$

This generic transfer function can be particularised for various converters as shown in Table 4.2 for the ideal lossless case, and as shown in Table 4.3 when including resistive loss. This leads to the particular form of the transfer function in Table 4.4 [6]. It can be seen that the major difference between

TABLE 4.2
Details of Transfer Functions for Various Converters, with Ideal Output Capacitor

Converter	G_{do}	ω_0	Q	ω_z
Buck	$\frac{V_{in}}{D}$	$\frac{1}{\sqrt{L \cdot C}}$	$\frac{R}{\sqrt{L/C}}$	∞
Boost	$\frac{V_{in}}{1-D}$	$\frac{1-D}{\sqrt{L \cdot C}}$	$\frac{(1-D) \cdot R}{\sqrt{L/C}}$	$\frac{(1-D)^2 \cdot R}{L}$
Buck–boost	$\frac{V_{in}}{D \cdot (1-D)^2}$	$\frac{1-D}{\sqrt{L \cdot C}}$	$\frac{(1-D) \cdot R}{\sqrt{L/C}}$	$\frac{(1-D)^2 \cdot R}{D \cdot L}$

TABLE 4.3

Details of Transfer Functions for Various Converters, with Ideal Output Capacitor: R = Load Resistance, (L, r_L) = Inductance, (C, r_{ESR}) = Capacitance

Converter	G_{do}	ω_0	Q	ω_z
Buck	$\frac{V_{in}}{D}\cdot\frac{R}{R+r_L}$	$\frac{1}{\sqrt{L\cdot C\cdot\frac{R+r_{ESR}}{R+r_L}}}$	$\frac{1}{\frac{\sqrt{L/C}}{r_L+R}+\frac{r_C+(r_L\,\|\,R)}{\sqrt{L/C}}}$	$\frac{1}{r_{ESR}\cdot C}$
Boost	$\frac{V_{in}}{1-D}$	$\frac{1}{\sqrt{L\cdot C}}\cdot(1-D)$	$\frac{1}{\frac{r_L}{L}+\frac{1}{(r_{ESR}+R)\cdot C}}\cdot\frac{1}{\sqrt{L\cdot C}}\cdot(1-D)$	$\frac{1}{r_{ESR}\cdot C}$, ω_p
Buck–boost	$\frac{V_o}{1-D}\cdot\frac{1}{R+r_{ESR}}$	$\sqrt{\frac{r_L+(1-D)^2\cdot R}{L\cdot C\cdot(R+r_{ESR})}}$	$\frac{\sqrt{L\cdot C\cdot(R+r_{ESR})\cdot\left(r_L+(1-D)^2\cdot R\right)}}{C\cdot\left(r_L\cdot(R+r_{ESR})+(1-D)^2\cdot R\cdot r_{ESR}\right)+L}$	$\frac{1}{r_{ESR}\cdot C}$, ω_p

TABLE 4.4

Actual Converter Transfer Functions

Boost Converter	Buck Converter
$G_{d-v}=G_{d0}\cdot\frac{\left(1+\frac{s}{\omega_z}\right)\cdot\left(1-\frac{s}{\omega_p}\right)}{1+\frac{s}{Q\cdot\omega_0}+\left(\frac{s}{\omega_0}\right)^2}$	$G_{d-v}=G_{d0}\cdot\frac{1+\frac{s}{\omega_z}}{1+\frac{s}{Q\cdot\omega_0}+\left(\frac{s}{\omega_0}\right)^2}$

the buck converter and the boost converter is the presence of a positive real number zero located within the right-half plane (RHP) of the complex plane.

The significance of a zero within the RHP is illustrated in Figure 4.28. The step response starts off in the wrong direction, then recovers towards the desired steady-state value. An RHP zero may produce phase reversal at higher frequencies, which is a possible issue for system stability.

Compensation is achieved by adding phase at the frequency of interest through the compensation law. A structured presentation of the compensation options for analogue implementation is included next.

4.5 Analogue-Mode Feedback Control Solutions

Within DC/DC power supplies, control is implemented with linear operational amplifiers and a passive compensation network. In order to simplify the controller structure, only certain control topologies (compensation laws) are selected for implementation with linear analogue circuits. These are repetitively used within various power supplies; therefore,

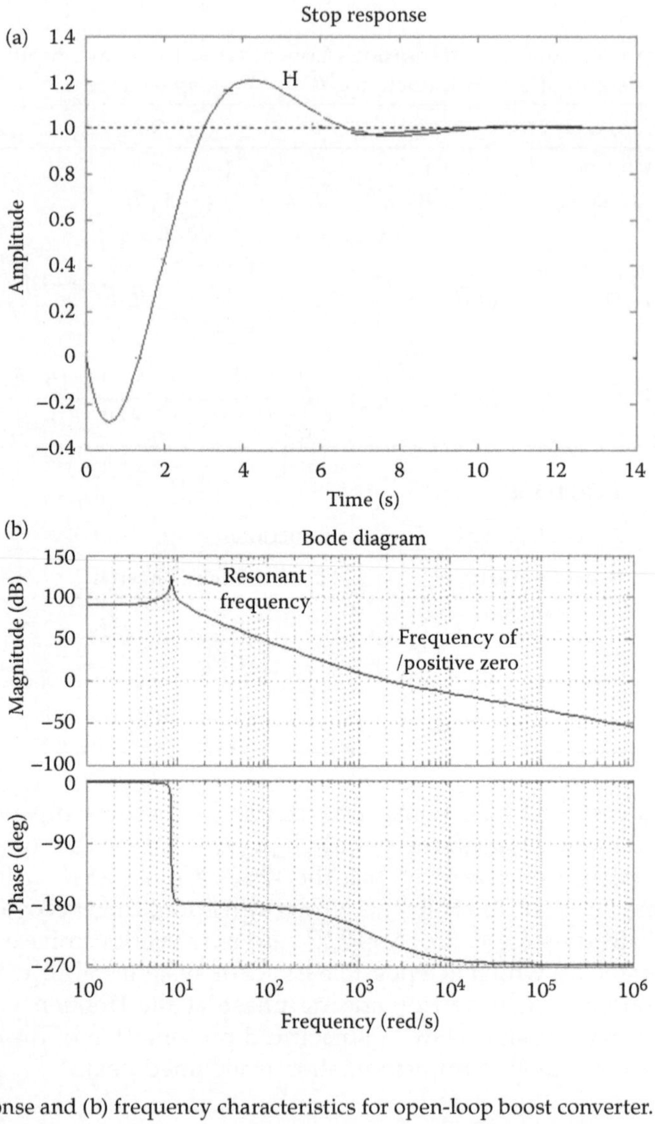

FIGURE 4.28
(a) Step response and (b) frequency characteristics for open-loop boost converter.

classifications and a structured approach are recommended for understanding them. Their classification and names come from the number of capacitors within the passive feedback network.

4.5.1 Type I Compensation

The circuitry is shown in Figure 4.29, and the characteristics of this compensation are shown in Figure 4.30. The transfer function for the compensation network is:

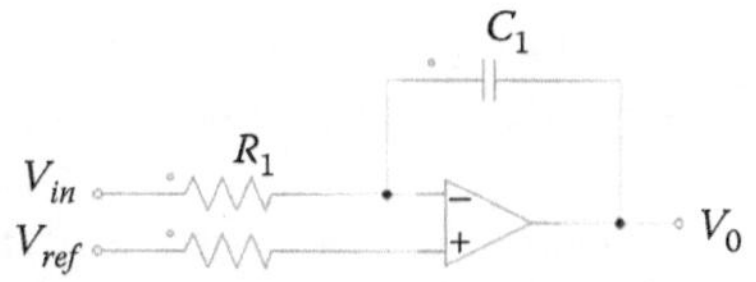

FIGURE 4.29
Type I compensation.

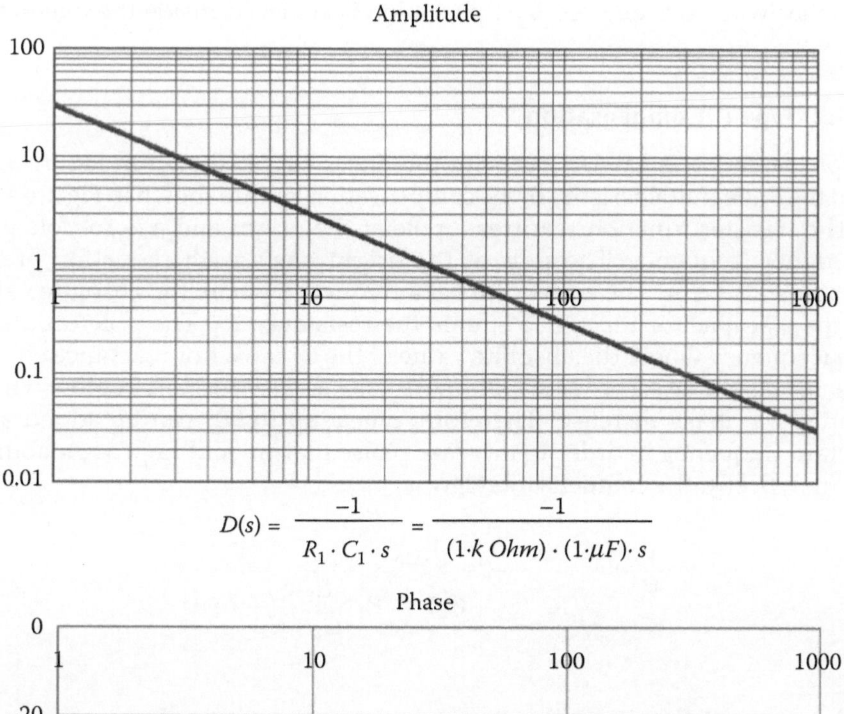

$$D(s) = \frac{-1}{R_1 \cdot C_1 \cdot s} = \frac{-1}{(1{\cdot}k\ Ohm) \cdot (1{\cdot}\mu F) \cdot s}$$

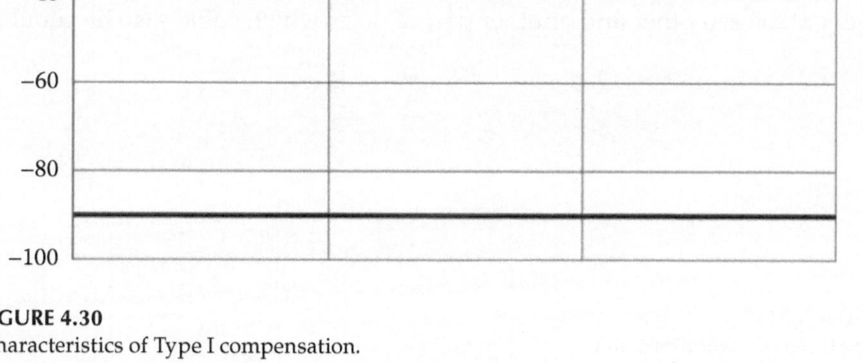

FIGURE 4.30
Characteristics of Type I compensation.

$$\frac{V_{out}}{V_{in}} = -\frac{1}{R_1 \cdot C_1 \cdot s} \tag{4.19}$$

The controller has an integral action with advantages in steady-state error reduction. The magnitude characteristics fall by 20 dB/decade, and the added phase equals 90°. The magnitude characteristics cross unity at a frequency where the absolute value of the C1 capacitor's reactance equals the R1 resistance's value. Type I compensation is often used within power supplies with pass devices. Compensation is achieved inside the integrated circuit.

4.5.2 Type II Compensation

Implementation is carried out with the linear circuit shown in Figure 4.31, and the Bode characteristics of the compensation law are shown in Figure 4.32.

The transfer function features a pole at the origin and a zero-pole pair at higher frequency. The pole at the origin helps with reduction of the steady-state error. The zero occurs at a frequency where the absolute value of the C_2 capacitor reactance equals the resistance R_2. The pole occurs at the frequency where the absolute value of the C_1 capacitor reactance equals the resistance R_1. The zero-pole pair creates a zero-gain region which features a phase increase. Therefore, phase up to 90° can be added at a certain frequency in order to increase phase margin and improve stability.

Analytically, the compensation law is:

$$\frac{V_o}{V_{in}} = -\frac{R_2 \cdot C_2 \cdot s + 1}{R_1 \cdot (C_1 + C_2) \cdot s \cdot \left(R_2 \cdot \dfrac{C_1 \cdot C_2}{C_1 + C_2} \cdot s + 1 \right)} \tag{4.20}$$

4.5.3 Type III Compensation

Implementation is carried out with the linear circuit shown in Figure 4.33. The transfer function features a pole at the origin, a pair of zeros which could be equal to each other and another pair of poles which could also be equal to

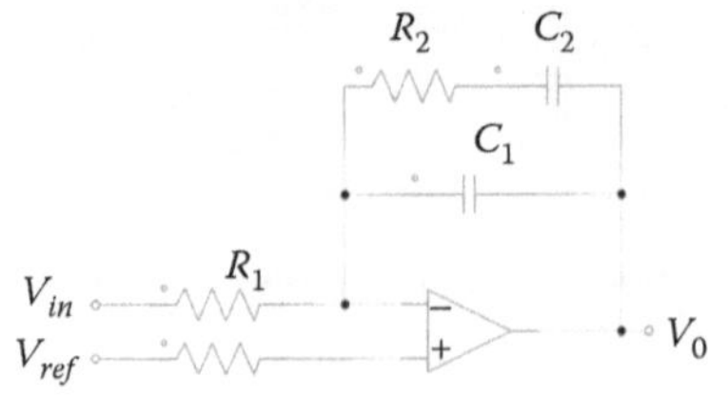

FIGURE 4.31
Type II compensation circuitry.

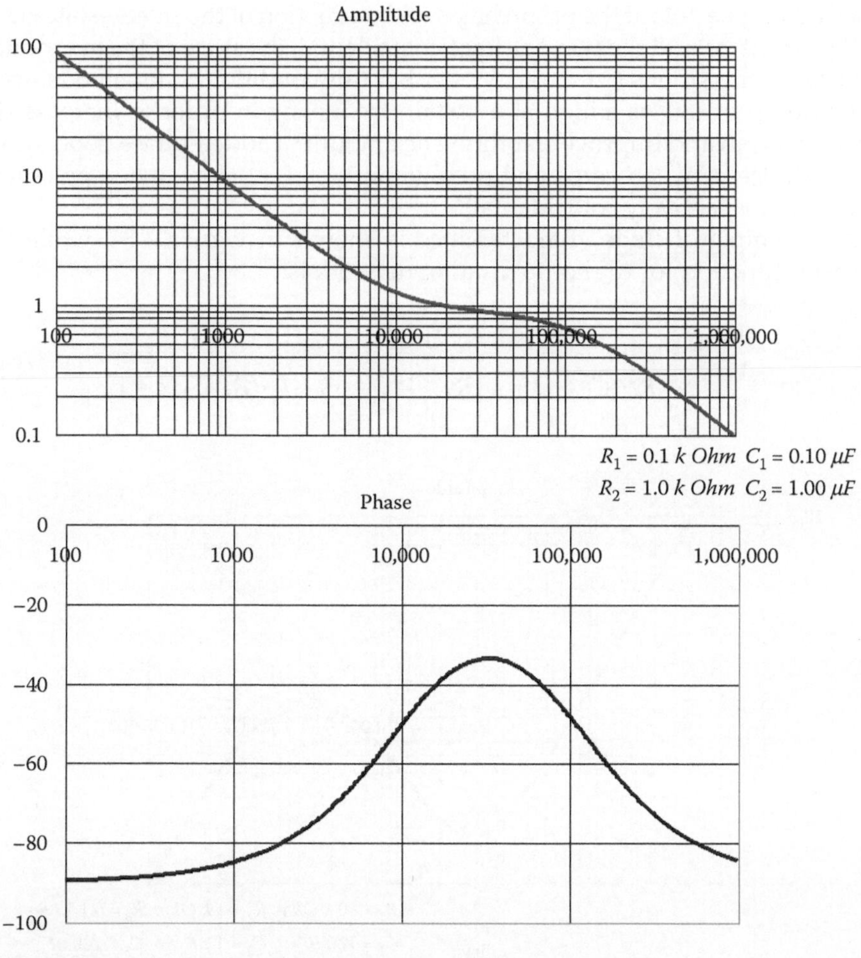

FIGURE 4.32
Bode characteristics for Type II compensation.

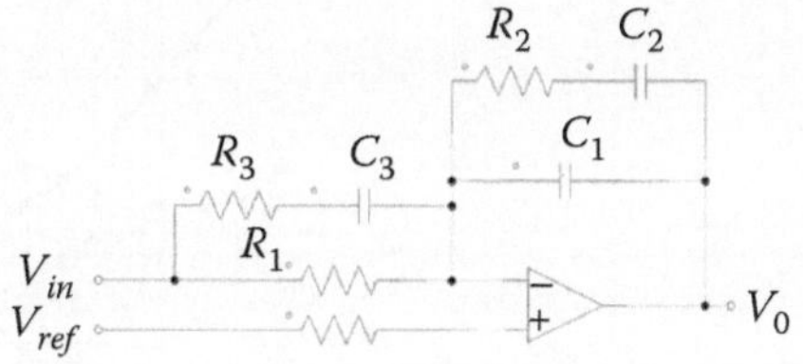

FIGURE 4.33
Type III compensation.

each other. The pole at the origin helps with reduction of the steady-state error. If the two zeros and the two poles are coincidental, the slope of the magnitude plot easily transitions ($-1 \rightarrow +1 \rightarrow -1$), but this condition is not mandatory.

Phase up to 180° is added at a certain frequency in order to increase the phase margin and improve stability. The amount of added phase depends on the distance between zeros and poles (a region of +1 slope when poles and zeros are respectively coincidental).

An example of Bode characteristics is shown in Figure 4.34, while the transfer function for the compensation network is:

$$\frac{V_o}{V_{in}} = -\frac{(R_2 \cdot C_2 \cdot s + 1) \cdot [(R_1 + R_3) \cdot C_3 \cdot s + 1]}{R_1 \cdot (C_1 + C_2) \cdot s \cdot \left(R_2 \cdot \frac{C_1 \cdot C_2}{C_1 + C_2} \cdot s + 1 \right) \cdot (R_3 \cdot C_3 \cdot s + 1)} \qquad (4.21)$$

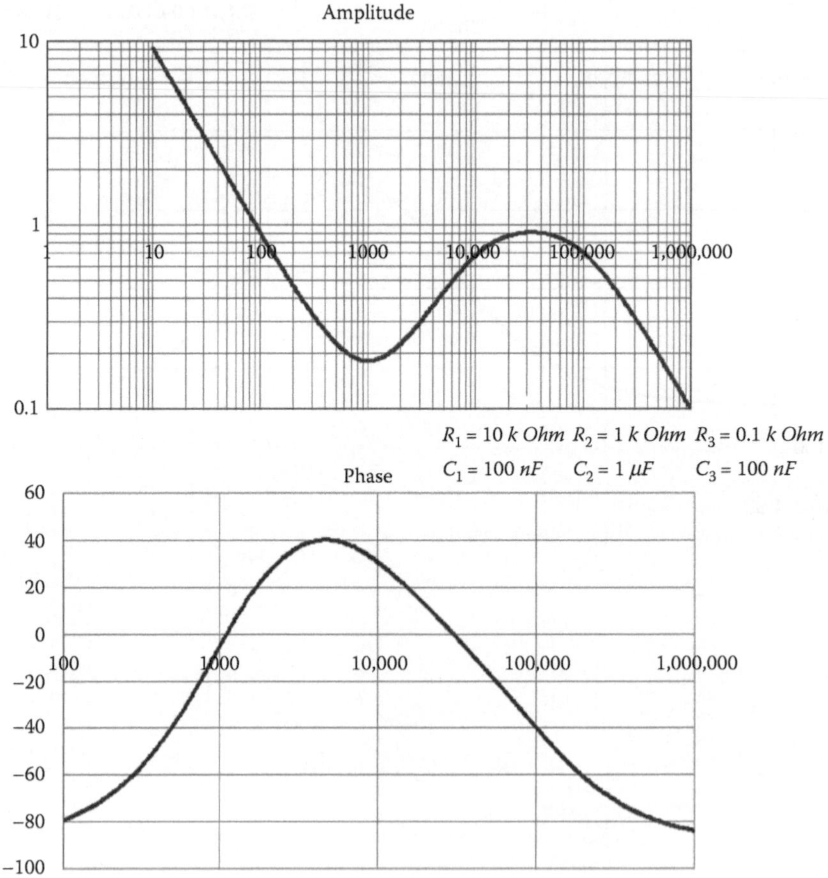

FIGURE 4.34
Bode characteristics for a Type III compensation network able to add 130° of phase (−90°–40°).

4.6 Design Process from Constraints to Component Selection

4.6.1 Pole-Zero Empirical Selection Based on Converter Model

Finally, Figure 4.35 shows an actual complete example of usage for a Type III compensation network. The example illustrates a boost converter with a switching frequency at 140 kHz. The power stage, the PWM circuitry and the compensation law are included and shown as blocks.

The converter has the following characteristics:

- $V_{in} = 12$ V DC
- $V_{out} = 20$ V DC
- $D = 0.60$
- $R = 30\ \Omega$ (load)
- $L = 22\ \mu H$, $R_L = 20\ m\Omega$
- $C = 220\ \mu F$, $ESR = 20\ m\Omega$

While analytical design methods are mostly known, the design of the compensation law is achieved herein with empirical remarks able to create a meaning for the entire process. The compensation law is designed to provide a crossover frequency of 25 kHz and a phase margin around 45°.

The core remarks are:

- A pole at the origin is considered for reduction of the steady-state error.

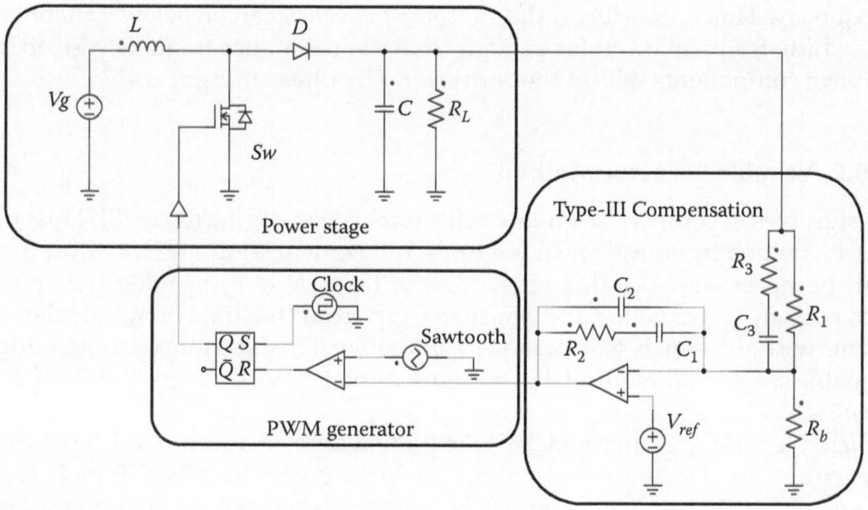

FIGURE 4.35
Example of a boost converter controlled with Type III compensation.

- A double zero occurs at a frequency equal to the resonance frequency within the converter model (resonance from converter inductance and output capacitor). Note that the resonance frequency is around 3 kHz for the given numerical example.
- A second pole equals the zero produced by the output capacitor's ESR.
- A third pole equals the RHP zero as absolute value for the boost converters.

As can be seen, the principle in this design method consists of direct compensation of the poles in the small-signal model of the power converter (in control systems terms, 'the plant') with zeros introduced with the compensation law. Finally, poles are appropriately added at higher frequencies to create multiple lead compensation terms. If the value of any of these poles or zeros is above the switching frequency, its value is limited to the switching frequency.

Bode characteristics for the compensation, plant and overall system are shown in Figure 4.36.

Application of this design to a Type III compensation network produces the control law from Figure 4.36, while the entire system (compensation plus converter) has the transfer function illustrated in Figure 4.37.

The gain of the converter model at the required crossover frequency is calculated by substitution, and its inverted value is used as the desired gain of the compensation law at that frequency. This process is illustrated in Table 4.5.

The stability of the power supply is ensured by adding phase at a desired frequency. This is possible if the *crossover frequency* can be below 1/10 of the switching frequency and far enough from the resonance frequency given by passive components within the converter. The phase margin is 45°.

4.6.2 Venable *K*-Factor Method

Design of the compensation law with either lead-lag terms or PI/D terms can be helped by computer-aided tools, but some trial-and-error effort may still be necessary. For this reason, algorithms able to provide the specified crossover frequency and phase margin from the first try by calculation alone without bench tests are very valuable. This section will introduce Venable's *K*-factor method [7], while Chapter 5 will discuss state-space design.

Venable's *K*-factor method [1] takes advantage of three fixed hardware structures for the compensation law, the previously described Type I, Type II and Type III compensations. It is demonstrated that a unique relationship exists between the components within these networks and the desired performance for any given converter model transfer function.

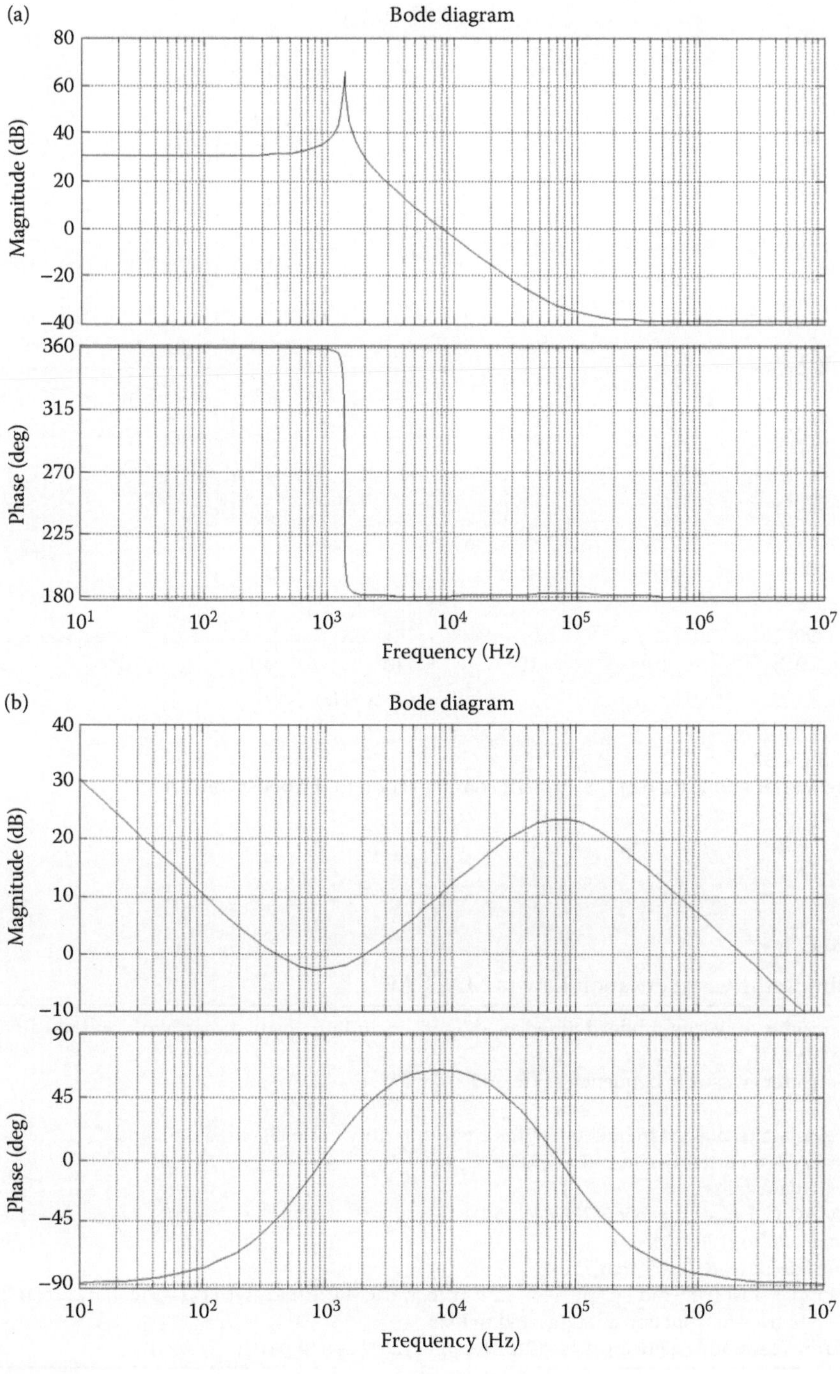

FIGURE 4.36
Design with Bode characteristics for (a) plant model and (b) compensation law.

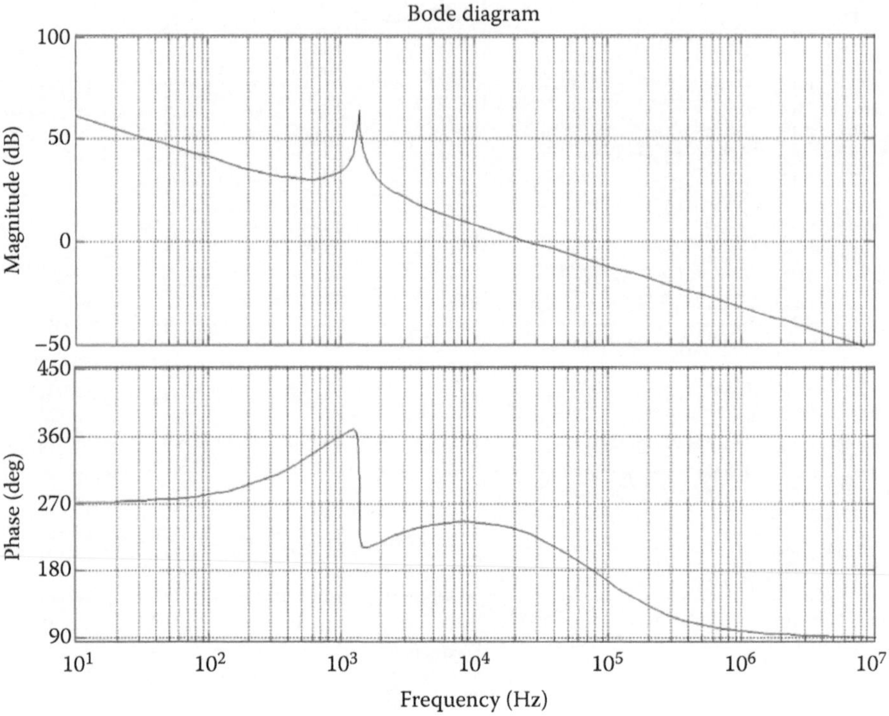

FIGURE 4.37
System after application of the Type III compensation to the boost converter.

TABLE 4.5
Defining the Compensation Law in MATLAB

```
% Consider converter model defined as "d2Out", including positive zero "wp1" and negative
  zero "wz1"
% Consider crossover frequency as "fc_rad"
%------------
% Calculate resonance frequency (with or without duty cycle D) based on L and C
fres=1/(2*pi*sqrt(L*C)); fc_rad = 25,000*2*pi;
% Calculate gains
[G_MAG,G_PHASE] = boe(d2Out,fc_rad)
G_mag=20*log10(G_MAG)
gainmod=power(10,(-G_mag/20));
gaincmp = abs( (i*fc_rad+(2*pi*fres))⊥2 / (i*fc_rad*(1+i*fc_rad/wp1)*(1+i*fc_rad/wz1)) );
% Define transfer function as suggested in text
G_emp=(gainmod/gaincmp)*((s+(2*pi*fres))⊥2)/(s*(1+s/wp1)*(1+s/wz1));
```

The *K*-factor is defined and calculated as

$$K = \sqrt{\frac{pole\_frequency}{zero\_frequency}} \quad (4.22)$$

where the pole and zero correspond to a Type II compensation or to a Type III compensation with double zeros and double poles. Figure 4.38 illustrates the *K*-factor atop the gain characteristics (numerical values are for illustration purposes only). The frequency marked with *f* on the figure can be chosen as the crossover frequency, and it will be the geometrical mean of the pole(s) and zero(s).

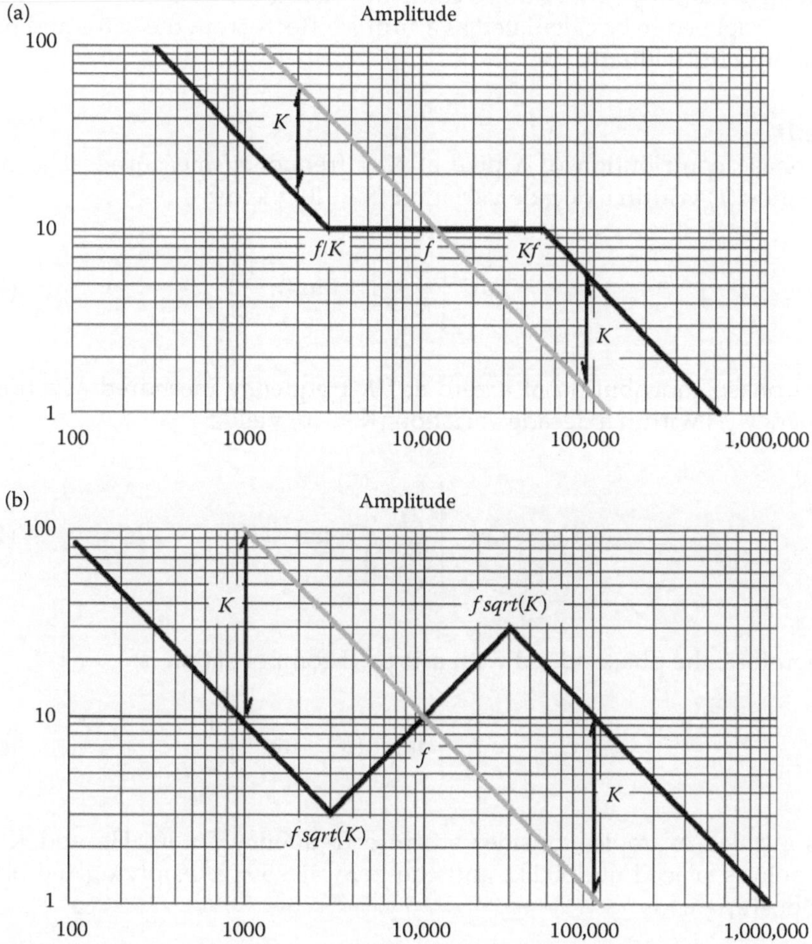

FIGURE 4.38
Venable *K*-factor definitions for (a) Type II and (b) Type III compensation network with double zeros and double poles, where numerical values are for illustration only.

The maximum added phase occurs at this frequency, so the only task for the designer is to adjust the gain to make this frequency indeed the crossover frequency of the entire system. The larger the *K* factor, the larger the phase added to the loop. This is because the distance between pole(s) and zero(s) is larger in such cases.

The characteristics in gray for both Type II and Type III cases represent the ideal integrator characteristics. It can be observed that the actual compensation decreases the gain at low frequencies and increases the gain at high frequencies. Both these effects of the applied Type II or Type III compensation are not desirable, and this is a drawback of the proposed compensation method.

This representation does not yet provide information about the amount of phase generated by any of these compensation laws. The amount of phase shift is proposed to be calculated as a sum of effects from the zeros and poles considered individually.

Type II

The phase contribution of a pole at $K \cdot f$ frequency measured at a lower frequency (*f*) within a decade variation ($K < 10$) yields:

$$\alpha_{pole} = \tan^{-1}\left(\frac{f}{K \cdot f}\right) = \tan^{-1}\left(\frac{1}{K}\right) \tag{4.23}$$

The phase contribution of a zero at *f*/*K* frequency measured at a higher frequency (*f*) within a decade variation (K < 10) yields:

$$\alpha_{zero} = \tan^{-1}\left(\frac{f}{\frac{1}{K} \cdot f}\right) = \tan^{-1}(K) \tag{4.24}$$

Therefore, the phase added with a Type II compensation is:

$$\alpha = \alpha_{zero} - \alpha_{pole} = \tan^{-1}(K) - \tan^{-1}\left(\frac{1}{K}\right) \tag{4.25}$$

As a quick example, consider a centre frequency $f = 100$ Hz and $K = 4$. The pole is placed at 400 Hz and the zero at 25 Hz. Applying the above relationship yields

$$\alpha = \alpha_{zero} - \alpha_{pole} = 75.96° - 14.03° = 61.93° \tag{4.26}$$

which can be further observed within Figure 4.39.

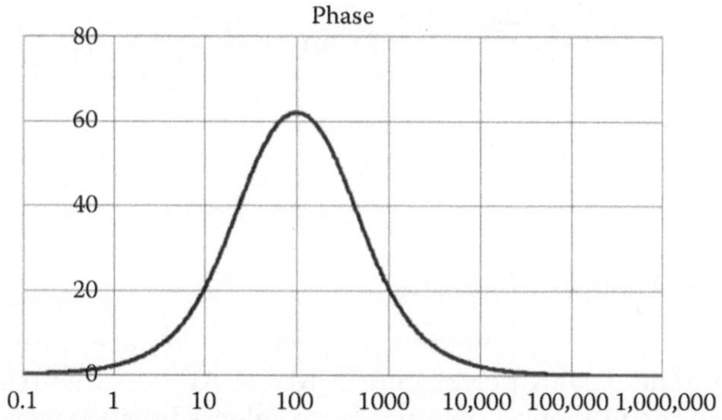

FIGURE 4.39
The added phase with a pole at 400 Hz and a zero at 25 Hz, for a centre frequency of 100 Hz.

The design problem is reversed in practice. An investigation of the power converter model (in control systems terms, 'plant') reveals the phase at the desired crossover frequency. Meanwhile, the requirements for stability, steady-state error and transient response define the desired crossover frequency and phase margin. Usually, the design of a switched-mode power supply leads to a desired phase margin of 45°, which is mostly uprated to 60°. The compensation law has to provide the difference between this desired phase margin and the phase currently residing in the converter model. The designer needs to come up with a value of the *K* factor able to generate the additional phase.

Some calculation follows:

$$\frac{\pi}{2} = \tan^{-1}(K) + \tan^{-1}\left(\frac{1}{K}\right) \Rightarrow \alpha = 2\cdot\tan^{-1}(K) - \frac{\pi}{2} \Rightarrow \tan^{-1}(K) = \frac{\alpha+\pi}{2} \Rightarrow K = \tan\left(\frac{\alpha+\pi}{2}\right) \tag{4.27}$$

Using this provides the exact location of poles and zeros.

Type III

The phase added by a compensation law with two coincidental poles and two coincidental zeros yields the double of a single zero-pole pair:

$$\alpha = 2\cdot(\alpha_{zero} - \alpha_{pole}) = 2\cdot\left(\tan^{-1}\left(\sqrt{K}\right) - \tan^{-1}\left(\frac{1}{\sqrt{K}}\right)\right) \tag{4.28}$$

Using the previous trigonometric identity leads to:

$$\alpha = 2\cdot\left(2\cdot\tan^{-1}\left(\sqrt{K}\right)-\frac{\pi}{2}\right) = 4\cdot\tan^{-1}\left(\sqrt{K}\right)-\pi \Rightarrow \tan^{-1}\left(\sqrt{K}\right)$$
$$= \frac{\alpha+\pi}{4} \Rightarrow K = \left[\tan\left(\frac{\alpha+\pi}{4}\right)\right]^2 \tag{4.29}$$

4.6.3 MATLAB Analysis of Results

The precise *K*-factor design method proposed by Venable can be illustrated with an automatic design performed in MATLAB®. In this respect, a buck converter with the following data has been selected for analysis:

- $V_{in} = 5$ V DC = input voltage
- $V_{out} = 1.2$ V DC = output voltage
- $f_{SW} = 500$ kHz = switching frequency
- $L = 10$ μH = inductance within buck converter (loss resistance of 30 mΩ)
- $C_{out} = 310$ μF = output capacitance (ESR = 5 mΩ)
- $I_{out} = 20$ A = nominal load current

A self-resonance frequency of 2,858 Hz can be calculated based on converter dataset. Design requirements for a buck converter are usually similar to the following:

- $f_c = 25$ kHz = design-imposed crossover frequency
- $\phi = 60°$ = design-imposed phase margin
- Zero steady-state error, which means the compensation law contains an integrator

The Laplace transfer function for the small-signal model of the buck converter and modulator yields (see Table 4.2 and numerical data):

$$G(s) = \frac{2.114\cdot10^{-6}\cdot s + 1.515}{3.1\cdot10^{-9}\cdot s^2 + 1.317\cdot10^{-5}\cdot s + 1} \tag{4.30}$$

This has the Bode plot shown in Figure 4.40. It can be seen that the phase is at –166.07° and the phase margin is around 13°, with a crossover frequency of around 30 krad/sec; that is, around 5 kHz. The phase margin should be augmented to 60°, and the crossover frequency moved to a higher frequency of 25 kHz as per the design requirements. This can be achieved with a double lag compensator and by considering an integrator for steady-state error

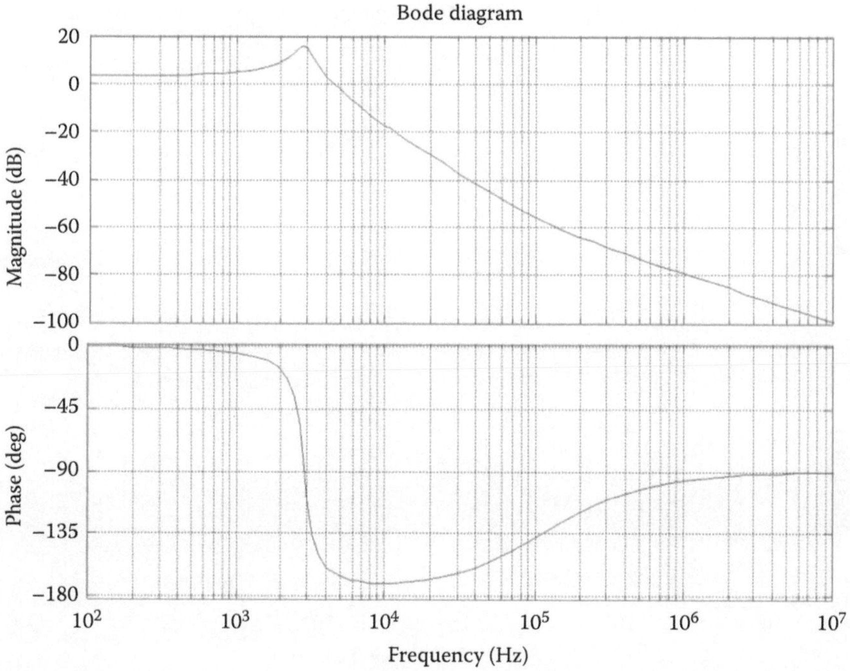

FIGURE 4.40
Bode characteristics for the considered dataset.

reduction. The lead compensator can further be designed with the Venable K-factor method.

Applying the Venable K-factor method determines $K = 26.5418$ and produces a compensation law with the Laplace form:

$$D(s) = \frac{1.953 \cdot 10^8 \cdot s^2 + 1.214 \cdot 10^{13} \cdot s^3 + 1.887 \cdot 10^{17}}{s^3 + 1.588 \cdot 10^6 \cdot s^2 + 6.302 \cdot 10^{11} \cdot s} \tag{4.31}$$

Figure 4.41 shows the Bode characteristics with the axis changed to Hz instead of rad/sec. A phase lead contribution of 135° can be noticed on top of the integrator loss of 90°. This produces 45° net phase contributions.

Considering numerical data, the open-loop transfer function is:

$$D(s) \cdot G(s) = \frac{412.8 \cdot s^3 + 3.216 \cdot 10^8 \cdot s^2 + 1.887 \cdot 10^{13} \cdot s + 2.859 \cdot 10^{17}}{3.100 \cdot 10^{-9} \cdot s^5 + 0.004936 \cdot s^4 + 1975 \cdot s^3 + 9.891 \cdot 10^6 \cdot s^2 + 6.302 \cdot 10^{11} \cdot s} \tag{4.32}$$

This has the Bode characteristics shown in Figure 4.42. The phase margin is around 60° at 25 kHz, as imposed by the design.

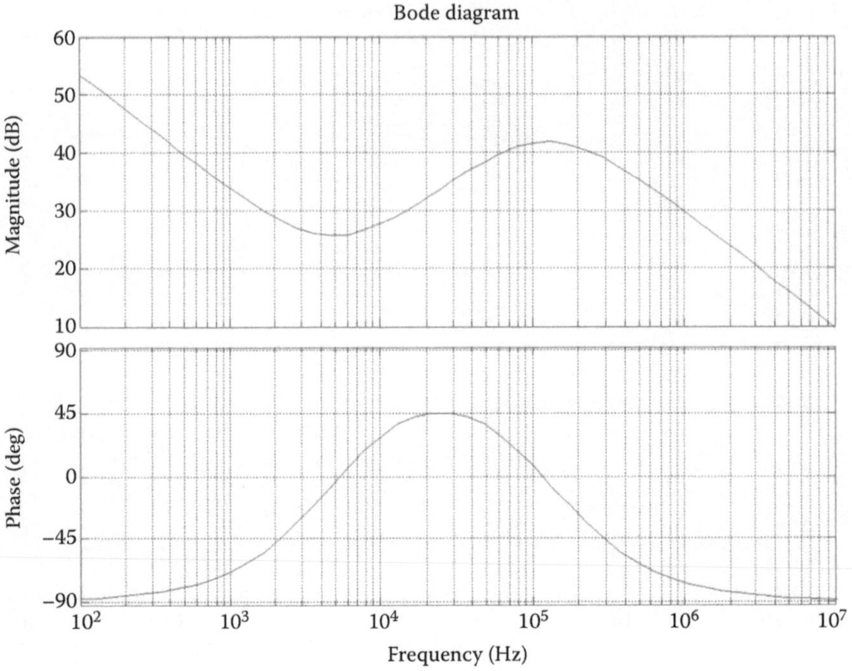

FIGURE 4.41
Compensation law for the considered converter dataset.

The compensation law can be implemented by either analogue or digital means. The conversion from *K*-factor terms and the Laplace form shown above to the compensation circuits illustrated in Figure 4.31 for Type II compensation or Figure 4.33 for Type III compensation is next presented.

Type II

If the required phase contribution is under 70°, that means the phase generated with a Type II compensator is enough, and the structure from Figure 4.31 is considered. The passive components are calculated with the following assumptions:

- The compensation gain G_{req} at the required crossover frequency is calculated as $1/G_{buck}$.
- The *K* factor is calculated from the *D*(*s*)*G*(*s*) transfer function and Equation 4.27.
- A fixed resistance R_1 is selected as the starting point for calculation.

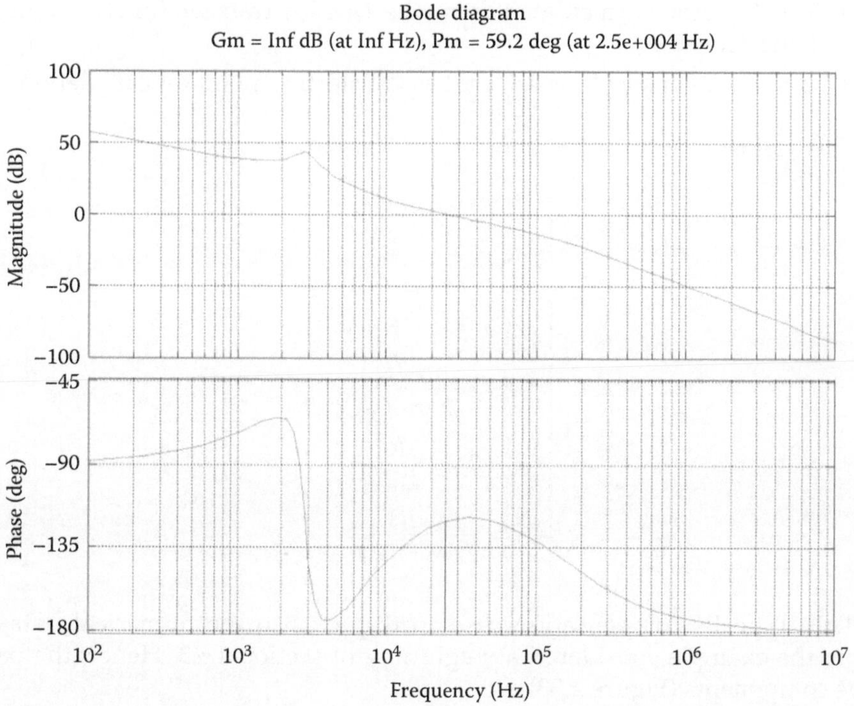

FIGURE 4.42
Bode plot for the open loop *D(s)G(s)*, calculated for a crossover frequency of 25 kHz.

This yields:

$$\begin{cases} R_1 = 10\,K\ Ohm \\ C_1 = \dfrac{1}{2\cdot\pi\cdot f_c\cdot G_{req}\cdot K\cdot R_1} \\ C_2 = C_1\cdot(K^2-1) \\ R_2 = \dfrac{K}{2\cdot\pi\cdot f_c\cdot C_2} \end{cases} \tag{4.33}$$

Type III

If the required phase contribution is above 70°, that means the phase generated with a Type III compensator is necessary, and the structure from Figure 4.33 is considered. The passive components are calculated with the following assumptions:

- The compensation gain G_{req} at the required crossover frequency is calculated as $1/G_{buck}$.

- The K factor is calculated from the $D(s)G(s)$ transfer function and Equation 4.29.
- A fixed resistance R_1 is selected as the starting point for calculation.

This yields:

$$\begin{cases} R_1 = 1kOhm \\ C_1 = \dfrac{1}{2\cdot\pi\cdot f_c \cdot G_{req} \cdot R_1} \\ C_2 = C_1 \cdot (K-1) \\ R_2 = \dfrac{\sqrt{K}}{2\cdot\pi\cdot f_c \cdot C_2} \\ R_3 = \dfrac{R_1}{K-1} \\ C_3 = \dfrac{1}{2\cdot\pi\cdot f_c \cdot \sqrt{K} \cdot R_3} \end{cases} \tag{4.34}$$

This Type III compensation case corresponds to the numerical dataset from the example considered at beginning of section 4.6.3. Hence, the passive components (Figure 4.33) are:

$R_1 = 1\ \text{k}\Omega$
$R_2 = 9.8197\ \text{k}\Omega$
$R_3 = 39.1515\ \text{k}\Omega$
$C_2 = 3.34\ \text{nF}$
$C_1 = 130.77\ \text{pF}$
$C_3 = 31.563\ \text{nF}$

Obviously, these values have to be adjusted to the standard values of passive components.

4.6.4 The Problem of Conditionally Stable Systems

The Bode plot for the compensated system shown in Figure 4.37 denotes a phase closer to 180° than planned for around 1.5 kHz, while the response shown in Figure 4.42 shows a very small phase margin at around 4 kHz. In both cases, a decrease of the loop gain could bring us closer to instability or at least show some oscillations.

This situation, when the compensated system is designed properly to deliver a good dynamic performance with enough phase margin at a certain crossover frequency while the phase is considerably decreased at frequencies under the crossover frequency, denotes a conditionally stable system. This can be identified only by the Bode response. The time response waveforms will not show anything close to instability for either step response to a load current change or to a small change in input voltage. Such a transient response does not oscillate. Thus, a frequency response plot is used to detect the conditional stability of the converter.

Unfortunately, a more dramatic change in operating conditions could lead to a change in loop gain and produce oscillations and instability. An example of unstable operation could correspond to operation of the synchronous buck converter with an input voltage considerably smaller than nominal. For instance, the buck converter used for the results from Figure 4.42 could aim at producing 1.2 V DC output from 3.3 V DC input rather than 5.0 V DC. This might show oscillations in time domain waveforms, denoting an unstable system. Such conditions may also happen during startup or during fault management procedures.

There are several ways to avoid this happening.

An advanced control system could monitor the converter operation and change gains adequately when general operation conditions change. This is studied in Section 5.5 when the design of the compensation law is achieved with a state-space approach.

As a much simpler solution, application engineers avoid conditionally stable designs by attempting not to allow the phase margin to drop below a certain level, such as 45°, for any frequency under the gain crossover frequency. The easiest way to achieve this goal is to lower the expectations for the crossover frequency. For instance, the buck converter considered in Section 6.4.4, with previous results in Figure 4.42, could moderate the crossover frequency expectations to 15 kHz instead of 25 kHz. This yields the Bode plot from Figure 4.43 for the open loop $D(S)G(s)$ calculated with the previous converter dataset. This is equivalent to a relocation of the zeros of the compensation network towards lower frequencies.

As a final solution to conditionally stable systems, some control integrated circuits, like IR3895, offer a feed-forward control scheme able to keep the loop gain constant. The feed-forward does not refer to the compensation network, but to a direct supply of the input voltage into the PWM generator. This allows the PWM ramp amplitude (V_{ramp}) to be proportionally changed with V_{in} supply voltage in order to maintain V_{in}/V_{ramp} almost constantly throughout the V_{in} variation range.

4.7 On Usage of Conventional Proportional–Integral–Derivative Controllers

Control systems using a PI/D compensation law are widely used in industry due to their simplicity and generality. First, a proportional gain is considered to

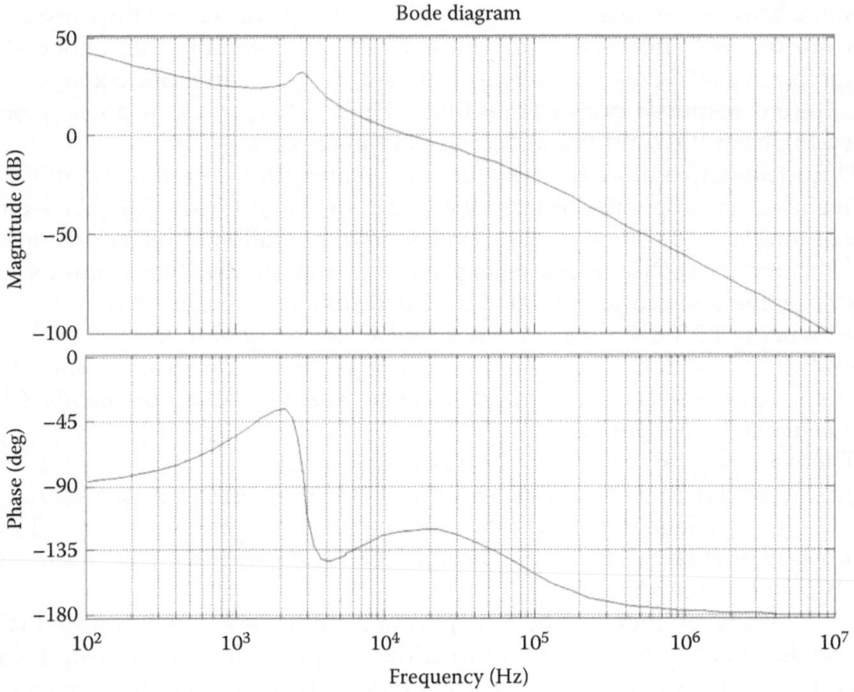

FIGURE 4.43
Modified Bode plot for the open loop $D(s)G(s)$ for 15-kHz crossover frequency.

act against the variations of the output measure from a reference. The possible steady-state error is reduced with an integrative term, and a PI is thus formed. In cases where fast dynamics or special requirements for the transient response are desired, the transient response is improved with a derivative term. The effect of the derivative term is more obvious with higher-order plant systems.

The compensation law is as follows:

$$u(t) = k_p \cdot e(t) + k_I \cdot \int_{t_o}^{t} e(\tau)d\tau + k_D \cdot \frac{de(t)}{dt} \tag{4.35}$$

where the first term corresponds to the proportional law, the second term represents an integrator and the last term shows the derivative term. Even if the PI/D control is inherently in the time domain, the compensation law can also be expressed in Laplace form for analysis of results. A standard form illustrates the gains for each term as:

$$\frac{U(s)}{E(s)} = k_p + \frac{k_I}{s} + k_D \cdot s \tag{4.36}$$

while the parallel form is more attractive for industrial control since it expresses each term with time constants (T_I, T_D), which are more easily understood and set in practical equipment.

$$\frac{U(s)}{E(s)} = k_P \cdot \left(1 + \frac{1}{T_I \cdot s} + T_D \cdot s\right) \tag{4.37}$$

4.8 Conversion of Analogue Control Law to Digital Solutions

After the compensation law has been decided with either lead-lag or PI/D compensators, it can be converted to a digital form and uploaded onto a microcontroller [8,9]. Figure 4.44 illustrates the practical setup.

The input signals are sampled at a constant sampling rate and acquired within the digital controller through an analogue–digital converter. The samples are called discrete signals, while the data acquired in digital form are called the digitised signal. The digitised signal is quantised on a number of bits which define the digital resolution.

A very large sampling frequency makes the system's response closed to its analogue equivalent. The sampling rate is, however, limited in a practical system due to physical implementation. A very low sampling frequency tends to slow down the system's response. In order to deal with this compromise, an empirical design rule suggests that the sampling rate be chosen so that during the rise time of a step response, there would be 5–10 samples of the input signal. Specific digital control methods can compensate for the sampling effect and further reduce the sampling frequency to 2–3 samples during the rising time.

The compensation law is implemented in software or hardware with a relationship using finite differences. The calculation result is converted back to analogue with a digital–analogue converter and sent to the real world for control of the power converter (in control systems terms, the 'plant').

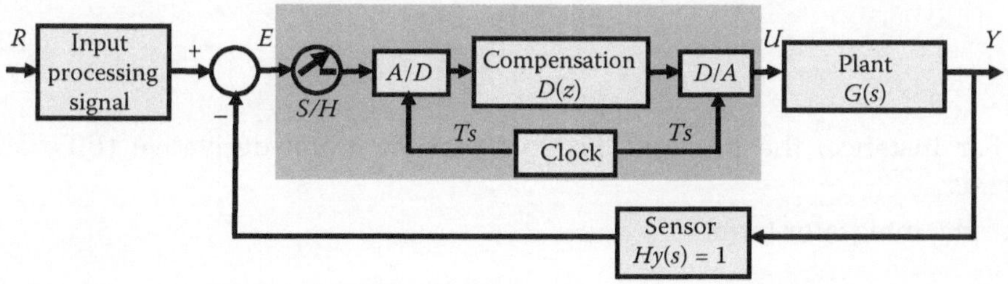

FIGURE 4.44
Implementation of a digital control law in practice.

While design methods specific to digital systems exist, in most cases design is first performed in analogue with the Laplace method. Then, the Laplace form of the compensation law is converted into the z-transform. As a last step, the z-transform is further converted to a relationship using finite differences, where succeeding samples of the input or output measures are used to compose a compensation law.

A first example for the well-known PI/D controller considers each term of the Laplace form for the compensation law and converts it into digital.

$$U(s) = \left(k_P + \frac{k_I}{s} + k_D \cdot s\right) \cdot E(s) \tag{4.38}$$

$$u(t) = k_P \cdot e(t) + k_I \cdot \int_0^t e(\tau)d\tau + k_D \cdot \frac{de}{dt} = u_P + u_I + u_D \tag{4.39}$$

Each term is written after sampling:

$$u_P(k \cdot T_s + T_s) = k_P \cdot e(k \cdot T_s + T_s) \tag{4.40}$$

$$\begin{aligned} u_I(k \cdot T_s + T_s) &= k_I \cdot \int_0^{k \cdot T_s + T_s} e(\tau)d\tau = k_I \cdot \left(\int_0^{kT_s} e(\tau)d\tau + \int_{kT_s}^{kT_s + T_s} e(\tau)d\tau \right) \\ &= u_I(k \cdot T_s) + [area\_under\_e(t)] \\ &\approx u_I(k \cdot T_s) + k_I \cdot \frac{T_s}{2} \cdot \{e(kT_s + T_s) + e(kT_s)\} \end{aligned} \tag{4.41}$$

$$\frac{T_s}{2} \cdot \{u_D(k \cdot T_s + T_s) + u_D(k \cdot T_s)\} = k_D \cdot \{e(k \cdot T_s + T_s) - e(k \cdot T_s)\} \tag{4.42}$$

When the compensation law is more complex, as in a Laplace form with multiple poles and zeros, an operator $z = e^{s \cdot Ts}$ is defined to facilitate the usage of the z-transform. This has the properties:

$$\begin{aligned} u(k \cdot T_s) &\leftrightarrow U(z) \\ u(k \cdot T_s + T_s) &\leftrightarrow z \cdot U(z) \end{aligned} \tag{4.43}$$

For instance, the previous proportional-integrator-derivative (PID) law yields:

For the integrator term:

$$z \cdot U_I(z) = U_I(z) + k_I \cdot \frac{T_s}{2} \cdot [z \cdot E(z) + E(z)] \Rightarrow U_I(z) = k_I \cdot \frac{T_s}{2} \cdot \frac{z+1}{z-1} \cdot E(z) \tag{4.44}$$

For the derivative term:

$$U_D(z) = k_D \cdot \frac{2}{T_s} \cdot \frac{z-1}{z+1} \cdot E(z) \tag{4.45}$$

For the control law:

$$U(z) = \left(k_P + k_I \cdot \frac{T_s}{2} \cdot \frac{z+1}{z-1} + k_D \cdot \frac{2}{T_s} \cdot \frac{z-1}{z+1} \cdot \right) E(z) \tag{4.46}$$

Similar results can be achieved when replacing the operator s with:

$$s \leftrightarrow \frac{2}{T_s} \cdot \frac{z-1}{z+1} \tag{4.47}$$

and this is called the Tustin method. Methods for conversion from the Laplace form to the *z*-transform include:

- Matched pole-zero method
- Zero-order hold method
- Linear interpolation of inputs method
- Other similar methods

It can also be observed that the digital form and all the coefficients from the digital form depend on the actual value of the sampling time T_s.

As an example, the digital form for the previous Venable *K*-factor design (4.31) was:

$$F(z) = \frac{64.52 \cdot z^3 - 56.74 \cdot z^2 - 64.29 \cdot z + 56.98}{z^3 - 1.23 \cdot z^2 + 0.2429 \cdot z - 0.01312} \tag{4.48}$$

when a sampling time of 2 µsec is considered. A change of the sampling period changes all the coefficients in Equation 4.48. The effect of sampling is shown with a Bode plot in Figure 4.45, to be compared with the *s*-domain form from Figure 4.41. The characteristics are curved down and then limited before a frequency of 250 kHz; that is, half of the sampling frequency.

The software implementation of the control law requires working with an equation with finite differences. The control action is herein denoted by *U*(*s*) in the *s*-domain and *U*(*z*) in the *z*-domain.

Equation 4.48 is rewritten as:

$$\begin{aligned} &U(z) \cdot (1 - 1.23 \cdot z^{-1} + 0.2429 \cdot z^{-2} - 0.01312 \cdot z^{-3}) \\ &\quad = E(z) \cdot (64.52 - 56.74 \cdot z^{-1} - 64.29 \cdot z^{-2} + 56.98 \cdot z^{-3}) \end{aligned} \tag{4.49}$$

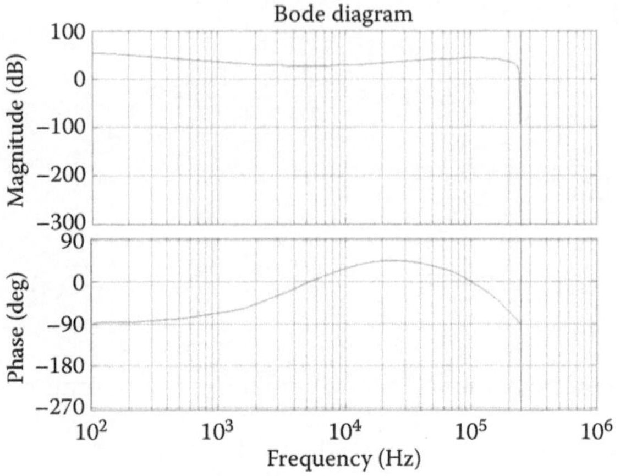

FIGURE 4.45
Bode plot for the *z*-domain transfer function.

where each term in *z* is replaced with a previous time instant. The current instant of time is denoted by *k*, while previous time instants are defined as $[k-1]$, $[k-2]$ and $[k-3]$. This form of the control law is easy to implement in software. It yields:

$$u[k]-1.23\cdot u[k-1]+0.2429\cdot u[k-2]-0.01312\cdot u[k-3]$$
$$=64.52\cdot e[k]-56.74\cdot e[k-1]-64.29\cdot e[k-2]+56.98e[k-3]$$

$$u[k]=1.23\cdot u[k-1]-0.2429\cdot u[k-2]+0.01312\cdot u[k-3]+64.52\cdot e[k]$$
$$-56.74\cdot e[k-1]-64.29\cdot e[k-2]+56.98e[k-3] \quad (4.50)$$

The actual implementation requires an important effort in scaling the input and output measures. Furthermore, many software systems use a fractional data (all data have subunitary values) format where each variable needs to previously be scaled within a subunity range (Figure 4.46).

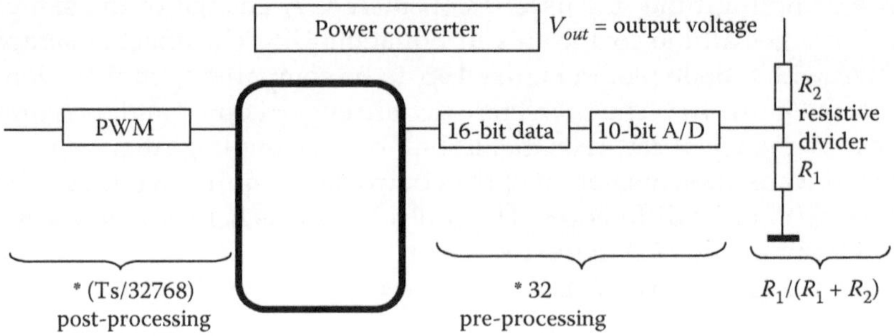

FIGURE 4.46
Scaling associated with microcontroller implementation.

Summary

Linear analogue circuits with amplifiers have been suggested for frequency compensation of DC/DC power supplies. Output voltage conditioning is done in a closed loop with feedback. Electronic circuits with continuous operation necessitate a pass device, which would dissipate a large amount of energy.

There is always a compromise between the usage of a low-dropout circuit and closed-loop system stability. The design of the low drop-out (LDO) converters using PNP bipolar transistors or p-type metal-oxide-semiconductor (pMOS) transistors depends on the selection of the load circuitry since we cannot interfere with compensation inside the integrated circuit.

Both power supplies with analogue operation and switched-mode power supplies may use analogue circuits with amplifiers and passive components for frequency compensation. In order to simplify the design, these control circuits have predefined structures – Type I, Type II or Type III – which allow the addition of passive components along with the existing integrated circuits. Design examples were briefly presented in this chapter.

The digital form for the compensation law is derived for a software implementation. Either a hardware or software implementation of the compensation law is currently very attractive for modern power supplies used in the telecom industry. This allows a correlation with digital management efforts in the telecom industry.

References

1. Franklin, G.F., Powell, J.D., Emami-Naemi, A., 2014, *Feedback Control of Dynamic Systems*, 7th edition, Prentice-Hall, Englewood Cliffs, NJ.
2. Ogata, K., 1995, *Discrete Time Control Systems*, Prentice-Hall, Englewood Cliffs, NJ.
3. Ghosh, A., Prakash, M., Pradhan, S., Banerjee, S., 2014, "A Comparison among PID, Sliding Mode, and Internal Model Control for a Buck Converter", *IEEE IECON 2014*, Dallas, TX, USA, pp. 1001–1007.
4. Kelly, A., Rinne, K., 2005, "Control of DC-DC Converters by Direct Pole Placement and Adaptive Feed-Forward Gain Adjustment", *IEEE APEC 2005*, 6–10 March 2005, Austin, TX, USA, pp. 1970–1975.
5. Neacşu, D.O., Bonnice, W., Holmansky, E., 2010, "On the Small-Signal Modeling of Parallel/Interleaved Buck/Boost Converters", *IEEE International Symposium in Industrial Electronics*, Bari, Italy, July 2010, pp. 2708–2713.
6. Zaitsu, R., 2009, "Voltage Mode Boost Converter Small Signal Control Loop Analysis Using the TPS61030", *Texas Instruments Application Report SLVA274A*, May 2007, Rev. January 2009.

7. Venable, D. H., 1983, "The K Factor: A New Mathematical Tool for Stability Analysis and Synthesis", *Proceedings of Powercon*, 22–24 March 1983, San Diego, CA, USA, pp. 1–12.
8. Maksimovic, D., 2013, "Digital Control of High-Frequency Switching Power Converters", *Tutorial Presented at The 14th IEEE Workshop on Control and Modeling for Power Electronics (COMPEL)*, Salt Lake City, UT, pp. 1–100.
9. Jones, M., 2011, "Digital Compensators Using Frequency Techniques", *EEWeb Documentation*, Posted November 2011.

5

State-Space Control of DC/DC Converters

5.1 Modern versus Classical Control Methods

An important development is currently being experienced with digital control of telecom power supplies. Most of this effort actually concerns a large set of functions for fault management and communication protocols with upper hierarchical levels. There have been very limited reports on feedback control–related references.

This chapter expands on the exciting new development of digital hardware for DC/DC converters with digital control systems based on state-space feedback control, also referred to as *modern control*. A complete design example is provided from a generic textbook definition [1] for a boost power converter.

All steps involved in a complete design can be achieved on a digital platform like a computer, and this represents the major advantage of using state-space controllers. Such steps include:

- State-space modelling and averaging
- Digital implementation of auxiliary functions like startup and precharging
- Feedback control setup
- Pole selection and actual state-space controller design
- Equivalence to an optimal PI controller
- Reference introduction to the state-space form
- Usage of a feed-forward component for improvement of the dynamic performance
- Full- and reduced-order state estimation with estimator pole selection
- Closing the feedback control loop after the estimated variable
- Evaluation of system performance

The step-by-step design procedure [1–5] was followed up with results derived with a comprehensive MATLAB-SIMULINK analysis and a microchip microcontroller platform implementation.

5.2 State-Space Modelling

While state-space control can be applied to any power converter structure, a DC/DC boost converter has been chosen herein as an example for a conversion from 12 V DC input to 20 V DC output (Figure 5.1).

The theoretical presentation is accompanied throughout this chapter by numerical examples for a converter with parameters:

- Input voltage $V_{in} = 12$ V DC
- Output voltage $V_{out} = 20$ V DC
- Duty cycle of 0.4; that is, $D = 0.600$ ideal or $D = 0.585$ real, per converter leg measurement
- Boost inductance $L = 22$ μH
- Output capacitance $C = 220$ μF

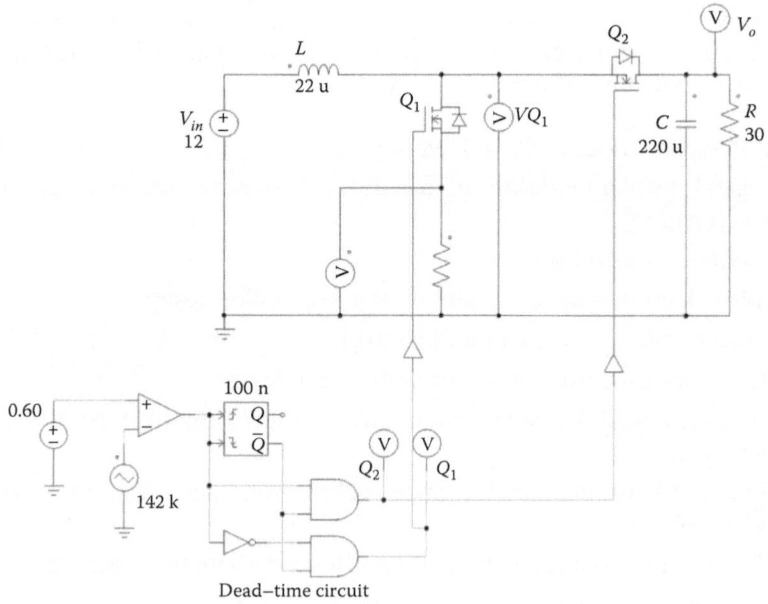

FIGURE 5.1
Circuit for the boost converter considered as an example, presented in open loop.

- Generic load resistance $R = 30\ \Omega$, used during main steady-state tests
- Switching frequency 142 kHz, selected to avoid stringent EMI requirements and standards above 150 kHz

As explained in Chapter 3, the duty cycle for a boost converter is defined with a relationship (Equation 3.7) that is true for operation in continuous conduction mode or for a synchronous converter.

The synchronous converter with digital control from a microcontroller is often redefined with a converter leg that is a pair of two MOSFET transistors. The new setup and definition are closer to the reasoning behind the microcontroller peripherals. Dedicated PWM peripherals define control for a generic pair of IGBT or MOSFET devices, which can be further used for either an AC/DC inverter or DC/DC converter leg. Hence, it is easier to redefine the control parameters as:

$$\frac{V_o}{V_d} = \frac{T_s}{t_{off}} = \frac{1}{1-\delta} \Rightarrow 1-\delta = D = \frac{V_d}{V_o} \tag{5.1}$$

This way, the outcome of the compensation law calculation can be directly loaded into the peripheral as its control parameter – the leg (branch) duty cycle. Another way to explain this relates to the synchronous converter operation with reversed power transfer from V_o to V_d.

Figure 5.2 illustrates the operation in an open loop in continuous conduction mode, with possible oscillation induced by the *L*–*C* pair of poles. Waveforms are shown for a switching frequency around 142 kHz and a duty cycle $D_0 = 0.585$, measured for the converter leg Q_1–Q_2, which means it is actually the duty cycle for the top device Q_2, mostly known for the diode role within a boost converter.

Modelling the power converter with averaged state equations is well known in the power electronics community since the plant model in the Laplace form is often derived with this approach. Operational states are identified, and equivalent circuits are derived for each state. Circuit equations for each state are next averaged with a weighing relationship over the duration of the sampling interval. The time-averaging method is suitable for the low-frequency analysis of power converters. This means it can be used when the system's time response lasts longer than the period of converter pulses. Observing the time intervals spent within each state allows us to define an averaged model over the states' repetition period.

There are two possible applications of the time-averaging method:

- Derive linear models able to describe transfer functions through manipulations of the equivalent circuit.
- Leave the circuit equations in the state form.

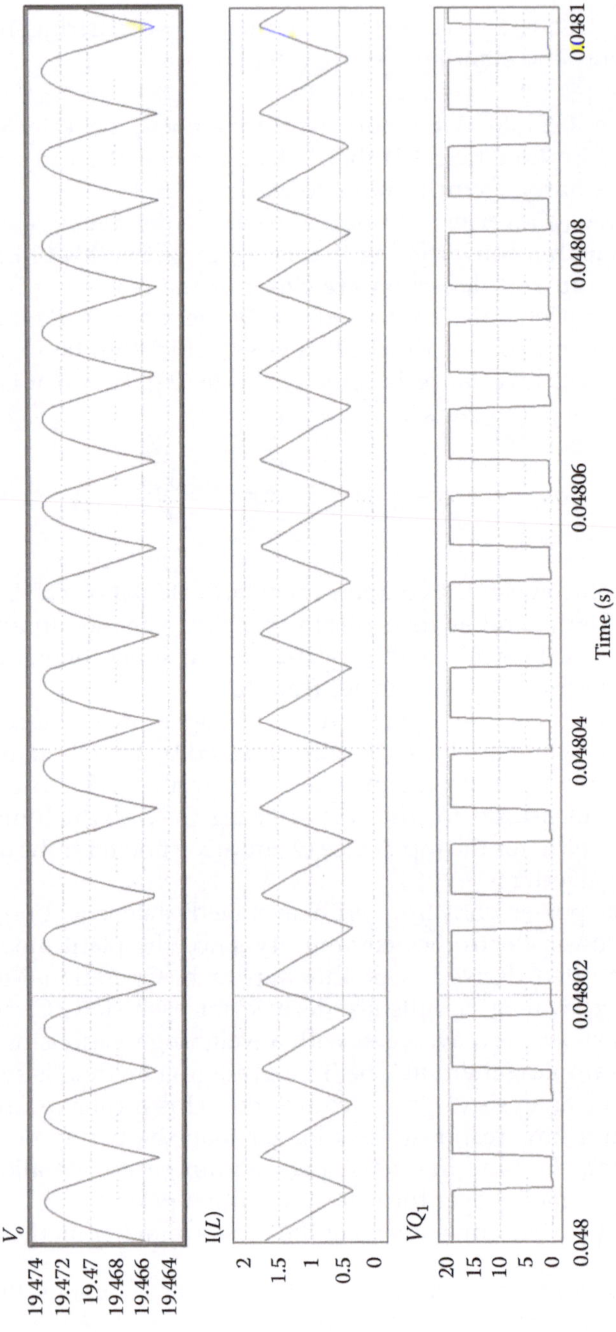

FIGURE 5.2
Output voltage and inductor current for an operation with fixed duty cycle in open loop for a load producing continuous conduction mode.

The latter is the basis of the state-space control method. Generally, a system can be associated with a state-space model defined with first-order differential equations organised in matrix form:

$$\begin{cases} \frac{d}{dt}X = F \cdot X + G \cdot U \\ Y = H \cdot X + K \cdot U \end{cases} \tag{5.2}$$

where *F*, *G*, *H* and *K* are constant coefficient matrices, *X* is a column matrix of state variables, *U* is the set of input measures and *Y* is the matrix of output variables.

A power converter is inherently a system with one input and one output. The operation of the boost converter recommends selection of two state variables as the inductor current and capacitor voltage:

$$[x_1\ x_2] = [i\ v_c]. \tag{5.3}$$

The output voltage is also considered the output variable: $y = V_o$.

For the boost converter in the example, the two operational states in continuous conduction mode are shown in Figure 5.3.

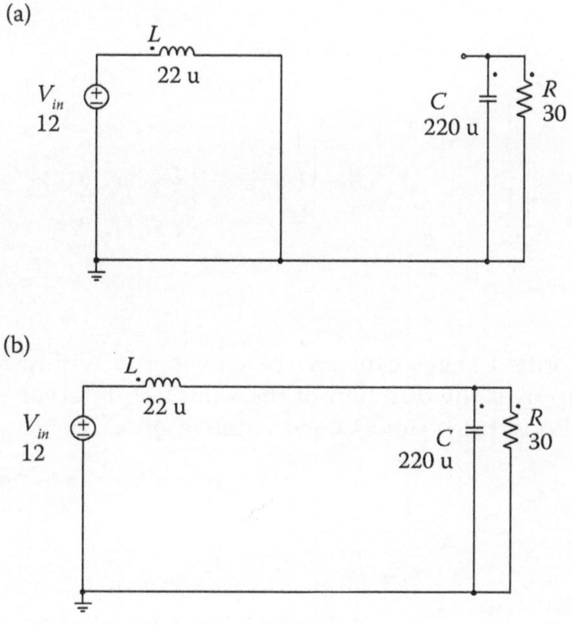

FIGURE 5.3
Equivalent circuits for the two states of a boost converter within continuous conduction mode.

Equations for the state shown with Figure 5.3a are:

$$\begin{cases} V_L = L\cdot\dfrac{di}{dt} = L\cdot\dot{X}_1 \\ \dfrac{V_{out}}{R} = -C\cdot\dfrac{dV_{out}}{dt} = -C\cdot\dot{X}_2 \end{cases} \Rightarrow \begin{cases} \dot{X}_1 = \dfrac{1}{L}\cdot V_L = \dfrac{1}{L}\cdot V_{in} \\ \dot{X}_2 = -\dfrac{1}{C\cdot R}\cdot X_2 \end{cases} \tag{5.4}$$

which can be rewritten in state-space form:

$$\begin{cases} \dot{X} = \begin{bmatrix} \dot{X}_1 \\ \dot{X}_2 \end{bmatrix} = \begin{bmatrix} 0 & 0 \\ 0 & -\dfrac{1}{R\cdot C} \end{bmatrix}\cdot\begin{bmatrix} X_1 \\ X_2 \end{bmatrix} + \begin{bmatrix} \dfrac{1}{L} \\ 0 \end{bmatrix}\cdot V_{in} = F_a\cdot X + G_a\cdot U \\ Y = \begin{bmatrix} 0 & 1 \end{bmatrix}\cdot\begin{bmatrix} X_1 \\ X_2 \end{bmatrix} = H_a\cdot X \end{cases} \tag{5.5}$$

The state shown with Figure 5.3b yields the circuit equations:

$$\begin{cases} V_{in} - V_{out} = L\cdot\dfrac{di}{dt} = L\cdot\dot{X}_1 \\ \dfrac{V_{out}}{R} = i - C\cdot\dfrac{dV_{out}}{dt} = X_1 - C\cdot\dot{X}_2 \end{cases} \Rightarrow \begin{cases} \dot{X}_1 = \dfrac{1}{L}\cdot V_{in} - \dfrac{1}{L}\cdot X_2 \\ \dot{X}_2 = \dfrac{1}{C}\cdot X_1 - \dfrac{1}{R\cdot C}\cdot X_2 \end{cases} \tag{5.6}$$

This can be rewritten in state-space form:

$$\begin{cases} \dot{X} = \begin{bmatrix} 0 & -\dfrac{1}{L} \\ \dfrac{1}{C} & -\dfrac{1}{R\cdot C} \end{bmatrix}\cdot X + \begin{bmatrix} \dfrac{1}{L} \\ 0 \end{bmatrix}\cdot V_{in} \\ Y = \begin{bmatrix} 0 & 1 \end{bmatrix}\cdot X \end{cases} \Rightarrow \begin{cases} \dot{X} = F_b\cdot X + G_b\cdot U \\ Y = H_b\cdot X \end{cases} \tag{5.7}$$

The two identified states can now be considered within a time-averaging relationship over the duration of the sampling interval T and keeping account of the boost transistor Q_1 conduction interval T_c:

$$\begin{aligned} F &= \frac{T_c}{T}\cdot F_a + \frac{T-T_c}{T}\cdot F_b \\ G &= \frac{T_c}{T}\cdot G_a + \frac{T-T_c}{T}\cdot G_b \\ H &= \frac{T_c}{T}\cdot H_a + \frac{T-T_c}{T}\cdot H_b \end{aligned} \tag{5.8}$$

This yields an average state-space model:

$$\begin{cases} \dot{X} = \begin{bmatrix} 0 & -\frac{1}{L}\cdot\frac{T-T_c}{T} \\ \frac{1}{C}\cdot\frac{T-T_c}{T} & -\frac{1}{R\cdot C} \end{bmatrix}\cdot X + \begin{bmatrix} \frac{1}{L} \\ 0 \end{bmatrix}\cdot U \\ Y = \begin{bmatrix} 0 & 1 \end{bmatrix}\cdot X \end{cases} \tag{5.9}$$

Considering the duty cycle definition for the converter leg (branch) as:

$$D = \frac{T-T_c}{T} \tag{5.10}$$

yields the voltage transfer relationship:

$$D = 1 - \left[1 - \frac{V_{in}}{V_{out}}\right] = \frac{V_{in}}{V_{out}} \tag{5.11}$$

The state-space model has its final form:

$$\begin{cases} \dot{X} = \begin{bmatrix} 0 & -\frac{1}{L}\cdot D \\ \frac{1}{C}\cdot D & -\frac{1}{R\cdot C} \end{bmatrix}\cdot X + \begin{bmatrix} \frac{1}{L} \\ 0 \end{bmatrix}\cdot U \\ Y = \begin{bmatrix} 0 & 1 \end{bmatrix}\cdot X \end{cases} \tag{5.12}$$

This set of equations should be considered further for small-signal variations around a known bias operation point. For instance, the model can be developed for small-signal variations around a bias operation point with an input voltage $V_{in} = 12$ V DC, output voltage $V_{out} = 20$ V DC and leg duty cycle $D_0 = 0.6$. Such a model is different from a model derived with any other operation settings, for instance, the case with $V_{in} = 9$ V DC.

To depict the small-signal model, the following notations are used to illustrate both the bias (uppercase letters) and the small-signal variation (lowercase letters).

$$\begin{aligned} D &= D_o + d \\ i_L &= I_L + i \\ v_C &= V_{out} + v_c \\ i_{OUT} &= I_{out} + i_{out} \\ v_{IN} &= V_{in} + v_{in} \end{aligned} \tag{5.13}$$

It yields:

$$\begin{cases} \begin{bmatrix} \dfrac{d(I_L+i)}{dt} \\ \dfrac{d(V_C+v_c)}{dt} \end{bmatrix} = \begin{bmatrix} 0 & -\dfrac{1}{L}\cdot(D_o+d) \\ \dfrac{1}{C}\cdot(D_o+d) & -\dfrac{1}{R\cdot C} \end{bmatrix} \cdot \begin{bmatrix} I_L+i \\ V_C+v_c \end{bmatrix} + \begin{bmatrix} \dfrac{1}{L} \\ 0 \end{bmatrix} \cdot (V_{in}+v_{in}) \\ V_{out}+v_{out} = [0 \quad 1]\cdot \begin{bmatrix} I_L+i \\ V_C+v_c \end{bmatrix} \end{cases} \quad (5.14)$$

After the terms corresponding to the bias operation point or the product of two small-signal variations, like $(d \cdot i)$ or $(d \cdot v)$, are eliminated, the small-signal model equations yield:

$$\begin{bmatrix} \dfrac{d(i)}{dt} \\ \dfrac{d(v_c)}{dt} \end{bmatrix} = \begin{bmatrix} 0 & -\dfrac{1}{L}\cdot(D_o) \\ \dfrac{1}{C}\cdot(D_o) & -\dfrac{1}{R\cdot C} \end{bmatrix} \cdot \begin{bmatrix} i \\ v_c \end{bmatrix} + \begin{bmatrix} 0 & -\dfrac{1}{L}\cdot(d) \\ \dfrac{1}{C}\cdot(d) & 0 \end{bmatrix} \cdot \begin{bmatrix} I_L \\ V_C \end{bmatrix} + \begin{bmatrix} \dfrac{1}{L} \\ 0 \end{bmatrix} \cdot (v_{in}) \quad (5.15)$$

These equations can be arranged in a more generic state-space form as:

$$\begin{cases} \dot{X} = \begin{bmatrix} 0 & -\dfrac{1}{L}\cdot D_0 \\ \dfrac{1}{C}\cdot D_0 & -\dfrac{1}{R\cdot C} \end{bmatrix} \cdot X + \begin{bmatrix} \dfrac{1}{L} \\ 0 \end{bmatrix} \cdot V_{in} + \left\{ \begin{bmatrix} 0 & -\dfrac{1}{L} \\ \dfrac{1}{C} & 0 \end{bmatrix} \cdot \begin{bmatrix} \dfrac{D_0}{R} \\ 1 \end{bmatrix} \cdot V_{out} \right\} \cdot d \\ Y = [0 \quad 1]\cdot X \end{cases} \quad (5.16)$$

The usage of the state-space theory usually stops here. The model based on state equations is generally used to derive the Laplace transfer function from a change in the duty cycle to a change in output voltage, which subsequently helps the design of the control (compensation) law. To differentiate from the actual control of transistors through the gate drivers, the term 'compensation' is preferred for the equation which defines the adjustment of the control pulses, and it ensures a compensation of the frequency effects of the plant model. This plant model incorporates all information about the dynamics of the boost converter and can be used for analytical investigation of the stability and dynamic transients of the system.

The main drawback for the state-space approach consists of the need for prior knowledge of the time intervals T_c and $T - T_c$, or, in another form, the duty cycle D_0. The model strongly depends on the bias operation point.

For instance, the converter in Figure 5.1, operated with a bias duty cycle of $D_0 = 0.6$, has a small-signal model in the Laplace form:

$$H_{conv}(s) = \frac{1818 \cdot s - 2.47 \cdot 10^6}{s^2 + 0.1515 \cdot s + 74.3} \tag{5.17}$$

If the bias duty cycle is changed to $D_0 = 0.3$, the small-signal model in the Laplace form changes to:

$$H_{conv}(s) = \frac{909.1 \cdot s - 1.24 \cdot 10^6}{s^2 + 0.1515 \cdot s + 18.6} \tag{5.18}$$

Figure 5.4 shows the Bode characteristics for the Laplace transfer function from small variations of the duty cycle to small variations in the output voltage for any bias duty cycle within the range 0.1–0.9. The small-signal model displays a relatively important change from the bias duty cycle, and it is less sensitive to a change in load resistance.

Any of these model transfer functions for the dynamics of the boost converter are defined with a pair of complex poles and a zero within the right

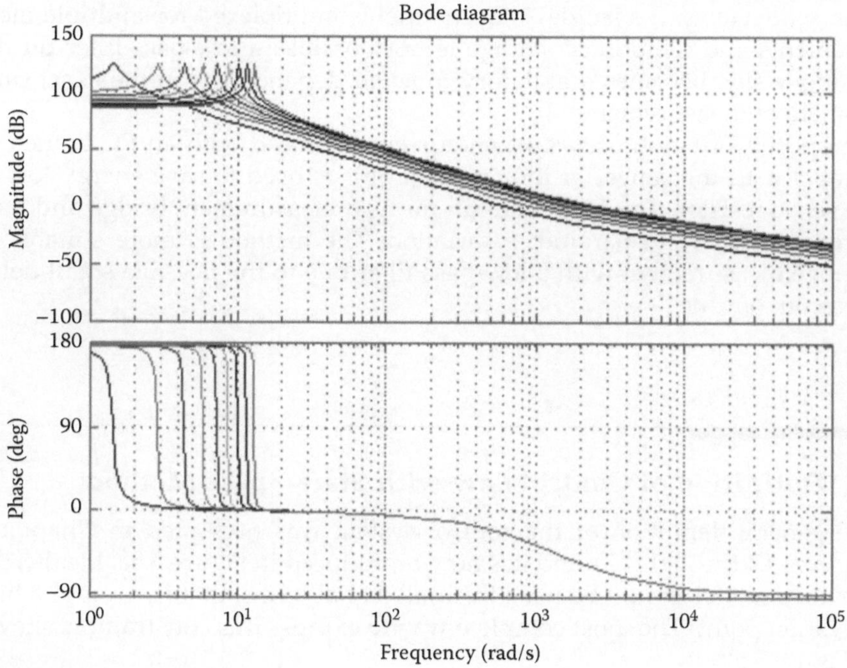

FIGURE 5.4
Bode characteristics for the converter model, with duty cycle variation between 0.1 and 0.9.

side of the complex plane. A more complete model, including inductor and capacitor loss, was provided in Tables 4.2 through 4.4 of Chapter 4.

Since the plant dynamic model has been herein depicted with a time-averaging method, it is important to understand that measured variables or parameters, like input voltage, output voltage, boost inductor current or the like, need to be time-averaged before being used within the control or compensation law. This would eliminate aliasing or instabilities produced by sampling in the wrong moment. This aspect relates to the digital implementation of the compensation law and may be seen as different from the previous analogue implementation. Hence, secondary effects related to analogue implementation, like slope compensation, are also avoided in digital implementation.

The time-averaging of the measurement signals is achieved with small low-pass filters on the analogue side of the A/D converters, tuned to average over the sampling interval. Such an implementation would introduce relatively large delay times due to low-pass filtering, which are sometimes not allowed in the control loop. Alternatively, advanced design can consider sampling in the middle of the main transistor conduction interval (Q_1 in the example) in order to assess the average value closer to the middle of the ripple trip from peak to peak. This is possible, especially when the digital platform allows for simultaneous acquisition at exactly the same time. Conversely, if a single A/D channel is multiplexed for multiple measurements and temporary storage is not possible, a low-pass filter on the analogue side becomes mandatory to avoid dependency on the exact time moment of measurement.

A special situation occurs when current is sensed with an RC branch in parallel with the converter inductor [6]. This is used to save energy lost in the sense resistor. The RC||RL circuit forms a measurement bridge and may become sensitive to parameter variation. The method is more suitable to average measurement with a low-pass filter due to the possible small delay when out of tune.

5.3 Definition of Control Law with State-Space Method

The generic definition of the control system was presented in Chapter 4 (Figures 4.4 through 4.7), and it is now reproduced in Figure 5.5a. Ideally, the previous model should be used in small-signal terms; that is, around a bias operation point. The most complete way to express this constraint is shown in Figure 5.5b.

Equation 5.16 can be used further for design of the compensation law directly in state-space form without the need for Laplace transforms. This

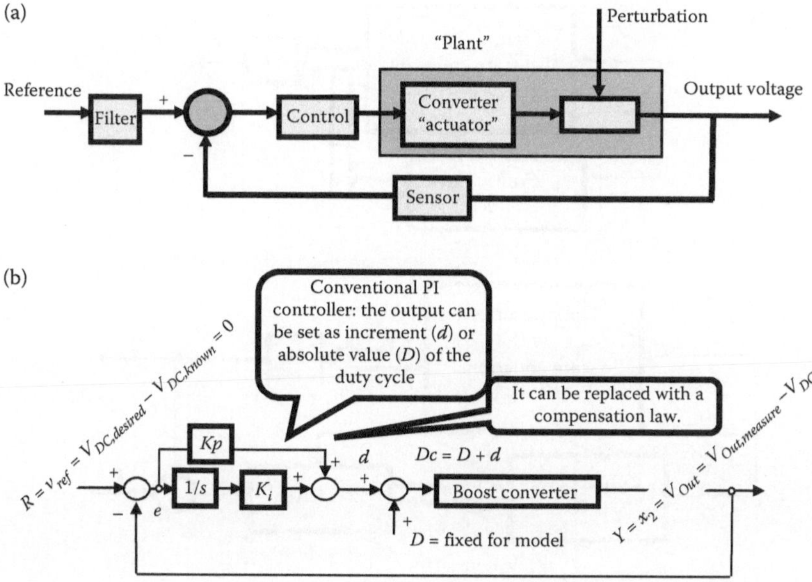

FIGURE 5.5
Control system setup around a power converter: (a) theoretical diagram, (b) usage of small signal concept.

type of control system design is called modern control, emphasising the complete digital (computer)-based design. This is a very promising field for the future, with the move of all development steps to a computer.

An attractive characteristic of design based on state variables consists of the structured design of the compensation law through a set of steps independent of the actual plant system. This represents a major advantage for the computer-based design of the control. The following standard steps to the design of a full controller can be observed through the representation in Figure 5.6:

- Establish the state equations and decide on the need for estimation of state variables. Note that the feedback is always after the state variables. If we cannot measure all state variables to complete the feedback loop, these can be calculated or estimated.
- The compensation law is always introduced as a linear combination of state variables.
- The compensation law plus the estimation represents the system compensation.
- Introduce the control reference through a linear relationship to state variables.

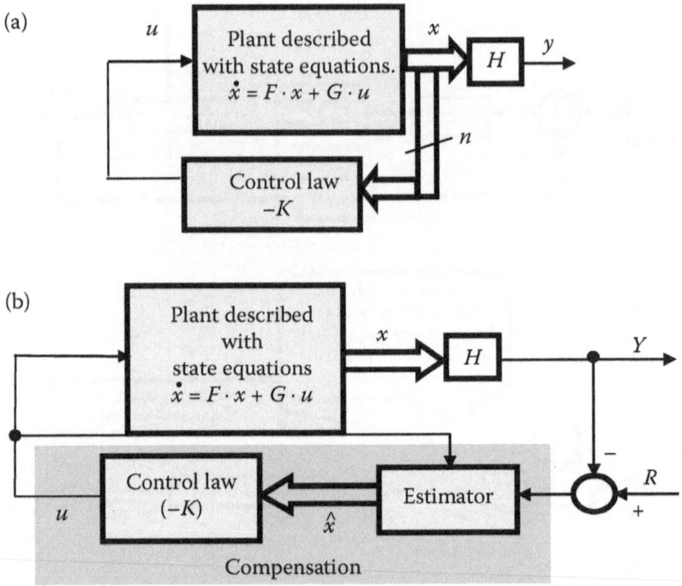

FIGURE 5.6
Introducing the control law based on state-space equations: (a) definition of control, (b) deciding between measurement and estimation.

The goal of compensation law design is to change the location of the system poles to satisfy certain dynamic performance requirements. After the design criteria have been processed, the desired location of the system poles is established. The number of state variables corresponds to the number of roots of the characteristic equation for the converter model. The location of each root can be achieved independently through a coefficient in the control law which acts to move an existing pole. Design of the compensation law with state variables assumes all state variables are known, either through direct measurement or by estimation. The compensation law as a linear combination of the state variables [x] is:

$$u = -K \cdot x = -[K_1 \quad K_2 \quad \ldots \quad K_n] \cdot \begin{bmatrix} x_1 \\ x_2 \\ \ldots \\ x_n \end{bmatrix} \tag{5.19}$$

Replacing (5.19) in previous state-space equations of form (5.2) yields:

$$\begin{aligned} \dot{x} &= F \cdot x - G \cdot K \cdot x \Rightarrow \dot{x} = (F - G \cdot K) \cdot x \Rightarrow s \cdot I \cdot x \\ &= (F - G \cdot K) \cdot x \Rightarrow (s \cdot I - F + G \cdot K) \cdot x = 0 \end{aligned} \tag{5.20}$$

The characteristic equation of the closed-loop system:

$$\det[s \cdot I - F + G \cdot K] = 0 \tag{5.21}$$

leads to a n-order polynomial with coefficients $K_1, \ldots K_n$.

The coefficients (K) are calculated so that the roots of the characteristic equations are moved to the desired locations. For n poles, we have n coefficients in K. These gains can be very easily calculated with generic computer programs suitable for any control system designed within state-space equations. A possibility is MATLAB.

In MATLAB, the gains K are calculated with either the *acker* or *place* instruction:

- K = *acker(F,G,Pd)* = Recommended for systems of order less than 10
- K = *place(F,G,Pd)* = Does not work well for multiple poles

Applying this theory to the converter small-signal model defined in Equation 5.16 means to consider the output voltage the system's output and the duty cycle variation the input, while the input voltage v_{in} and output current i_{out} are considered perturbation.

$$\begin{cases} \begin{bmatrix} \dot{x}_1 \\ \dot{x}_2 \end{bmatrix} = F \cdot \begin{bmatrix} x_1 \\ x_2 \end{bmatrix} + G_w \cdot v_{in} + G_o \cdot d \\ y = H \cdot \begin{bmatrix} x_1 \\ x_2 \end{bmatrix} = [0 \quad 1] \cdot \begin{bmatrix} i_L \\ v_C \end{bmatrix} \end{cases} \leftarrow \begin{bmatrix} x_1 \\ x_2 \end{bmatrix} = \begin{bmatrix} i_L \\ v_C \end{bmatrix} \tag{5.22}$$

The system poles can be moved into the desired locations through the gains K, helping improve transient response. Unfortunately, such a solution will depend upon the parameter variation; each parameter variation or wrong assessment would produce a steady-state error.

A general method to alleviate steady-state errors consists of adding an integral action against the error between reference and measured feedback. The error integral can now be seen as an additional state-space variable x_I. This is another major advantage of the state-space theory since it can easily modify the 'plant' model when composing the complete system from subsystems with known, yet different, state variables.

The new set of state variables is:

$$x = \begin{bmatrix} x_I \\ x_1 \\ x_2 \end{bmatrix} \tag{5.23}$$

and the state-space equations for the small-signal model yield:

$$\begin{cases} \begin{bmatrix} \dot{x}_I \\ \dot{x} \end{bmatrix} = \begin{bmatrix} 0 & H \\ 0 & F_s \end{bmatrix} \cdot \begin{bmatrix} x_I \\ x \end{bmatrix} + \begin{bmatrix} 0 \\ G_{sw} \end{bmatrix} \cdot V_{in} + \begin{bmatrix} 0 \\ G_{s0} \end{bmatrix} \cdot d - \begin{bmatrix} 1 \\ 0 \end{bmatrix} \cdot r \\ v_{out} = y = [0 \quad H] \cdot \begin{bmatrix} x_I \\ x \end{bmatrix} = [0 \quad 0 \quad 1] \cdot \begin{bmatrix} x_I \\ i_L \\ v_C \end{bmatrix} \end{cases} \leftarrow x = \begin{bmatrix} x_1 \\ x_2 \end{bmatrix} = \begin{bmatrix} i_L \\ v_C \end{bmatrix} \tag{5.24}$$

or, specifically, the state matrices are:

$$\begin{aligned} F_I &= \begin{bmatrix} 0 & H \\ 0 & F_s \end{bmatrix} = \begin{bmatrix} 0 & 0 & 1 \\ 0 & 0 & -\frac{1}{L} \cdot D_0 \\ 0 & \frac{1}{C} \cdot D_0 & -\frac{1}{R \cdot C} \end{bmatrix} \\ G_I &= \begin{bmatrix} 0 \\ G_s \end{bmatrix} = \begin{bmatrix} 0 \\ \frac{V_{out}}{L} \\ \frac{D_0 \cdot V_{out}}{R \cdot C} \end{bmatrix} \\ H_I &= [0 \quad H] = [0 \quad 0 \quad 1] \end{aligned} \tag{5.25}$$

Using Equation 5.24 in Equation 5.19 yields:

$$\begin{cases} \begin{bmatrix} \dot{x}_I \\ [x] \end{bmatrix} = FI \cdot \begin{bmatrix} x_I \\ [x] \end{bmatrix} - GI \cdot K \cdot \begin{bmatrix} x_I \\ [x] \end{bmatrix} - \begin{bmatrix} 1 \\ [0] \end{bmatrix} \cdot r = (FI - GI \cdot K) \cdot \begin{bmatrix} x_I \\ [x] \end{bmatrix} - \begin{bmatrix} 1 \\ [0] \end{bmatrix} \cdot r \\ v_{out} = y = [0 \quad H] \cdot \begin{bmatrix} x_I \\ x \end{bmatrix} = [0 \quad 0 \quad 1] \cdot \begin{bmatrix} x_I \\ i_L \\ v_C \end{bmatrix} \end{cases} \leftarrow [x] = \begin{bmatrix} x_1 \\ x_2 \end{bmatrix} = \begin{bmatrix} i_L \\ v_C \end{bmatrix} \tag{5.26}$$

Figure 5.7 shows a schematic representation of the resulting control system.

The design of the compensation law consists of defining the gains $K_x = [K_i \ [K]]$ able to move the system poles into a newly desired location. The problem really becomes the choice (selection) of the new position of the poles. Two important observations occur herein:

- Moving poles over a large distance means an important gain *K* (large control effort).
- It is very difficult to move poles near a system's zero-s since this also implies a large control effort.

Various methods link the desired dynamic performance requirements to the choice of the location of the poles. There are numerous methods able to define the best locations of the system poles [5,7–10]:

- Using heuristics, with a direct equivalence to a dominant pair of complex poles [7] based on a selection of the system's resonant frequency from system delays after sampling.
- Solving for a Bessel polynomial is used when the dynamic performance requirements are not clearly specified for easier identification of the pole location [11], and Bessel-derived poles tend to provide a slower response solution.
- Matching an *n*th-order polynomial that was designed to minimise the integral time absolute error (ITAE). ITAE represents the integral of the time multiplied by the absolute value of the error, calculated in response to a step function [11]. ITAE poles are not very damped; they usually lead to overshoot and faster rise times. Note that both ITAE and Bessel solutions are tabulated for 1 sec in technical literature like [11], and the designer need only map these results into any desired frequency range.
- A pure optimal criterion is sought with the symmetrical root locus [12].
- Sometimes, an inverse Hilbert transform (IHT) method is used.

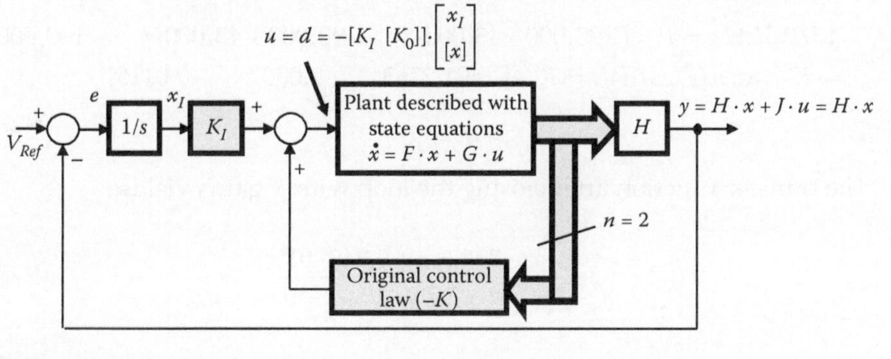

FIGURE 5.7
Control system with error integral term, showing the converter (plant) model and the state-space–based controller with previously defined gains *K*.

Any of these methods can be implemented offline on computers, and the design becomes a computer-based optimisation problem. As an alternative to computer optimisation, the pole location can also be imposed empirically from designer experience. This approach is considered herein to provide more insight into the process of defining the compensation law for the boost converter. A fully optimised process may lose the first-time reader with complex equations.

The switching frequency has been already selected at 142 kHz (7.04 μs). The effect of digitisation allows a proper signal reconstruction up to half the sampling (or switching) period.

The converter's small-signal model was depicted through averaging over a sampling cycle. This implies the introduction of a delay of half of $T_s = 7.04$ μs. The control system theory says that any time delay reduces phase reserve and pushes the system towards instability.

For instance, the phase delays introduced at various frequencies yield:

$$f_o = 5.0 \text{ kHz} \rightarrow \theta = 4.5°$$

$$f_o = 7.5 \text{ kHz} \rightarrow \theta = 9.0°$$

$$f_o = 15.0 \text{ kHz} \rightarrow \theta = 18.0°$$

The resonant frequency for the pair of complex poles can be selected as $f_o = 15.915$ kHz; that is, 100 krad/sec. The two complex poles can be empirically desired at $-90{,}000 \pm 43{,}000{*}i$ [rad/sec] at a location set with a radius equal to the system's desired resonant frequency. The pole at the origin may be moved to a location beyond –100 krad/sec, straight on the real axis.

MATLAB can calculate gains $K = [Ki \quad [Ko]]$ with the *acker* instruction. This provides the gains:

$$\begin{aligned} f_0 = 15.915\,\text{kHz} &\Rightarrow P = [-90{,}000 - 43{,}000 \cdot j \quad -90{,}000 + 43{,}000 \cdot j \quad -100{,}000] \\ &\Rightarrow K = acker(FI, GI, Pd) \Rightarrow K = [-401.2763 \quad -0.0003 \quad -0.0115] \end{aligned} \tag{5.27}$$

The transfer function after closing the loop with K gains yields:

$$H(s) = \frac{-729.6 \cdot s + 9.949 \cdot 10^5}{s^3 + 280 \cdot s^2 + 2.795 \cdot 10^4 \cdot s + 9.949 \cdot 10^5} \tag{5.28}$$

This transfer function demonstrates that the zero on the right-hand side of the complex plane is preserved, and the poles are moved towards higher frequencies.

The novel location of poles is shown with Figure 5.8. These results are close to a selection based on the Bessel solution, with the difference of a larger attenuation. This choice for pole location is intended with regard to the possible increased overshoot due to the digital implementation, in equivalence to an analogue controller.

An additional pole has been introduced as a state variable in the system in order to reduce the steady-state error with an integral of the reference error. Unfortunately, this pole has a negative effect on the dynamic behaviour, as it slows down the transient response. In order to compensate for this negative effect, a feed-forward term (*N*) can be added from the reference to the boost the converter controller's input. The value of the gain *N* is chosen to produce a zero at the same frequency as the location of the third pole of the closed-loop system.

The previous decision was to have this third pole at a value of 15.915 kHz or 100,000 rad/sec in Equation 5.26, adopted from empirical reasons concerning phase delay. This now means adopting a feed-forward term of:

$$N = -\frac{K_{1[1]}}{P} \tag{5.29}$$

where $P = -100{,}000$ [rad/sec] equals the third single pole (real term only) in the adopted solution.

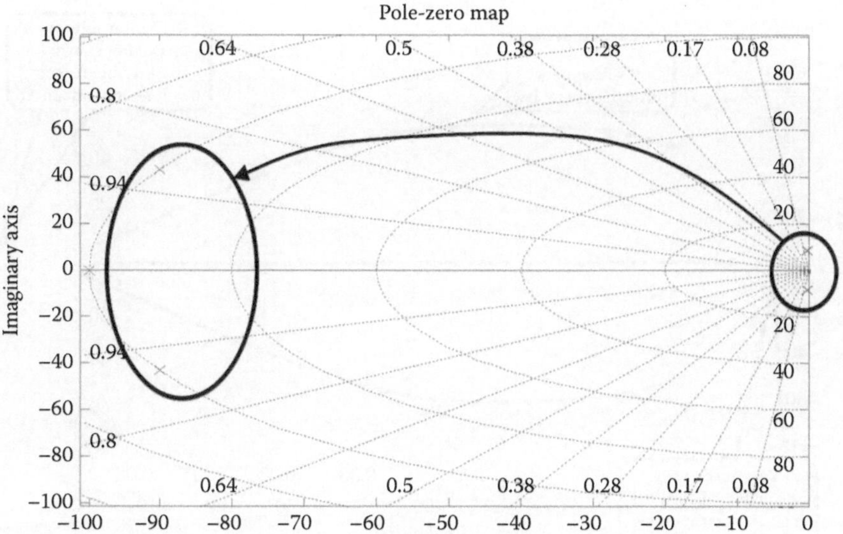

FIGURE 5.8
Displacement of the converter poles after *K*-law control.

For completeness of presentation, all poles and zeros of the system in the negative half of the complex plane are shown in Figure 5.9. The addition of the feed-forward term produces the system structure in Figure 5.10.

The state-space equations change into:

$$\begin{cases} \begin{bmatrix} \dot{x}_I \\ [x] \end{bmatrix} = FI \cdot \begin{bmatrix} x_I \\ [x] \end{bmatrix} + GI \cdot \left\{ -K \cdot \begin{bmatrix} x_I \\ [x] \end{bmatrix} + N \cdot r \right\} - \begin{bmatrix} 1 \\ [0] \end{bmatrix} \cdot r \\ \quad = (FI - GI \cdot K) \cdot \begin{bmatrix} x_I \\ [x] \end{bmatrix} + \left\{ GI \cdot N - \begin{bmatrix} 1 \\ [0] \end{bmatrix} \right\} \cdot r \qquad \leftarrow [x] = \begin{bmatrix} x_1 \\ x_2 \end{bmatrix} = \begin{bmatrix} i_L \\ v_C \end{bmatrix} \\ v_{out} = y = [0 \quad H] \cdot \begin{bmatrix} x_I \\ x \end{bmatrix} = [0 \quad 0 \quad 1] \cdot \begin{bmatrix} x_I \\ i_L \\ v_C \end{bmatrix} \end{cases} \tag{5.30}$$

Numerical data for the boost converter yield $N = -4.0128$ and a transfer function:

$$H(s) = \frac{-7.296 \cdot s^2 + 9219 \cdot s + 9.949 \cdot 10^5}{s^3 + 280 \cdot s^2 + 2.79510^4 + 9.949 \cdot 10^5} \tag{5.31}$$

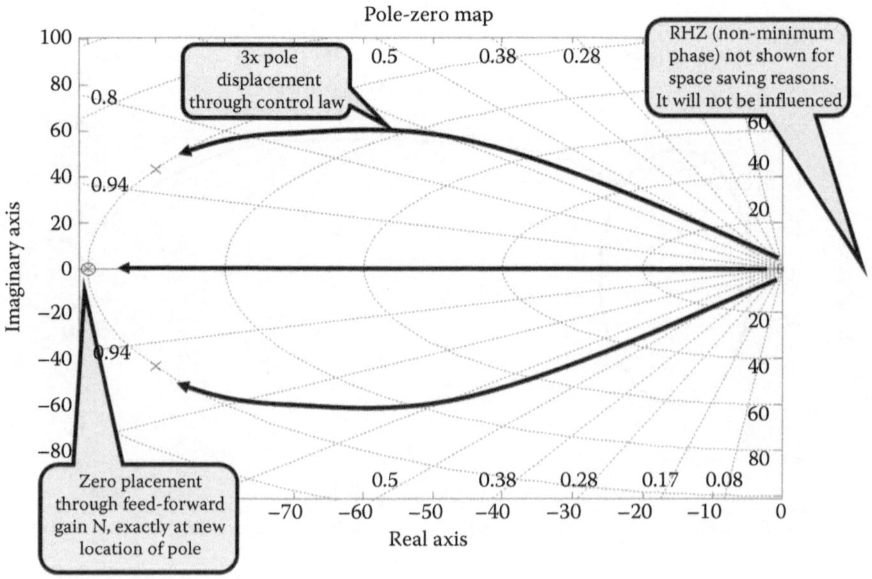

FIGURE 5.9
Final (desired) location of poles and zero for the boost converter system in the example.

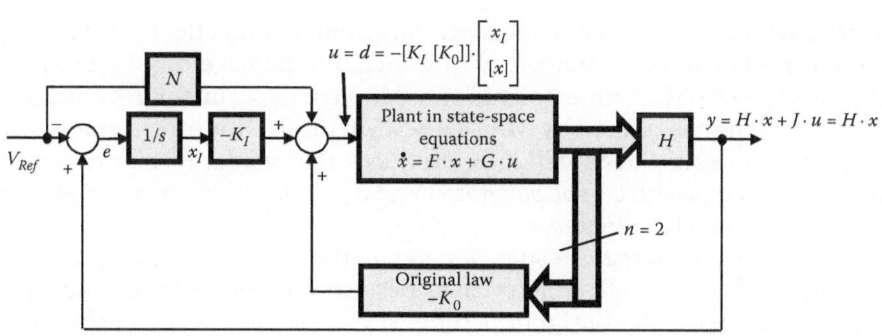

FIGURE 5.10
Boost converter system (converter plus compensation) with a state-space design of the K gains and feed-forward term.

A comprehensive analysis of results includes the visualisation of the new location of poles in Figure 5.9, the theoretical step response of the converter small-signal model in Figure 5.11 and the Bode characteristics for the system transfer function in Figure 5.12.

All these graphics are design and analysis tools applied to the converter's small-signal model. Therefore, they do not assess the operation of the actual hardware, which depends on real-time operation that is subject to sampling, delays, parasitic components and other possible nonlinearities. For instance, the previous design with a selection of the poles' location around 15 kHz is acceptable from an analytical point of view when analysis is performed with Laplace in the s-domain. However, it may produce instabilities when used in

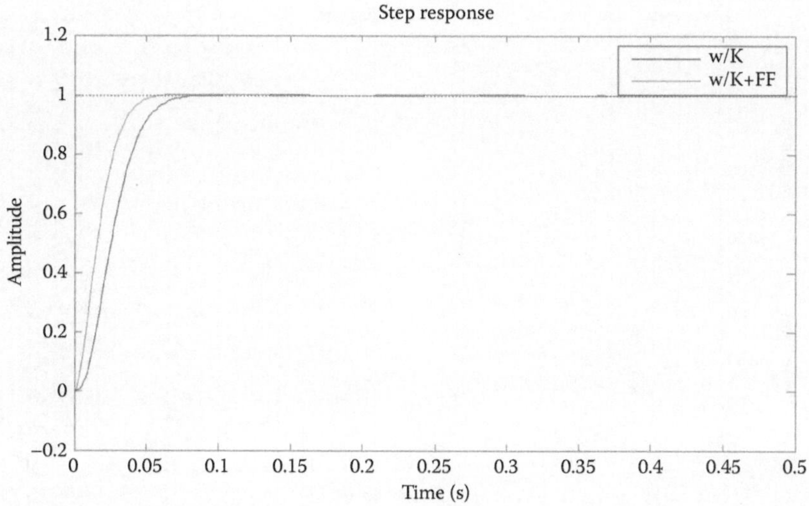

FIGURE 5.11
Step response of the converter's small-signal model, before (slower) and after (faster) adding the feed-forward term.

practice with digital systems due to the large delays and effects of sampling. These aspects can be contained within a comprehensive digital analytical study and proven with time domain simulation or experiments. A quick way to prove a design is offered by MATLAB-SIMULINK. Running a model with the previous design in SIMULINK produces the results in Figure 5.13. In such a case, the design can be reiterated and a pole location corresponding to 5.0 or 7.5 kHz can be adopted.

Furthermore, all of these results depend on the bias operational point, and a change in input voltage, duty cycle or passive components may affect performance. Usually, the designer accepts such a small loss in performance and the same gains are used for any load. The most important performance difference consists of the steady-state error, which is corrected with the introduced integral action. Small differences in response time, overshoot or stabilisation time are considered acceptable. Hence, it is very rare that an adaptive controller is used to change the gains with variation in the bias input voltage or load current.

The control structure presented in Figure 5.10 can be rearranged as shown in Figure 5.14. This emphasises the structure of the control system as a PI-equivalent module and an additional linear sum-of-products compensation term. Both terms, the PI equivalent and the linear sum-of-products compensation, can easily be implemented in software with library functions like PI or sum-of-products algorithms. This arrangement is considered the fastest implementation of the state-space controller within a digital structure. This is important because the state-space control is usually avoided for industrial usage due to its apparent complexity.

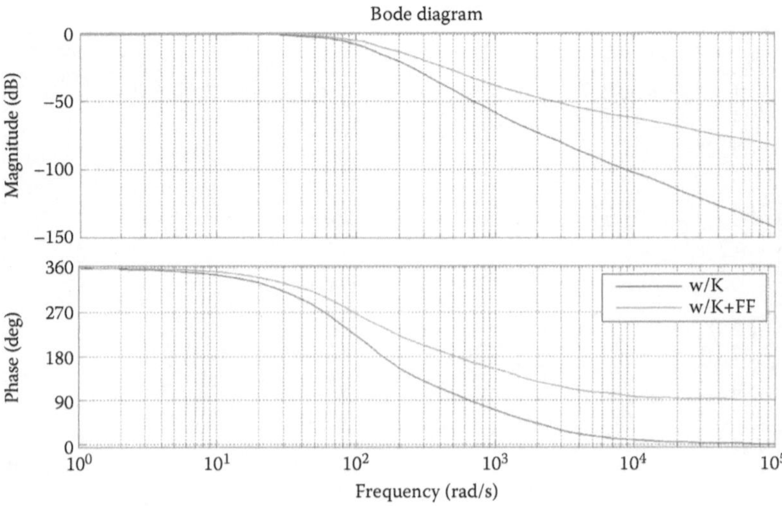

FIGURE 5.12
Bode characteristics for the converter's small-signal model, before (bottom) and after (top) adding the feed-forward term.

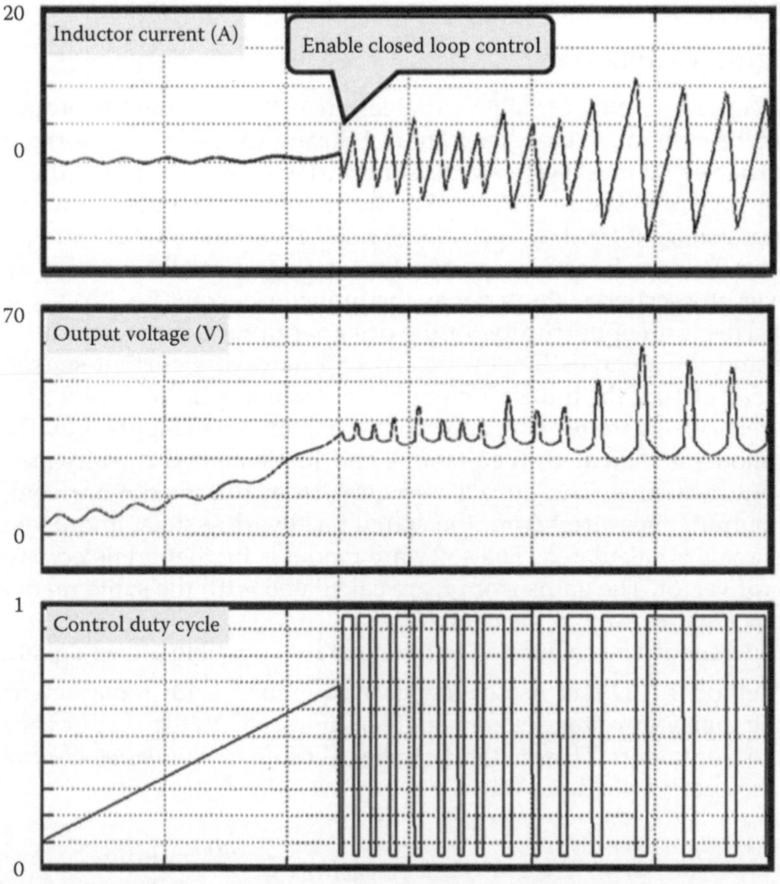

FIGURE 5.13
Instabilities proven with MATLAB-SIMULINK for a system pole location selected at 15 kHz and a system sampling at 142 kHz.

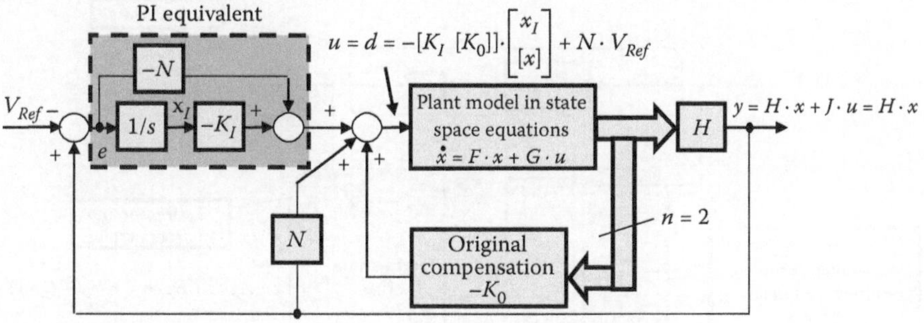

FIGURE 5.14
Equivalence of two well-known terms, a PI equivalent term and a sum-of-products compensation term, provides a fast implementation on a digital platform.

5.4 Using Estimators

It has been shown that the state variables are different from the output variables of the system, and feedback control through K gains is performed with state variables. In most systems, the state variables are internal to the system and not easy to measure. This is also the case with the boost converter: the capacitor voltage (first state variable) coincides with the output voltage of the converter. This can be easily measured for setting up feedback in any system. However, the second state variable, the inductor current, requires a second sensor. This is an opportunity for the designer to save money on the second sensor and to use an estimator instead of a hardware current sensor. This means calculating the inductor current from equations.

The generic scheme for using an estimator is shown in Figure 5.15. The converter model is recreated in equations and implemented in software using the input measures. The output measures from this model are compared to the outputs measured from the actual hardware system, and a vector of error signals is calculated. The software model is further adjusted based on this error vector. The gains from E are calculated with the same math as the previous K gains used in the control.

The current can be calculated with one of two possible sets of equations:

- Method 1 = Direct use of the state equations, with replacement of the capacitor voltage (state variable) samples with the output voltage measurements. This method can be called *direct calculation of current*. The current yields:

$$i_L(k+1) = A_1 \cdot v_{out}(k+1) + B_1 \cdot v_{out}(k) + C_1 \cdot \frac{d(k+1)+d(k)}{2} \tag{5.32}$$

where the coefficients are provided herein without demonstration.

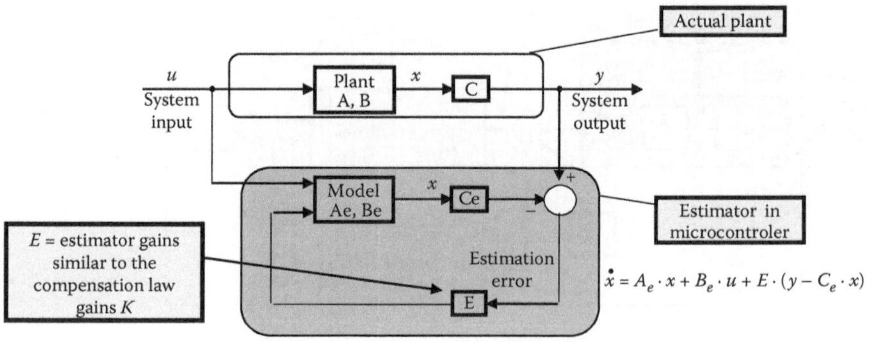

FIGURE 5.15
General structure for using an estimator.

$$A_1 = \frac{\left[1+\frac{1}{2}\cdot\frac{1}{R\cdot C}\cdot T\right]}{\frac{D_0}{C}\cdot T} - \frac{\left[\frac{D_0}{L}\cdot T\right]}{4}$$

$$B_1 = -\frac{\left[1-\frac{1}{2}\cdot\frac{1}{R\cdot C}\cdot T\right]}{\left[\frac{D_0}{C}\cdot T\right]} - \frac{\left[\frac{D_0}{L}\cdot T\right]}{4} \tag{5.33}$$

$$C_1 = \frac{\left[\frac{V_{out}}{L}\cdot T + \frac{2\cdot V_{out}}{R}\right]}{2}$$

- Method 2 = Using a reduced-order estimator with the usage of an intermediate error between the output voltage and estimated voltage. This method can be called *reduced order estimator.* The current yields:

$$i_L(k+1) = A_1\cdot i_L(k) + B_1\cdot v_{out}(k+1) + C_1\cdot v_{out}(k) + D_1\cdot\frac{d(k+1)+d(k)}{2} \tag{5.34}$$

where the coefficients are provided herein without demonstration.

$$A_1 = \frac{1-\left(\frac{E}{C}\cdot D_0\cdot\frac{T}{2}\right)}{1+\left(\frac{E}{C}\cdot D_0\cdot\frac{T}{2}\right)}$$

$$B_1 = \frac{E+\left(-\frac{1}{L}\cdot D_0 + \frac{E}{R\cdot C}\right)\cdot\frac{T}{2}}{1+\left(\frac{E}{C}\cdot D_0\right)\cdot\frac{T}{2}}$$

$$C_1 = \frac{-E+\left(-\frac{1}{L}\cdot D_0 + \frac{E}{R\cdot C}\right)\cdot\frac{T}{2}}{1+\left(\frac{E}{C}\cdot D_0\right)\cdot\frac{T}{2}} \tag{5.35}$$

$$D_1 = \frac{\left(-\frac{V_{out}}{L} - \frac{E}{C}\cdot\frac{D_0}{R}\cdot V_{out}\right)\cdot T}{1+\left(\frac{E}{C}\cdot D_0\right)\cdot\frac{T}{2}}$$

The second method; that is, the reduced order estimator, requires a definition of the gain E based on the specification of the estimator pole. There is a single estimator pole for the estimation of a single state variable, and it can be defined for a response as quickly as possible. Hence, an estimator pole with a frequency as high as possible is desirable. Furthermore, the estimator pole needs to be at least 2 … 8 times higher (in absolute terms) than the other system poles so it will not influence the system behaviour. That is, any estimator dynamics are done before the main compensation law can react. For the presented numerical data, since the system poles have been selected in the range 5–15 kHz, the estimator pole can be adopted at 35–70 kHz.

The principle of using the two current estimation methods is provided with Figure 5.16, where the gains K are calculated for the compensation law as discussed before.

The estimator design considers a gain E without any integral action. The steady-state error is not important in this case since the estimated current is used within a closed-loop control system for output voltage, which will take care of any internal steady-state error. A larger E gain corresponds to a higher estimator pole frequency. This cuts the steady-state error and also somewhat improves the transient. As a drawback, a higher estimator pole frequency will make the system more sensitive to noise.

If better performance is sought, an integrator action should also be considered for the estimator error. This should complicate the equations and slow down the response of the system.

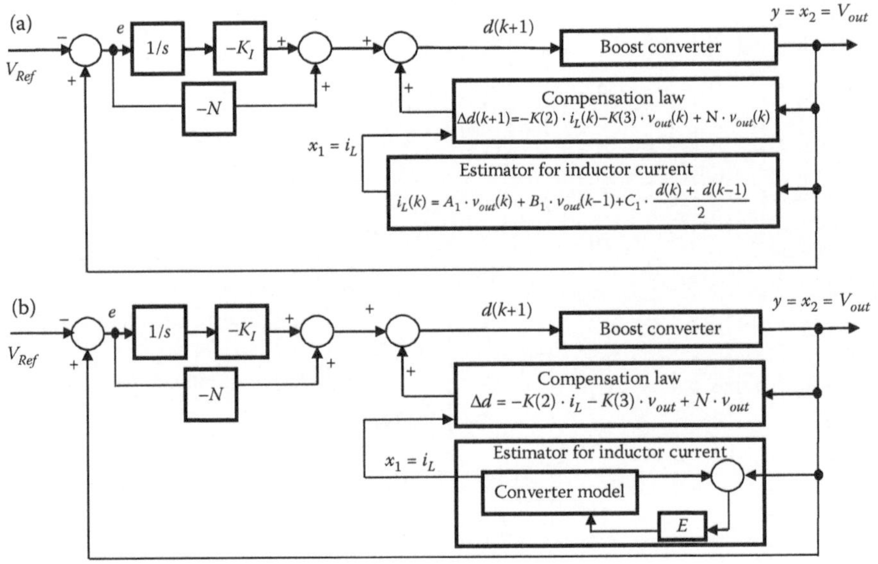

FIGURE 5.16
Usage of the two methods for current calculation, without current sensor: (a) direct calculation from circuit equations, (b) using a high gain against the error for correction of the internal model.

5.5 Change of Gains with Load and Dwell Time

It has been demonstrated that the gains suitable to a compensation law are not optimal for any set of converter parameters, and a change in input voltage or load current influences dynamic results.

In an extreme case, such a change in the plant model can lead to instabilities. The best-known case in practice occurs when a lower-than-expected input voltage produces instabilities of the power supply set to work with gains for a higher input voltage. A quick example is achieved by running the compensation law designed in this chapter for the 12–20 V DC boost converter at a lower input voltage in a closed loop. Results from a MATLAB-SIMULINK analysis of the system are provided in Figure 5.17. Decreasing the input voltage to 5 V DC produces instability in the system, while a small increase from 12 to 16 V DC is tolerated by the system.

Digital control of power supplies opens up many possibilities to address this matter.

The simplest solution is to adjust the control duty cycle based on direct measurement of the input voltage through a feed-forward scaling term. The change with the load current is addressed in the simplest form, with two sets of gains, one optimised for high current and the other for weak load current. Such a simple solution is implemented in numerous integrated circuits. For instance, Linear Technology's LTC3880 uses the power management bus for operation management. It designates the information from a 'bit' [7] of MFR_PWM_MODE_LTC3880 (command code 0 × D4) to adjust gains in two steps for low and high current. Changing this bit value changes the PWM loop gain and compensation: 0 – low current range, 1 – high current range. This is further explained in Chapter 8, along with the power management bus.

Other high-performance systems consider an adaptive change of the compensation law gains with the load current. The most intuitive approach monitors the load current and changes the gains at large changes in the load current. A gain scheduling technique is generally optimally introduced to improve the transient performances. The optimal settings for the control parameters at different load currents and input voltages, an adaptive mechanism based on machine learning algorithms, can be used [13]. When properly implemented, this approach can feature an improvement of up to 20%. Due to the need for optimal settings and online adaptation, such an approach may be difficult in practice.

An alternative can be provided with a selective change of gains without continuous action of the search and adaptation algorithm. The change is therefore done with a certain *dwell time* that is herein optimised for the case of either a voltage reference or a major perturbation change (either input voltage or output current) [14].

Switched linear state-space control is therefore used whenever the system operates over a wide range of input voltage and load current variations. All

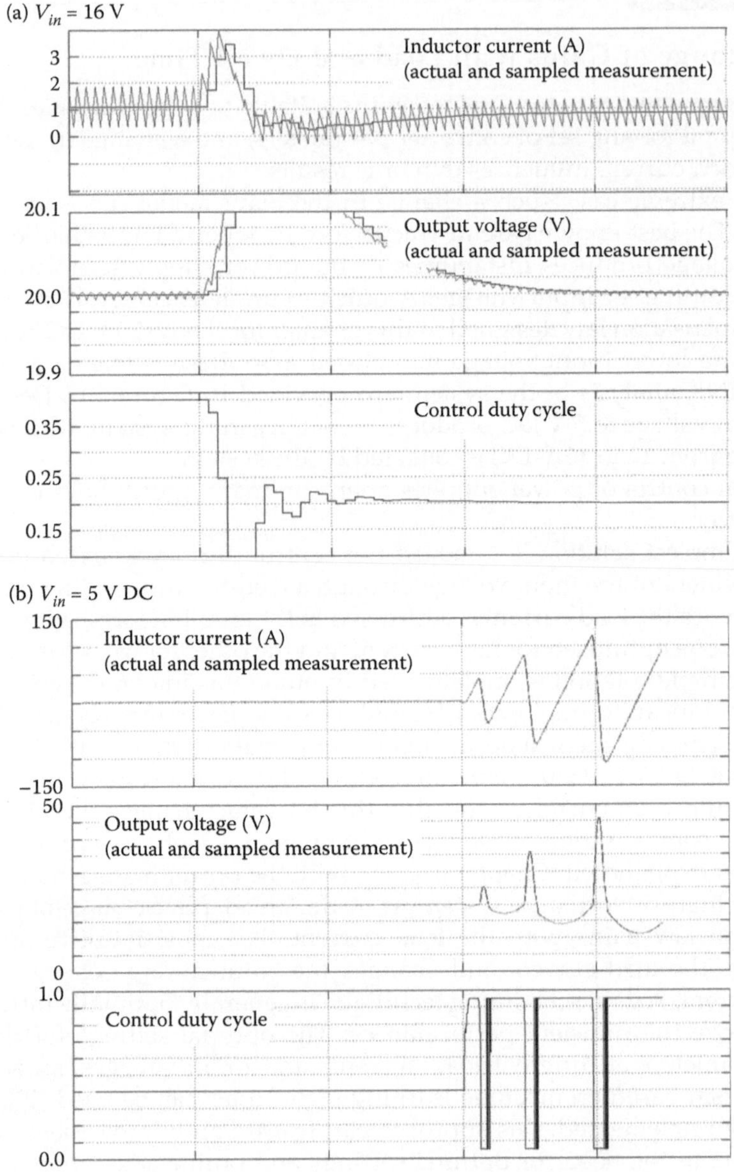

FIGURE 5.17
Main waveforms for the control with estimation, control poles at 5 kHz and estimator pole at 35 kHz.

control gains (K_i, $K_{(1)}$, $K_{(2)}$, N) are precalculated offline and stored. A minimal solution can consider 10 values of bias duty cycle, in steps of 0.1 from 0.1 to 0.9. Optimisation can further be performed for the best dwell time and the best switching constraint while operating across the same table of stored gains.

The small-signal output of the compensation law is continuously monitored. After this small-signal component of the duty cycle is maintained within a fixed interval (like 10% of nominal) for 24 sampling cycles at a PWM frequency of 142 kHz (that is, a total of 169 μsec), the bias duty cycle (D_0) is reassessed (Figure 5.18).

Reassessing the bias duty cycle switches the model between one and another of the 10 available data sets. Obviously, the table of stored gains can be designed with more than 10 data sets; this is just an example. The resolution is chosen herein based on accuracy of monitoring changes in controller output from inherent noise or ripple.

The effect of the dwell time is to limit the transient trip when gains are changed by changing them in smaller steps. For instance, a change from a duty cycle of 0.6 to a duty cycle of 0.3 is done over three steps, each of 169 μsec, in the sequence 0.6 → 0.5 → 0.4 → 0.3. At each change of gains, there is a minimal distortion of the output voltage. A more severe distortion could be seen if the gains were changed suddenly from 0.6 to 0.3. Over all operation, the output voltage stays within the desired specification, such as 50 mV peak–peak maximum variation. The process is slower but with less distortion. After a set of gains is changed, the system's response time becomes shorter, with less overshoot and shorter stabilisation time, all with advantages in efficiency and allowed speed of the digital load.

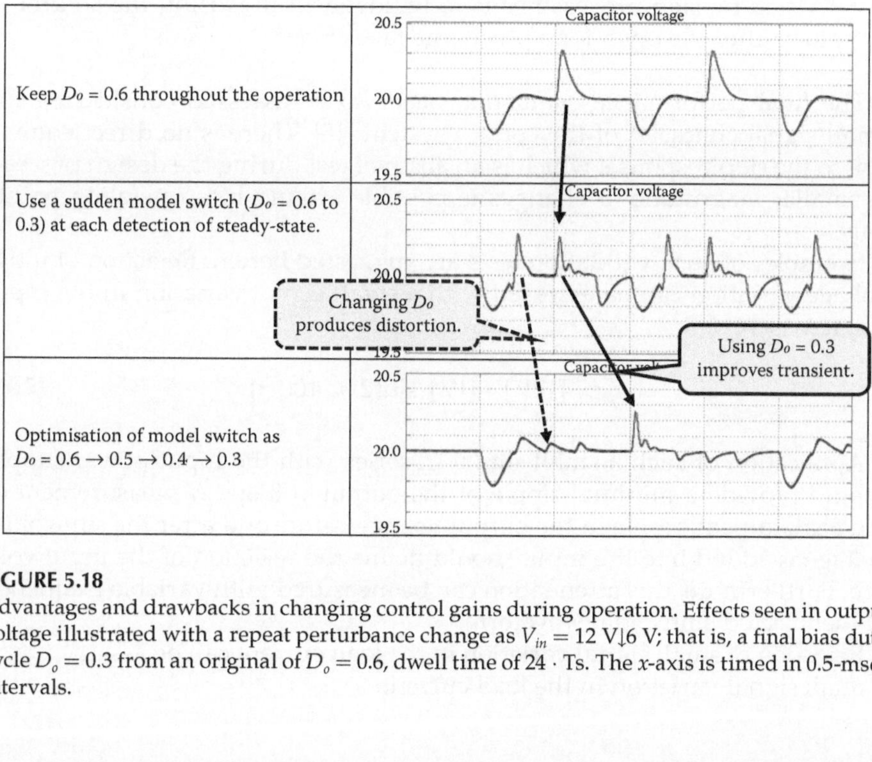

FIGURE 5.18
Advantages and drawbacks in changing control gains during operation. Effects seen in output voltage illustrated with a repeat perturbance change as $V_{in} = 12\ \text{V}{\downarrow}6\ \text{V}$; that is, a final bias duty cycle $D_o = 0.3$ from an original of $D_o = 0.6$, dwell time of $24 \cdot$ Ts. The x-axis is timed in 0.5-msec intervals.

Any of the algorithms presented for changing the compensation gains to keep up with input voltage and load current are not recommended when such dramatic changes occur frequently.

5.6 Practical Aspects: Implementation Problem

The previous development of the state-space control law works with small-signal models and does not make the implementation obvious. For implementation within a digital platform, some additional steps need to be considered:

- The charging of the output capacitor at startup has to be done with a specialized circuit with relay decoupling or be left alone to the series diode.
- An open-loop ramp-up algorithm needs to precharge the output capacitor at limited current smoothly from the input DC voltage level closer to the desired higher-output DC level through a slowly paced change of the duty cycle.
- A set of protection features needs to cease operation when a fault occurs in the system.
- A step-by-step process needs to be involved in closing the control loop after the open-loop precharge takes place.

The final performance evaluation needs to be assessed considering the small-signal character of the control system [15]. There is no direct equivalent to the step response, which is an abstract test during the design process. A smaller magnitude of changes is possible around bias operation points only.

Examples of such validation tests are suggested herein. Rejection of input voltage variation can be assessed with a small signal variation in the input voltage:

$$v_{in} = [12V] + [1V] \cdot \sin(2 \cdot \pi \cdot 100 \cdot t) \qquad (5.36)$$

Application of such a small signal in series with the input power supply should provide a minimal ripple of the output voltage. A measurement of the peak-to-peak ripple in the output voltage before and after the sinusoidal voltage is added into the input should define the rejection of the input voltage. Furthermore, this attenuation can be measured with variable frequency of the injected sinusoidal waveform.

Rejection of small signal variation in the load current can be assessed with a small signal variation in the load current.

$$i_{load} = \left[\frac{V_{out}}{R = 30\ \Omega}\right] + [0.55] \cdot \sin(2 \cdot \pi \cdot 1000 \cdot t) \tag{5.37}$$

Analogously, the ripple of the output voltage should quantify the attenuation at various frequencies of the injected sinusoidal waveform.

Considering a large signal step in the input voltage or output current would actually assess the dependency of the model on the bias operational point rather than the quality of the control system through a step response. It can be an assessment of performance if properly understood.

While an optimisation of performance was not the complete goal herein, this study proves the possibility of a complete design and implementation of the control system using computers without actual bench tests or tune-ups and provides the steps of the process.

Summary

The state-space–based control, also known as modern control, presents a series of advantages with the intensive usage of computers. The entire design and analysis can be performed offline on computers, minimising the tune-up times. All control gains can be calculated offline with conventional library functions. The control code is also standardised and can be automatically generated for further implementation on microcontrollers.

The state-space control can be seen as two components: a PI equivalent module and a sum-of-products linear compensation term.

The overall system can easily be optimised with an estimation of the inductor current, which saves a current sensor in the overall hardware.

The main concern with the adoption of digital control by the power supply industry relies on the complex mathematical development and the nonstandard form for code implementation of these equations. The state-space control method provides an arrangement able to keep the same software structure independently of the offline optimisation problem (pole selection for the control law or estimation of certain state variables). The software structure can thus be validated and the code can be implemented while having several parameters as key development parameters. The code can be tested and certified for fault-free operation without uncertainty independently of the actual dynamic performance of the power supply. Offline setting of the parameters can be made under the same real-time code.

The actual use of state-space–based control does not make the software more difficult than a conventional PI controller.

Finally, this chapter discussed steps taken in the digital implementation of the compensation law, with all the novel complexity introduced by this process.

References

1. Franklin, G.F., Powell, J.D., Emami-Naemi, A., 2014, *Feedback Control of Dynamic Systems*, 7th edition, Prentice-Hall, Upper Saddle River.
2. Maksimovic, D., 2014, "Digital Control of High-Frequency Switching Power Converters", in *Tutorial presented at The 14th IEEE Workshop on Control and Modeling for Power Electronics (COMPEL)*, Salt Lake City, UT, pp. 1–100.
3. Jones, M. 2011, Digital Compensators Using State Feedback Techniques, EEWeb Documentation.
4. Jones, M., 2011, Digital Compensators Using Frequency Techniques, EEWeb Documentation.
5. Neacsu, D., 2016, A State-Space Design Approach to Digital Feedback Control of DC/DC Converters, Tutorial IEEE APEC, Long Beach, CA, USA, pp. 1–130.
6. Anon, 1998, A Simple Current-Sense Technique Eliminating a Sense Resistor, *Linfinity Application Note AN-7*, pp. 1–6.
7. Neacsu, D.O., 2015, "A Simplified Approach to Implementation of State-Space Control of DC/DC Converters on Low-Cost Microcontrollers", in *Proceedings of IEEE IECON Conference*, Yokohama, Japan, pp. 631–636.
8. Kelly, A., Rinne, K., 2005, "Control of DC-DC Converters by Direct Pole Placement and Adaptive Feed-Forward Gain Adjustment", in *Proceedings of IEEE APEC*, 6–10 March 2005, Austin, TX, USA, pp. 1970–1975.
9. Bae, H.S., Yang, J.H., Lee, J.H., Cho, B.H., 2008, "Digital State Feedback Control and Feed-Forward Compensation for a Parallel Module DC-DC Converter using the Pole Placement", in *Proceedings of IEEE APEC*, 24–28 February 2008, Austin, TX, USA, pp. 1722–1725.
10. Usman Iftikhar, M., Godoy, E., Lefranc, P., Sadarnac, D., Karimi, C., 2008, "A Control Strategy to Stabilize PWM DC-DC Converters with Input Filters Using State-Feedback and Pole-Placement", in *Proceedings of IEEE INTELEC*, 14–18 February 2008, San Diego, CA, USA, pp. 1–5.
11. How, J., 2007, Course materials for 16.31 Feedback Control Systems – Fall 2007, *MIT OpenCourseWare, Massachusetts Institute of Technology*.
12. Ghosh, A., Prakash, M., Pradhan, S., Banerjee, S., 2014, "A Comparison among PID, Sliding Mode, and Internal Model Control for a Buck Converter", in *Proceedings of IEEE IECON*, 29 October–1 November 2014, Dallas, TX, USA, pp. 1001–1006.
13. Andries, V.D., Goras, L., Buzo, A., Pelz, G., 2017, Automatic Tuning for a DC-DC Buck Converter with Adaptive Controller, *IEEE ISSCS 2017*, Iasi, Romania, pp.
14. Neacsu, D., 2016, "Switched Linear State-Space Control of DC/DC Converters with Optimal Dwell-Time", in *IEEE IECON Conference*, Florence, Italy, October 2016, pp. 1423–1428.
15. Kurucso, B., Peschka, A., Stumpf, A., Nagy, I., Vajk, I., 2015, "State Space Control of Quadratic Boost Converter using LQR and LQG approaches", in *IEEE OPTIM Conference*, Side, Turkey, 2015, pp. 642–648.

6

Isolated DC/DC Converters: Flyback, Forward and Push-Pull

6.1 Role of Direct Current/Direct Current Converters with Isolation

Certain applications require galvanic isolation between the load circuit and the power supply. This is usually achieved with a power converter which includes a transformer. Usually, the isolation through the transformer is also responsible for any voltage difference between the circuitry on primary and secondary windings.

A good example consists of the measurement of current through a high-voltage line used to power a street car or electric train. Possible voltage systems for an electric train include 600–3 kV DC, 15 kV AC, 16.7 Hz, 25 kV AC, 50 Hz (EN 50163) and 60 Hz (IEC 60850). The ground of the measurement system therefore has a large voltage difference from other electric circuits and requires high-voltage insulation. Obviously, most telecom systems do not see such large voltage differences since they are primarily supplied from national grid voltages at 120 V AC or 220 V AC.

There are standards which define the protection or safety level; usually, they refer to the maximum isolation voltage. Examples of such standards applicable to low-voltage equipment include:

- IEC 60076-3, Power transformers – Part 3: Insulation levels, dielectric tests and external clearances in air, from 2013
- UL 5085-1, Low-voltage transformers – Part 1: General requirements, edition of 2006-04-17, ANSI approved: 2013-12-06
- National Electrical Code (NEC), or NFPA 70, which is a regionally adoptable standard for the safe installation of electrical wiring and equipment in the United States
- More general safety testing within European Standard with TUV certification EN 61558

The isolation voltage is defined as the maximum voltage tolerated without breakdown of the insulation when applied between the ground circuit of the primary side and ground circuit of the secondary side.

Several power converter topologies which offer galvanic isolation are analysed in detail for energy conversion with a change in DC voltage level.

- *Forward converter* with isolation (presented in this chapter)
- *Flyback converter,* derived from previous boost converters (this chapter)
- *Push-pull converter* (this chapter)
- *H-bridge converter with phase-shift control* (Chapter 7), which is the most-used topology for brick structures
- *LLC H-bridge converter* (Chapter 7)

The operation of any of these converter topologies depends on magnetic devices. The magnetic materials and technologies of fabrication have evolved so much over the last decade that magnetic devices have today become a commodity product that can be bought from elite vendors, rather than a custom design for each application, made up in one's own yard.

A brief description of magnetic devices follows from a physical-principle point of view.

6.2 Physics of the Pulse Transformer

Converters offer galvanic isolation through transformers able to work with high-frequency pulses. The most important characteristic of a transformer is the B-H characteristic (Figure 6.1a, where B = magnetic field induction, H = magnetic field intensity). The transformer schematic and principal

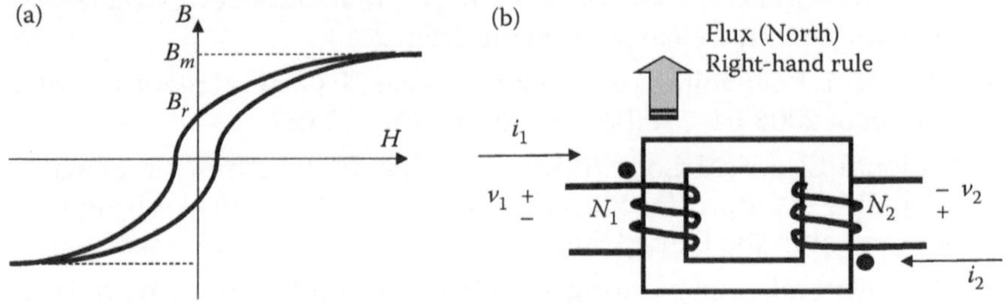

FIGURE 6.1
Physics of transformer operation: (a) B-H characteristic, (b) transformer in electric circuit.

measures are shown in Figure 6.1b. The currents entering a terminal denoted with '*' determine the flux direction.

Neglecting the flux leakage means the same flux within a coil is found within the other coils.

The voltages induced within each winding are:

$$\begin{cases} v_1 = \dfrac{d\lambda_1}{dt} = N_1 \cdot \dfrac{d\varphi}{dt} \\ v_2 = \dfrac{d\lambda_2}{dt} = N_2 \cdot \dfrac{d\varphi}{dt} \end{cases} \Rightarrow v_2 = \frac{N_2}{N_1} \cdot v_1 \tag{6.1}$$

where:

Φ = Magnetic flux within the magnetic circuit produced in a coil

$\lambda = N\Phi$ = Magnetic flux produced by the entire winding

The flux produced by a current $\boldsymbol{i}$ within each coil (winding turn) can be estimated for the ideal case ($\mu \to \infty$) and a transformer with a generic gap of length $\boldsymbol{l_g}$ (Figure 6.2):

$$\phi = \underset{\text{Reluctance}}{\frac{i}{R_c}} = \frac{i}{\underbrace{\dfrac{l_c}{\mu_c \cdot A_c}}_{\substack{\text{Geometry of}\\ \text{the section}\\ \text{through the coil}\\ \text{section}}} + \underbrace{\dfrac{l_c}{\mu_g \cdot A_g}}_{\substack{\text{Geometry of}\\ \text{the gap}\\ \text{[when present]}}}} \tag{6.2}$$

where:

l_c = Length of the winding (considered larger than the section radius).

A_c = Coil section area (usually circular).

μ_g = Permeability of the material used to build the transformer core, which represents the material capacity for building the magnetic field (ferrite, Cool-Mu or even air). For instance, $\mu_0 = 4\pi \times 10^{-7}\ \mathrm{H \cdot m^{-1}} \asymp 1.2566370614 \ldots \times 10^{-6}\ \mathrm{H \cdot m^{-1}}$ or $\mathrm{N \cdot A^{-2}}$.

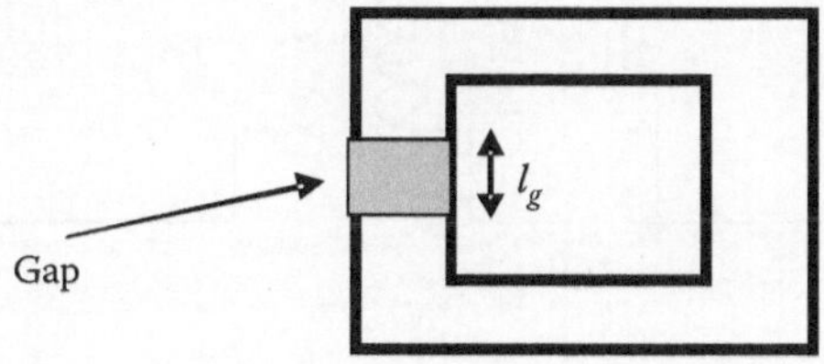

FIGURE 6.2
Core section with gap.

The ideal case, taking into account the magnetisation ($\mu \to \infty$), becomes:

$$\begin{cases} \varphi = \dfrac{N_1 \cdot i_1 + N_2 \cdot i_2}{R_c} \\ \mu \to \infty \Rightarrow R_c \to 0 \end{cases} \Rightarrow N_1 \cdot i_1 + N_2 \cdot i_2 \to 0 \Rightarrow N_1 \cdot i_1 = -N_2 \cdot i_2 \quad (6.3)$$

wherefrom:

$$\Rightarrow N_1 \cdot i_1 + N_2 \cdot i_2 = R_c \cdot \varphi \neq 0 \quad (6.4)$$

The real case, taking into account magnetisation ($\mu = \textit{finite}$), can be seen as an error within the transferred current.

Considering an open circuit on the secondary side (Figure 6.3, zero current through the secondary, $i_2 = 0$), magnetic flux is produced by the primary current circulation only.

$$i_2 = 0 \Rightarrow \varphi = \frac{N_1 \cdot i_1}{R_c} \Rightarrow v_1 = \frac{d\lambda}{dt} = \frac{d(N_1 \cdot \varphi)}{dt} = \frac{N_1^2}{R_c} \cdot \frac{di_1}{dt} = L_m \cdot \frac{di_1}{dt}$$

$$\text{where} \quad L_{m(1)} = \frac{\mu \cdot N_1^2 \cdot A}{l} \quad (6.5)$$

Similar results can be achieved when analysing the second winding when the first winding is in open circuit. L_m represents the magnetisation inductance and shows energy stored within the magnetic field.

$$L_{m(2)} = \frac{\mu \cdot N_2^2 \cdot A}{l} \quad (6.6)$$

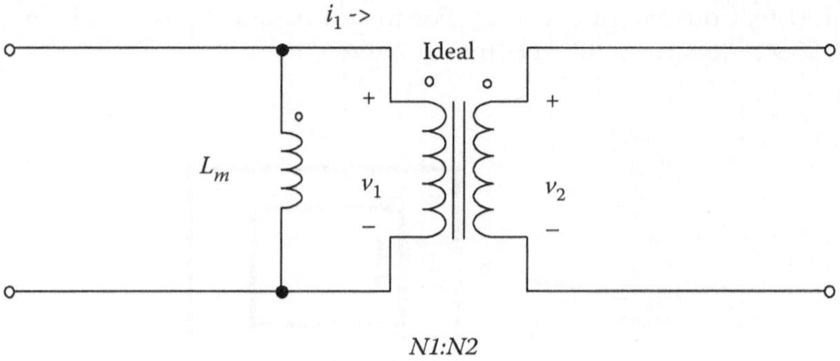

FIGURE 6.3
Simplified model of the transformer

Considering the model from Figure 6.3, one can place L_m on either side of transformer with the appropriate transformer ratio; that is, $n_2 = (N1/N2)^2$. For this reason, the transformer does not work in DC quantities. The flux would yield:

$$v_1 = \frac{d\lambda}{dt} = const \Rightarrow \lambda \Uparrow \tag{6.7}$$

In reality, the entire flux trajectory is closed from one winding to another, and there is a leakage flux (denoted with either l index or with μ) representing the nonideal character of the transformer and the energy leakage through windings (Figure 6.4).

The total flux within any winding is:

$$\begin{cases} \lambda_1 = N_1 \cdot (\phi_c + \phi_{l1}) = N_1 \cdot \dfrac{N_1 \cdot i_1 + N_2 \cdot i_2}{R_c} + N_1 \cdot \dfrac{N_1 \cdot i_1}{R_{l1}} = \underbrace{\dfrac{N_1^2 \cdot i_1}{R_{l1}}}_{L_{l1}} + \underbrace{\dfrac{N_1^2 \cdot i_1}{R_c}}_{L_m} + \dfrac{N_1 \cdot N_2 \cdot i_2}{R_c} \\ \lambda_2 = N_2 \cdot (\phi_c + \phi_{l2}) = N_2 \cdot \dfrac{N_1 \cdot i_1 + N_2 \cdot i_2}{R_c} + N_2 \cdot \dfrac{N_2 \cdot i_2}{R_{l1}} = \underbrace{\dfrac{N_2^2 \cdot i_2}{R_{l1}}}_{L_{l2}} + \dfrac{N_2^2 \cdot i_2}{R_c} + \dfrac{N_1 \cdot N_2 \cdot i_1}{R_c} \end{cases} \tag{6.8}$$

Neglecting the resistive energy loss (also called caloric energy) of the windings, a transformer can be represented as a magnetisation inductance additional to two leakage inductances. This means an increase of the magnetisation inductance is desired for temporary storage of energy. A reduction of the values of these two inductances is desired for the optimal transfer of energy.

The circuit model is shown in Figure 6.5.

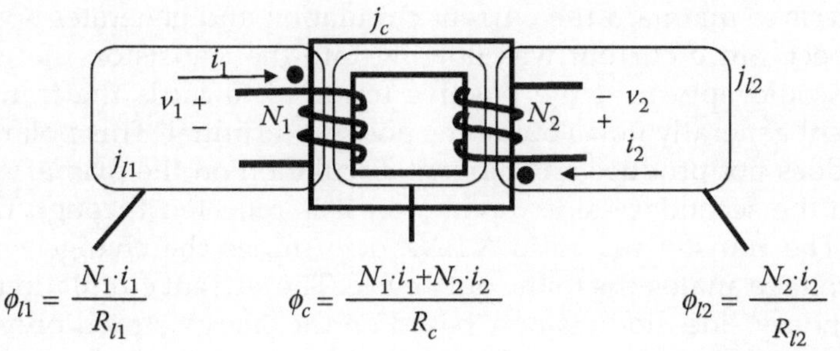

FIGURE 6.4
Flux components through a transformer.

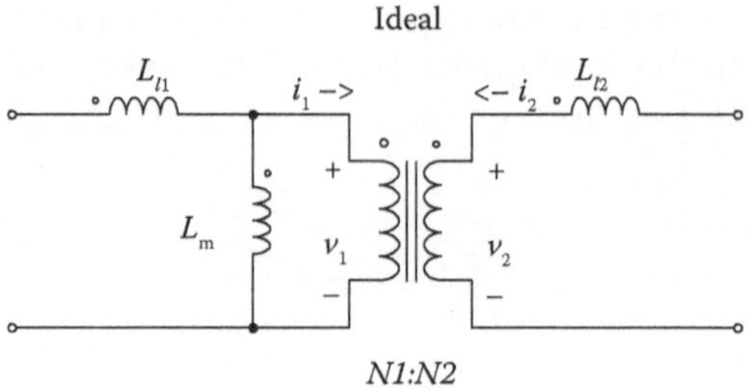

FIGURE 6.5
Circuit model for a transformer, including leakage inductances.

6.3 Flyback Converter

6.3.1 Operation

The operation of the flyback converter can be understood from Figure 6.6 through an analogy with a boost converter. When the transistor is on for the boost converter in Figure 6.6a, the supply voltage is applied to the inductance and the current increases linearly. When the transistor is off, the inductance current tries to maintain the current circulation through the generation of a counter-electro-motive-force (CEMF) voltage which turns on the diode *D*. The diode current charges the output capacitor. The output voltage can thus be controlled through the transistor's duty cycle. Figure 6.6b redraws the boost converter.

The boost inductance can be replaced with a pulse transformer, as further shown in Figure 6.6c. The primary winding offers a magnetisation inductance similar to the inductance from the boost converter. When transistor turns on, the entire supply voltage is applied to this boost inductance and current increases through primary winding. Energy is therefore stored within the magnetic field of the transformer.

When the transistor is turned off, the primary winding of the transformer tries to maintain the current circulation and generates a voltage in this respect. Since current was flowing into the transistor, the generated voltage would appear at the positive terminal towards the transistor, as the current generally flows out of the positive terminal. This polarity of the voltage does not produce any current circulation on the primary side, but turns on the secondary-side diode after it is reflected through the transformer. The transformer ratio ***N*1:*N*2** determines the change within the output voltage analogous to the duty cycle. The current circulation through the secondary-side diode is now based on the energy stored previously in the magnetic field of the transformer and looks similar to boost converter operation.

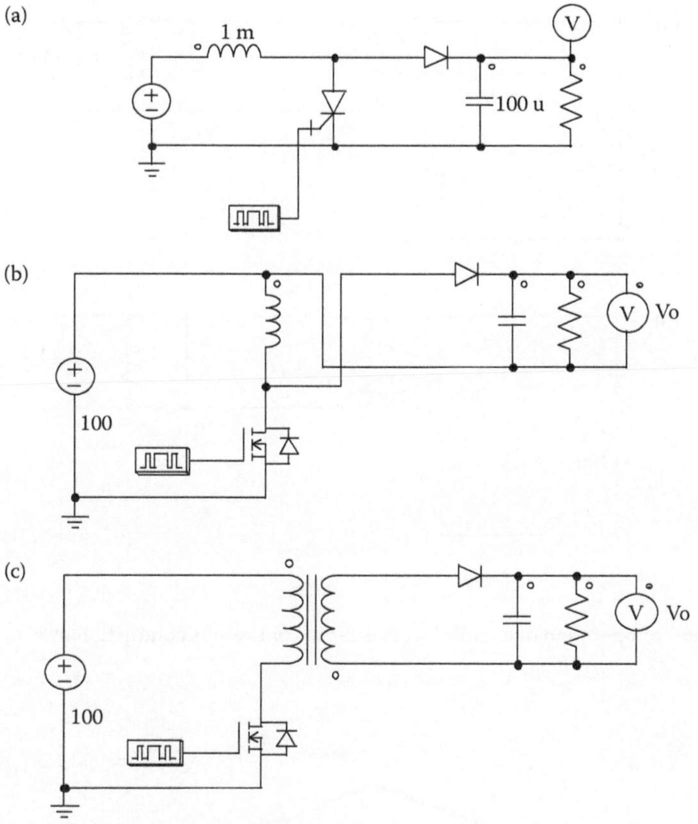

FIGURE 6.6
Understanding operation of the flyback converter.

The operation during the two conduction states for the transistor can be followed in Figure 6.7.

During the turn-on interval for the main transistor, equations similar to the boost converter can be written for the magnetic flux of the currents. The evolution of the magnetic flux through the transformer can be understood using Figure 6.8.

$$\Phi(t) = \Phi(0) + \frac{V_d}{N_1} \cdot t \tag{6.9}$$

$$\Phi_{peak} = \Phi(0) + \frac{V_d}{N_1} \cdot t_{on} \tag{6.10}$$

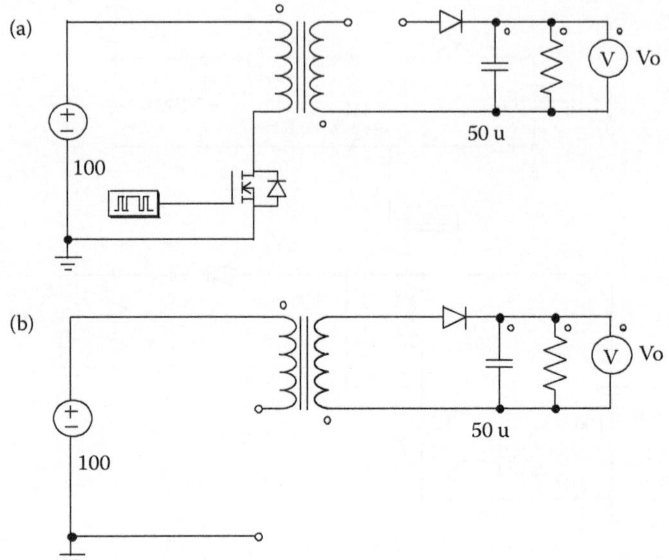

FIGURE 6.7
The two states of operation of a flyback converter in continuous conduction mode.

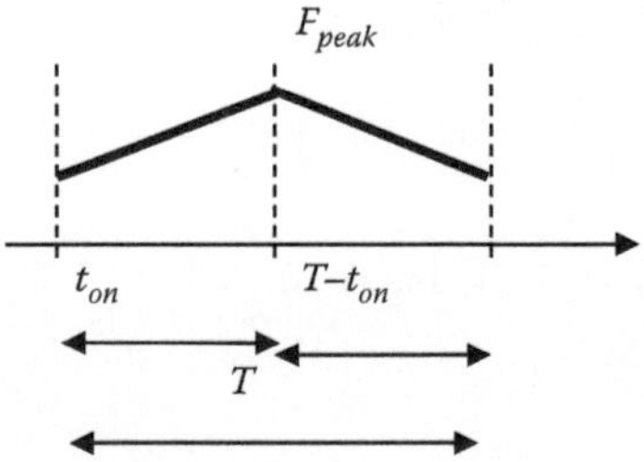

FIGURE 6.8
Magnetic flux inside the transformer.

The maximum flux can be achieved at the end of the conduction interval t_{on}. If the flux is not zero at the end of the period after the interval with the transistor in the off state, it yields:

$$\Phi(t) = \Phi_{peak} - \frac{V_o}{N_2} \cdot [t - t_{on}] \tag{6.11}$$

$$\Phi(T) = \Phi_{peak} - \frac{V_o}{N_2} \cdot [T - t_{on}] = \Phi(0) + \frac{V_d}{N_1} \cdot t_{on} - \frac{V_o}{N_2} \cdot [T - t_{on}] \tag{6.12}$$

In a steady-state regime of operation, it yields $\Phi(T) = \Phi(0)$. Hence:

$$\Phi(0) + \frac{V_d}{N_1} \cdot t_{on} - \frac{V_o}{N_2} \cdot [T - t_{on}] = \Phi(0) \Rightarrow \frac{V_d}{N_1} \cdot t_{on} = \frac{V_o}{N_2} \cdot [T - t_{on}]$$
$$\Rightarrow V_o = V_d \cdot \frac{N_2}{N_1} \cdot \frac{t_{on}}{T - t_{on}} \tag{6.13}$$

If we neglect the caloric loss and consider a continuous flux, the transfer characteristic yields:

$$\frac{V_o}{V_d} = \frac{N_2}{N_1} \cdot \frac{D}{1 - D} \tag{6.14}$$

Continuous conduction mode is defined when the transformer's magnetic flux does not reach zero, and it can be observed with waveforms from Figure 6.9. These results are achieved for an output voltage of 5 V DC, an input voltage of 100 V DC and a duty cycle of 0.385, following the conversion relationship (Equation 6.14):

$$V_o = 100 \cdot \frac{8}{100} \cdot 0.625 = 5\,\text{V} \tag{6.15}$$

Figure 6.9 also demonstrates that a load resistance of $R = 3$ Ohm corresponds approximately to the boundary (critical) conduction mode (often

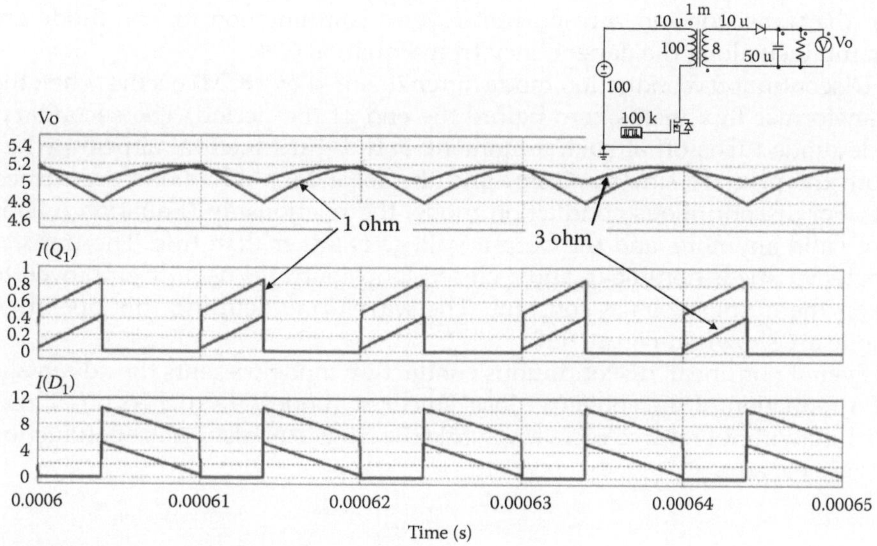

FIGURE 6.9
Waveforms and circuit for the flyback converter operated in continuous conduction mode at $R = 1$ Ohm and boundary conduction mode at $R = 3$ Ohm.

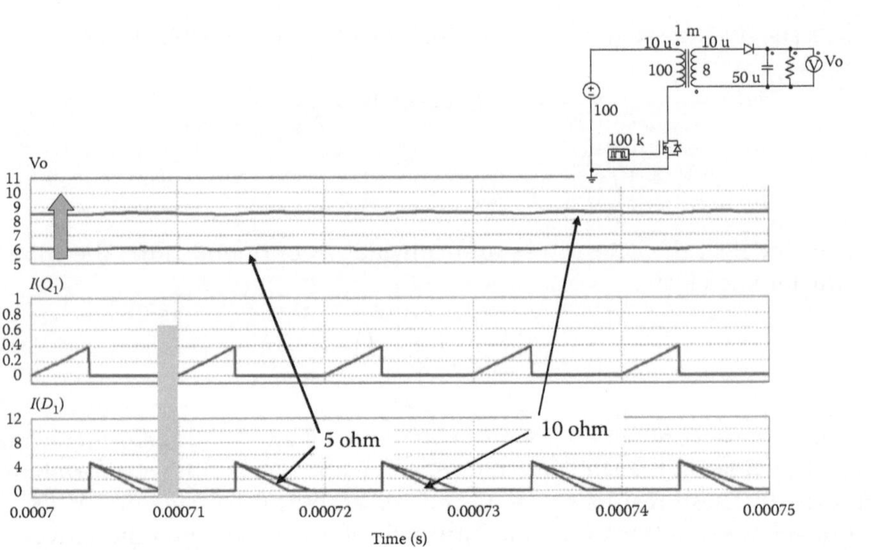

FIGURE 6.10
Waveform examples for operation in discontinuous conduction mode.

denoted by BCM) for the converter data in this example. This is defined as operation at the boundary between continuous and discontinuous conduction modes, where the current through the magnetic device touches the zero axis. This way, the efficiency-related advantages of zero-current turn-on for the transistor and zero-current reverse commutation for the diode are maintained along the dependency from Equation 6.14.

Discontinuous conduction mode (often denoted by DCM) occurs when the transformer flux yields zero before the end of the period. The secondary-side diode turns off at such a moment, and a third interval appears when both the primary-side transistor and the secondary-side diode are turned off. For discontinuous conduction mode, the relationship (Equation 6.14) is not valid anymore, and the output voltage is larger than this. The transfer characteristic is nonlinear, and a closed-loop control is usually employed to keep the output voltage constant. The waveforms defining this operation mode are shown in Figure 6.10.

Even if nonlinear, discontinuous conduction mode presents the advantage of a reduction of the converter loss, since the diodes are turned off at zero current without reverse recovery, while the transistor also enters conduction at zero current.

6.3.2 Control

The control of energy can be assured only during the on time of the primary-side MOSFET through conduction time of this transistor, while the energy is actually delivered towards the load during the off time. This means there is no control possible during the time the energy is transferred to the load.

If the converter operates in continuous conduction mode, the energy accumulated in the inductor will be discharged uncontrolled onto the load over several cycles before an action or any change can be done by the control. This behaviour is seen in the small-signal model (see Chapter 4) as a right-half-plane (RHP) zero, usually of a fairly large value.

This behaviour must be considered when defining the control-loop compensation, and it is similar to any other boost converter. The general design rule for converters with a right-half-plane zero is to design the compensation law at the lowest input voltage and maximum load current. An empirical rule of thumb restricts the bandwidth of the feedback loop to about one-fifth of the right-half-plane zero frequency. The right-half-plane zero frequency for continuous conduction mode (CCM) is:

$$f_{RHPZ} = \frac{(1-D)^2 \cdot V_o}{2 \cdot \pi \cdot L \cdot D \cdot I_{out} \cdot \left(\frac{N_2}{N_1}\right)^2} \tag{6.16}$$

The numerical values considered in the presented example yield:

$$f_{RHPZ} = \frac{(1-0.385)^2 \cdot 5}{2 \cdot \pi \cdot 0.001 \cdot 0.385 \cdot 5 \cdot \left(\frac{8}{100}\right)^2} = 24.43\,\text{kHz} \approx 140\,\text{krad/sec} \tag{6.17}$$

Figure 6.11 shows a possible location of the two resonant poles and the right-hand-plane zero in the converter model before and after the compensation. The compensation also considers an integral action able to address the steady-state error. The location of the right-hand-plane zero is not affected by the compensation through the control law.

The new location of the poles is selected so that the bandwidth of the control loop system is less than one-fifth of the right-hand-plane zero. On many occasions, the new location of the third pole is considered at a higher frequency than the new location of the resonant poles. The solution proposed herein is suggested as a minimal frequency constraint. This compensation law is just for exemplification, and more design detail can be found in Chapters 4 and 5.

The presence of the right-half-plane zero is also seen in DCM. However, this is usually not a problem anymore, as the frequency moves above half of the switching/sampling frequency. Furthermore, the DCM case is rarely seen today in simple buck/boost converters due to the usage of more and more synchronous rectification methods.

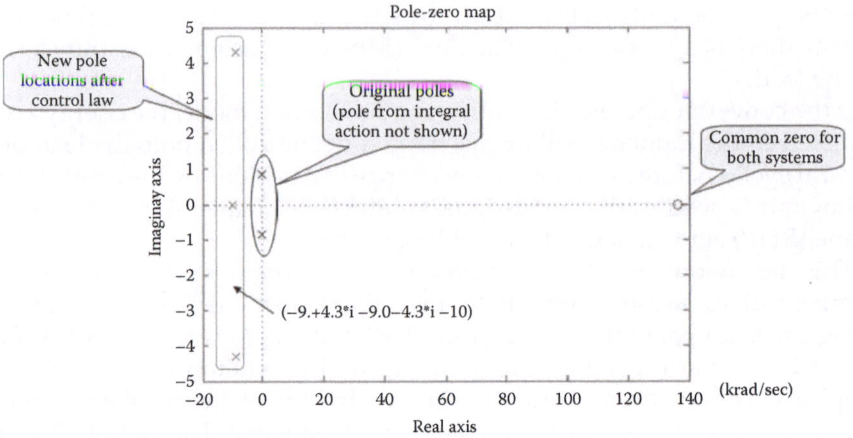

FIGURE 6.11
Example of pole-zero location for a flyback converter, before and after the compensation (real axis in krad/sec). New pole locations are at around 10 krad/sec (1.5 kHz bandwidth in control loop), while RHP zero is at ~140 krad/sec.

Feedback control is achieved with dedicated integrated circuits, somewhat similar to the circuits used for control of buck or boost converters presented in Chapter 4. The latest flyback control ICs benefit from the following possible features [1,2]:

- Operation with variable frequency to maintain discontinuous or boundary (also called transition mode) conduction modes, with the advantages of:
  - Efficiency optimisation (operation with diode turn-off by zero current is virtually lossless, with no diode reverse recovery, and it improves efficiency)
  - Smaller size of the flyback transformer given the requirement of a smaller inductance
  - Small-signal modelling, with the plant model reduced to a first-order model for either voltage-control mode or current-control mode
- Operation in 'green mode', where the PWM sequence is shut down at low load, with burst-mode operation
- Special protection features like overvoltage detection, maximum on-time programming, fast latch-up fault recovery and thermal shutdown

There is the disadvantage of a higher peak current when operating in DCM or BCM, with a negative influence on electromagnetic interference (EMI), MOSFET conduction loss and requirements for overvoltage protection.

However, the cost difference of using a higher-current MOSFET is minimal, and the problems are thus solved.

6.3.3 Multioutput Flyback Converters

Flyback converters are the preferred topology within circuits which necessitate multiple isolated supply voltages, which are achieved with multiple secondary windings [3]. This generally happens within the centralised architecture of the power supply systems explained in Chapter 1.

In such applications, there are multiple secondary windings of the flyback transformer required to supply multiple low-voltage circuits with galvanic isolation. These electronic loads require different currents, and it is virtually impossible to regulate all voltages simultaneously. A compromise is usually to set the feedback loop after the most restrictive secondary, usually the supply of digital circuitry. A digital circuit requires a smaller ripple of the supply voltage.

While the voltages in the other secondary windings cannot be adjusted or frequency compensated with the flyback converter's control loop, additional solutions require on-board three-pin voltage regulators on each secondary for precise regulation without isolation. However, the gate drivers and some current-, temperature- and voltage-sensing devices (connected in differential mode) do not require accurate regulation of voltage, and a single compensation circuit for the flyback is enough.

6.3.4 Protection

Finally, the reduction of transistor stress is achieved with snubber protection circuits. An example is shown in Figure 6.12. At turn-off of the MOSFET

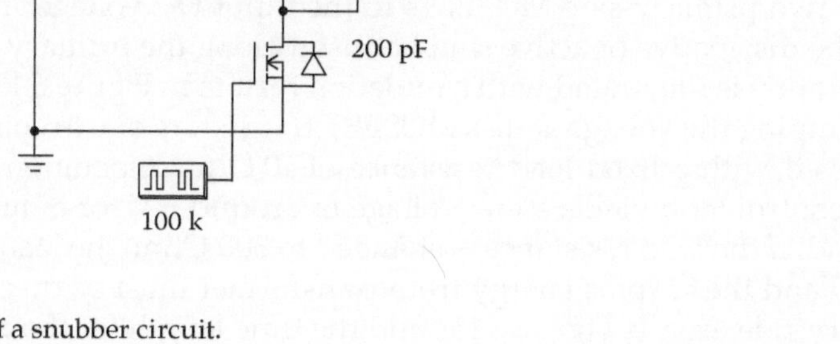

FIGURE 6.12
Utilisation of a snubber circuit.

transistor, a voltage appears at the drain/collector terminal due to the resonant circuit formed between the primary winding leakage inductance $L_{\mu 1}$ and the transistor output capacitance C_{OSS}.

The MOSFET transistor output capacitance C_{OSS} becomes, therefore, a very important design parameter. When MOSFET is hard switched within a converter, C_{OSS} is needed to calculate the additional power dissipation of the power MOSFET due to discharging this output capacitor every switching cycle. When MOSFET is soft switched within a converter, C_{OSS} is needed to calculate the resonant frequency or the transition time at state change. These are critical in establishing zero-voltage switching or zero-current switching conditions. As shown in Chapter 2, the value of C_{OSS} varies nonlinearly with the MOSFET's drain-source voltage V_{ds}.

The design problem is that the value of C_{OSS} specified in most data sheets is at a fixed voltage, which is not the information required in actual circuit calculation. Newer datasheets present the effective C_{OSS} as curves up to 80% of the rating voltage V_{dss}.

For the flyback converter, the voltage which appears at the drain/collector terminal due to the resonant circuit can produce MOSFET transistor breakdown due to overvoltage. Through the utilisation of snubber circuitry, the snubber diode (D_2) turns on when the main transistor ($Q_1 - D_1$) turns off, and the RC snubber circuit allows a limitation of the voltage in the transistor's collector. Waveforms characteristic to this operation are shown in Figure 6.13.

6.3.5 Two-Switch Flyback Converter

The requirement for proper voltage limitation across the power MOSFET within a flyback converter finds a solution with the two-switch flyback converter. The power MOSFET needs to be rated for half the DC input voltage, and the snubber circuitry can be avoided [4].

The converter is built up with two transistors and two diodes on the primary side (Figure 6.14). Both transistors are controlled with the same logic signal, and they determine the time interval when energy is stored in the magnetic field. After transistors are turned off, the primary-side diodes act as a clamp for the primary-side voltage: the primary side diodes turn on when the voltage on the secondary circuitry reflected into the primary exceeds the input DC voltage. This voltage clamping limits the voltage stress across the two primary-side MOSFETs to the input DC voltage. There is no need for the dissipative or active snubber used along the primary circuitry.

The operation is illustrated with simulation results in Figure 6.15. First, the case of clamping the voltage across MOSFET transistors due to commutation was achieved with a fixed load resistance of 10 Ohm. Operation without a feedback control loop yields a low voltage of around 5 V for a duty cycle of 0.32. Second, if the load resistance is changed to 300 Ohm, the load current is very small and the surplus energy from transformer must be discharged on the primary side (case b, Figure 6.15), and the time length for the conduction

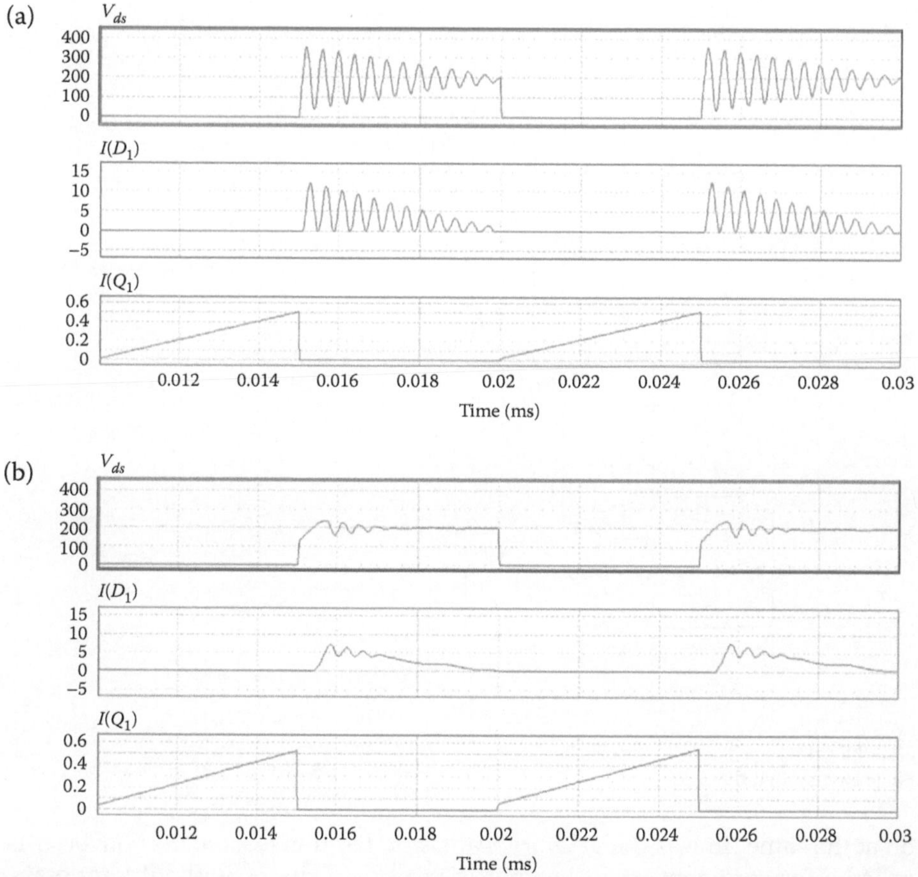

FIGURE 6.13
Waveforms for the flyback converter (a) without snubber, (b) with snubber. Converter data are shown in Figure 6.12.

intervals related to the primary-side diodes increases. Finally, a transient operation is imagined to generate a wider output voltage (case c).

6.4 Forward Converter

6.4.1 Operation

The operation of the forward converter is similar to the buck converter as a voltage step-down conversion. The topology is shown in Figure 6.13. The circuit looks like the flyback converter, with the difference of having the transformer's secondary winding reversely connected so that the secondary diode is turned

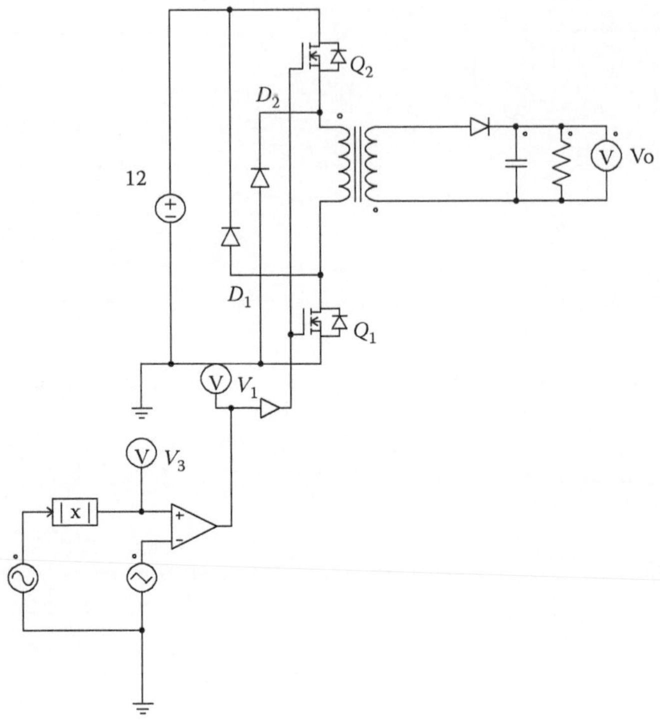

FIGURE 6.14
Basic two-switch flyback.

on at the same time as the primary transistor. For this reason, the converter is called a forward converter: the energy transfer is direct, without temporary storage, as has been seen in the flyback converter. The transformer used herein is as close to ideal as possible without finite magnetisation inductance. The circuit is completed with a diode D_3 connected in the output circuit to avoid the overvoltages induced by the filter inductance L_1. This diode D_3 works as a freewheeling diode in a buck converter. When the diode D_1 turns off, the filter inductance L_1 will try to maintain the current circulation through the generation of a voltage. Waveforms at ideal operation are shown in Figure 6.16.

Since the energy is transferred directly, it yields the numerical data from Figure 6.16:

$$\frac{V_o}{V_{in}} = \frac{N_2}{N_1} \cdot D \left(= \frac{100}{8} \cdot 0.625 = 0.05 \right) \tag{6.18}$$

The problem with this circuit relies on the possibility of the transformer magnetising due to the unipolar current pulse transferred through the forward converter. This problem limits the transferred energy through the direct

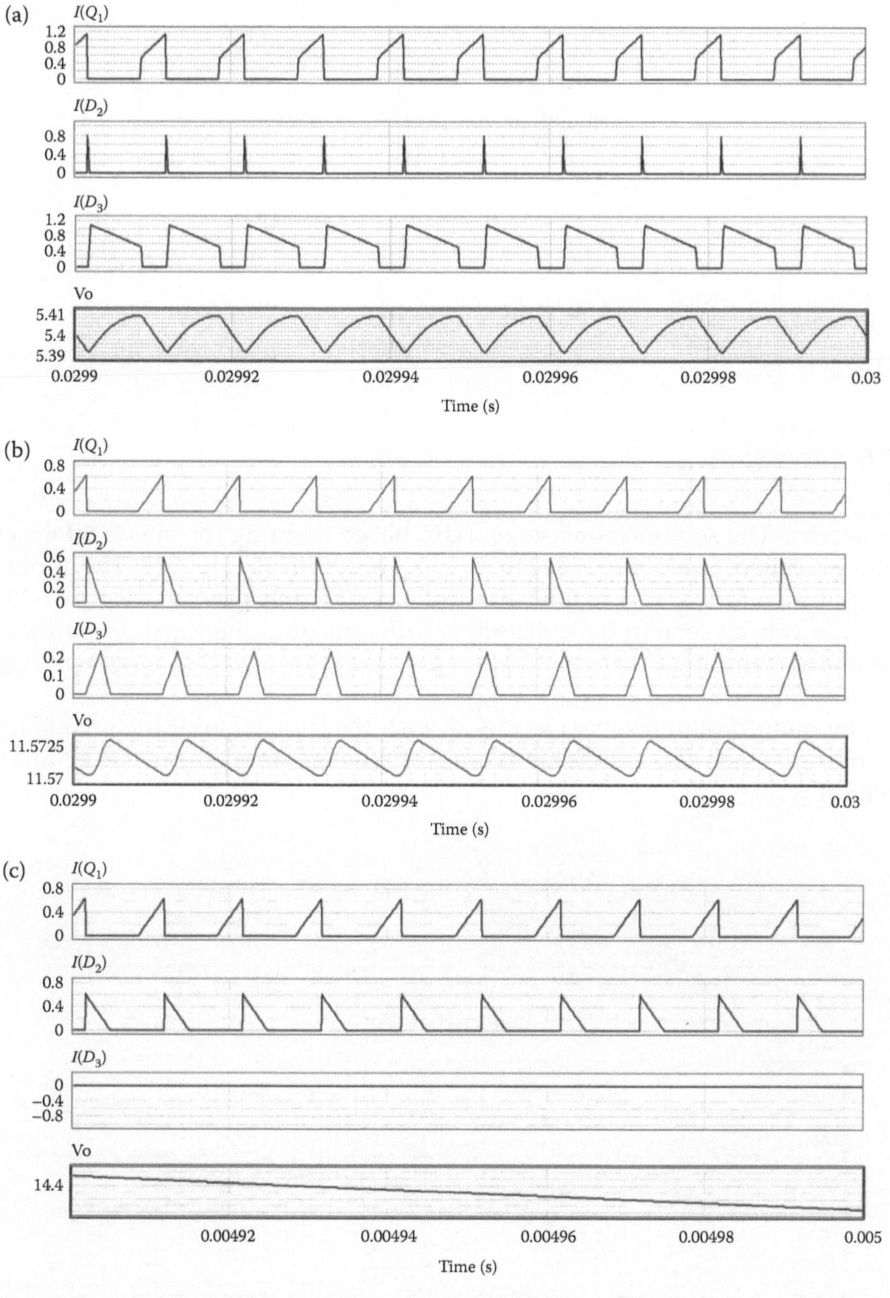

FIGURE 6.15
Operation of the two-switch converter: clamping diodes on the primary side turn-on for a short interval under heavy load (a), for a longer interval at light load (b), or when the output voltage exceeds the input, for instance due to transients (c).

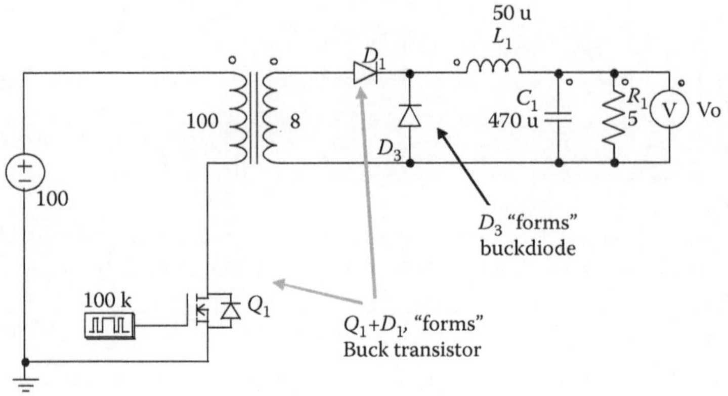

FIGURE 6.16
Forward converter.

converter. The solution consists of a discharge winding for the transformer energy called *transformer reset*, which is shown in Figure 6.17. The input–output waveforms (Figure 6.18) are similar to the previous case (Figure 6.19), but the voltage seen in the collector is different from the converter without discharge winding (Figure 6.20a) and the converter with discharge winding (Figure 6.20b).

The main design problem is raised with the protection of the secondary winding. When the transistor is on, the voltage applied on the primary winding is:

$$v_{IN} = V_d \tag{6.19}$$

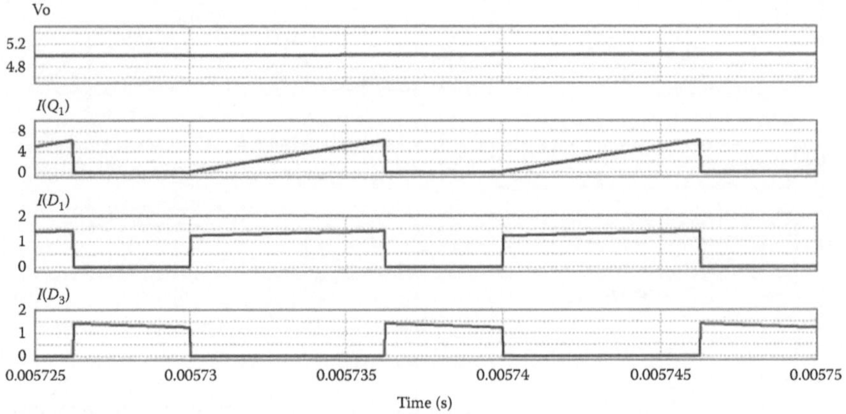

FIGURE 6.17
Forward converter with discharge winding.

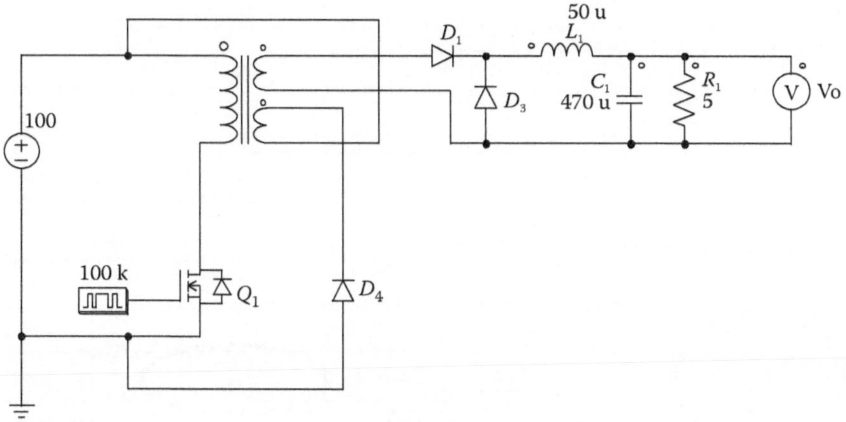

FIGURE 6.18
Waveforms for forward converter with discharge winding.

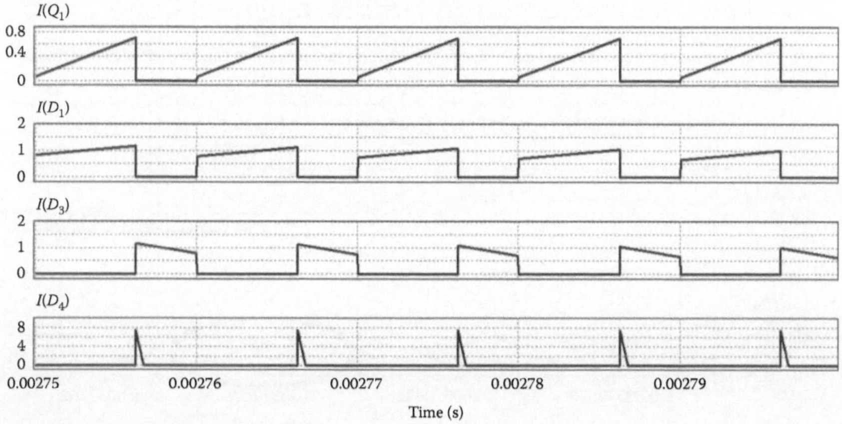

FIGURE 6.19
Waveforms for operation of a forward converter.

During conduction of the D_4 diode within the discharge winding, the voltage reflected in the primary winding becomes:

$$v_{IN} = -\frac{N_1}{N_3} \cdot V_d \tag{6.20}$$

The demagnetisation current through diode D_4 is:

$$i_3 = \frac{N_1}{N_3} \cdot i_m \tag{6.21}$$

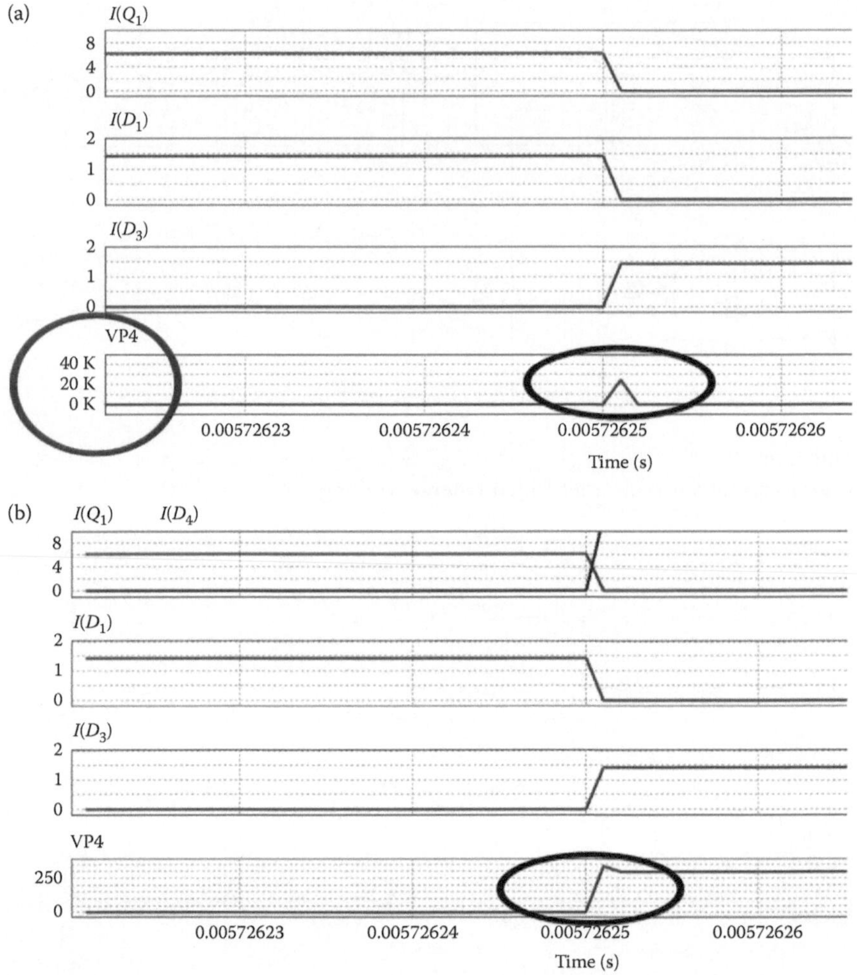

FIGURE 6.20
Operation for the forward converter without discharge winding (a) and with discharge winding (b).

The constraint can be expressed as the condition for the current/flux through the transformer to get to zero before the next conduction interval for the transistor (Figure 6.21).

$$t_m < (1 - D) \cdot T \tag{6.22}$$

It gives, for the conservation of the maximum (or peak) current:

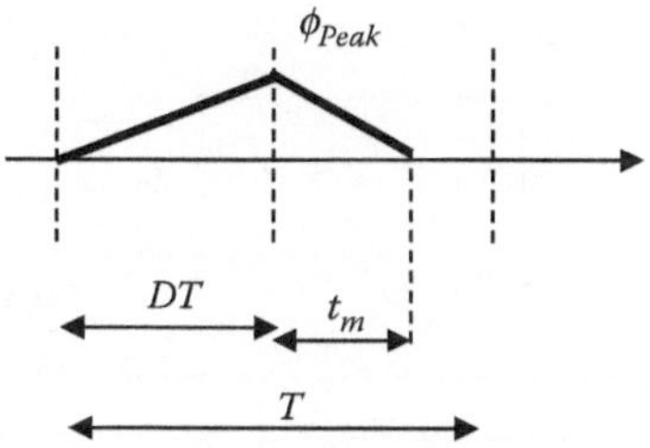

FIGURE 6.21
Transformer flux.

$$\begin{cases} Sw = ON \Rightarrow V_d = L_m \cdot \dfrac{di_m}{dt} = L_m \cdot \dfrac{I_{m,\max}}{D \cdot T} \Rightarrow I_{m,\max} = \dfrac{V_d \cdot D \cdot T}{L_m} \\ Sw = OFF \Rightarrow -\dfrac{N_1}{N_3} \cdot V_d = L_m \cdot \dfrac{di_m}{dt} = L_m \cdot \dfrac{I_{m,\max}}{t_m} \Rightarrow I_{m,\max} \qquad \Rightarrow D \cdot T = \left(\dfrac{N_1}{N_3}\right) \cdot t_m \\ = \dfrac{\left(\dfrac{N_1}{N_3} \cdot V_d\right) \cdot t_m}{L_m} \end{cases}$$

$$\Rightarrow t_m = \left(\frac{N_3}{N_1}\right) \cdot D \cdot T < (1-D) \cdot T \Rightarrow D < \left(\frac{N_1}{N_3}\right) \cdot (1-D) \Rightarrow D < D_{\max} = \frac{1}{1 + \dfrac{N_3}{N_1}} \tag{6.23}$$

$$D_{\max} = \frac{1}{1 + \dfrac{N_3}{N_1}} \tag{6.24}$$

For instance, when $N_3 = N_1$, then $D_{\max} = 0.5$.

Some practical notes:

- A very large isolation voltage is not needed between the primary winding and the demagnetisation winding since both connect to the same DC bus.
- The current through the demagnetisation winding is lower than the current through the primary or secondary winding.

6.4.2 Multioutput Forward Converter with Coupled Inductors

As discussed for the flyback converter, many applications require multiple isolated output voltages. This can also be accomplished with a forward

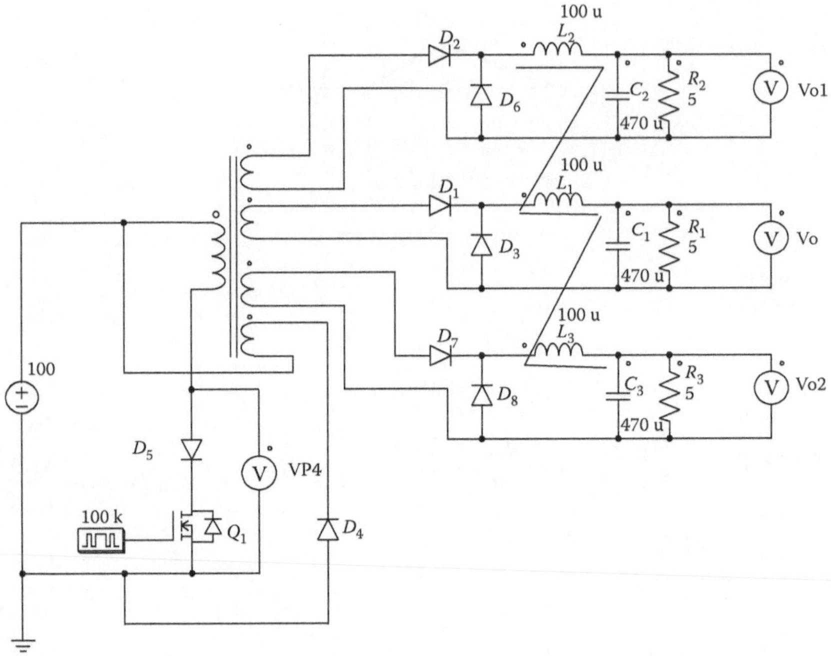

FIGURE 6.22
Multioutput forward converter with coupled inductors

converter. Actually, the forward converter with coupled filter inductors is very well accepted by the industry.

Figure 6.22 illustrates this principle. The output filter inductors are wound on the same core. This makes all outputs dynamically coupled: the inductors help filtering with energy stored while also accounting for cross-regulation through the transformer effect. The cross-regulation improves the output voltage regulation. This improves the overall performance and also saves costs through the use of a single core.

The inductor current and output voltage ripple are reduced when coupled inductors are used. This can be seen in Figure 6.23 when performance of the converter with coupled inductors is compared to the converter with individual inductances: the inductor current ripple is decreased from 0.5 to 0.2 Amp, and the output voltage ripple is also decreased from 2 to 0.4 mV. All operation conditions, open loop PWM control schemes and supply voltages are kept the same for the two cases.

This solution is very commonly used due to performance improvements and is almost exclusively preferred over individual output inductors [5]. A closed-loop operation can be formed with feedback from one of the output channels, while the other channels provide power with uncontrolled voltage. The output voltage regulation constraints therefore need to be relaxed. However, this solution has a good equivalence in complexity and

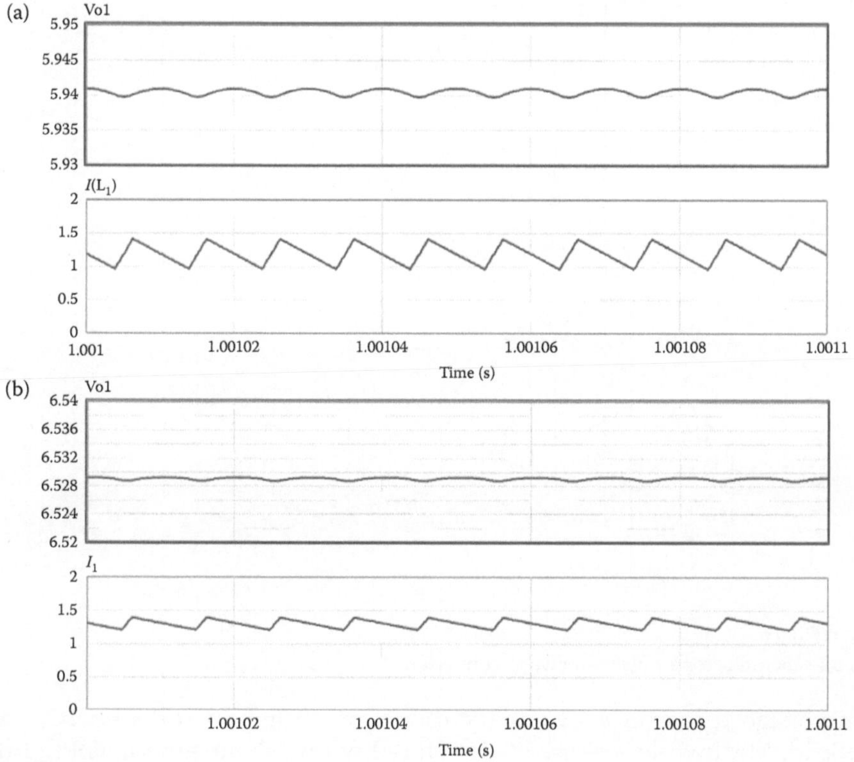

FIGURE 6.23
Waveforms for (a) individual inductors, (b) coupled inductors.

in the number of output point-of-load converters employed with the simple solution of using individual converters for each output. The multioutput converters with final point-of-load converter units cannot offer real cost savings. Adding cost advantage ambiguity to the improved reliability, the simpler solution of using multiple individual power supplies is more advantageous unless the power supply output specification is relaxed accordingly.

6.5 Forward-Flyback Converter

A new converter [6–8] has been widely accepted during recent years, and it can be seen as an extension of the operation described as a forward converter with a discharge branch (Figure 6.24). The primary-side transistor applies the input DC voltage across the primary winding of the transformer. A voltage is reflected into the secondary side. Unlike the flyback converter,

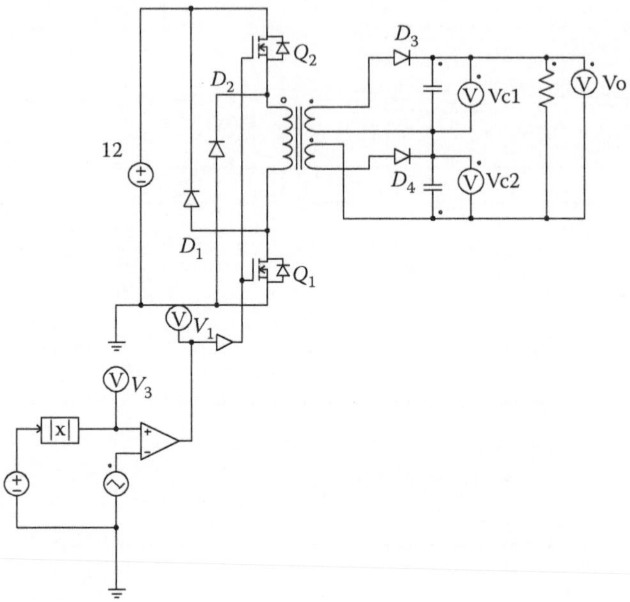

FIGURE 6.24
Circuit schematic for a flyback-forward converter.

this voltage turns on a secondary diode and transfers some energy into the load. Meanwhile, energy is also stored within the magnetic field of the transformer, as with a boost converter. When the transistor turns off, this energy is transferred to the load through the second secondary diode. This combination of a flyback and forward converter is possible when using different diodes for the two operation modes and two output capacitors. Since the transformer has two separate secondary windings for the two operation modes, the output voltages can be added up through a series connection of the two capacitors (Figure 6.24).

The operation can be understood with waveforms from Figure 6.25. The two different energy transfer modes lead to different current waveforms through diodes and different voltages across the output capacitors.

Each converter stage can be seen as a forward converter with reset winding. During the on state of the main switch, the peak of the magnetising current is calculated with:

$$I_1 = \frac{T_{on} \cdot V_{Bat}}{L} \tag{6.25}$$

where:

$$T_{on} = m \cdot T_s \Rightarrow I_1 = \frac{m \cdot T_s \cdot V_{Bat}}{L} \tag{6.26}$$

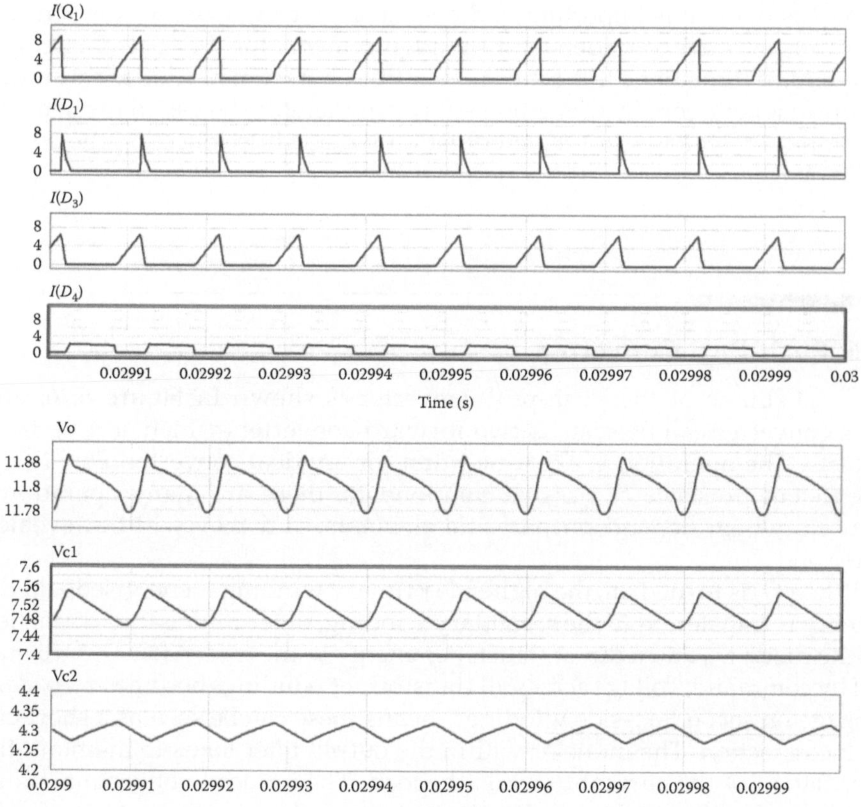

FIGURE 6.25
Waveforms describing the operation of a flyback-forward converter. Note different currents through secondary-side diodes and different voltages across the two output capacitors.

During this time interval, the energy is converted into the secondary side through a forward mode, according to the transformer turn ratio.

The peak of the secondary-side current for the flyback conversion in discontinuous conduction mode is:

$$I_2 = \frac{T_{off} \cdot V_{Out}}{n^2 \cdot L} \tag{6.27}$$

The DCM operation is guaranteed if (numerical data for simulation study include $V_{in} = 12$ V DC and $m = 0.32$, with open loop operation with a load of 10 Ohm):

$$\begin{cases} n \cdot V_{Bat} > V_{Out} \\ m < 0.5 \end{cases} \Rightarrow \begin{cases} n \cdot V_{Bat,Min} > V_{Out} \\ m < 0.5 \end{cases} \Rightarrow \begin{cases} n > \dfrac{11.8}{12} = 0.98 \\ m < 0.5 \end{cases} \tag{6.28}$$

At this point, it is important to draw static characteristics for the converter's gain in order to understand the requirements for the control circuitry and circuit limitations like saturation of this characteristic due to saturation of the magnetic core. A simulation study can therefore be useful in changing the input DC voltage and the modulation index while collecting the output DC voltage.

6.6 Push-Pull Converter

The schematic of the push-pull converter is shown in Figure 6.26, and this converter can be seen as two forward converters which add up their effects. The operation is explained first for an ideal converter. Due to the consistent presence of passive components, there are many operational aspects which depend on parasitic elements, and they will be revealed later on.

When Q_1 is turned on, the high-side primary winding is energised and the energy is transferred to the secondary winding, determining the conduction of the diode D_{s1} and a direct transfer of energy to the load. After Q_1 is turned off by command, voltage across all transformer windings becomes zero. Zero voltage on secondary-side windings means these can be seen as a short-circuit connection. The inductor within the output filter tends to maintain the circulation of the current towards the load. The induced voltage maintains both rectifier diodes, D_1 and D_2, in conduction. The conduction time for D_1 and D_2 can be until a novel command of transistors (continuous conduction mode) or shorter if their current is zero (discontinuous conduction mode).

In the second half of the operation period, Q_2 is turned on. The low-side primary winding is supplied with energy from input and through Q_2.

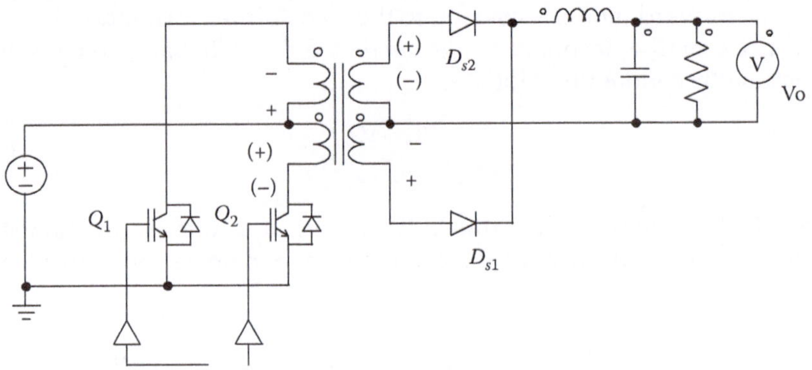

FIGURE 6.26
Push-pull converter.

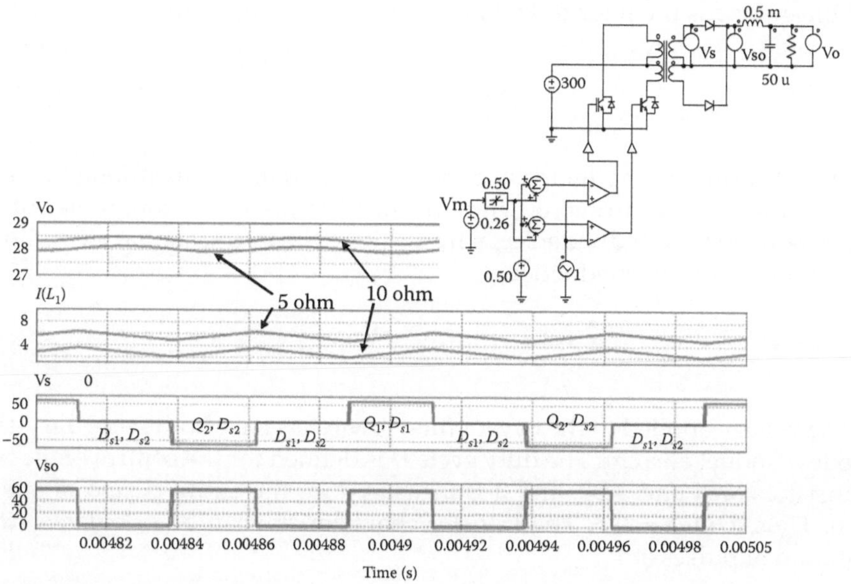

FIGURE 6.27
Waveforms for operation of the push-pull converter. Converter data: $V_{in} = 300$ V DC, $N = 5{:}1$, $D = 0.26$.

The energy is transferred to the secondary, determining the conduction of the diode D_{s2} and the transfer of energy to the load. During this interval, the transformer sees a flux of the opposite direction from the time interval with conduction of Q_1. This is very important since it avoids saturation of the magnetic core. The magnetic flux through the transformer has opposite signs on the two transistor conduction intervals.

The waveforms when considering an ideal transformer and continuous conduction mode are shown in Figure 6.27 for converter data $V_{in} = 300$ V DC, $N = 5{:}1$, $D = 0.26$.

This operation mode is not unique, and it depends in practice on the parasitic elements of the converter. Depending on the actual values of the transformer parameters and load current, the waveforms and operation are slightly different. The main difference may consist of the time intervals of actual conduction through the secondary-side diodes.

The ideal waveforms in Figure 6.27 can be used to derive the operation equations for the two conduction states. When transistor Q1 is turned on, the voltage across the secondary winding is transferred through diode ***Ds1*** to the load. The voltage relationship yields:

$$\frac{N_2}{N_1} \cdot V_d = v_0 \tag{6.29}$$

Voltage across the filter inductance can be expressed with:

$$v_L = \frac{N_2}{N_1} \cdot V_d - V_0 \tag{6.30}$$

This determines the increase of the load current. As mentioned before, when a transistor is turned off, the current is maintained through the output filter inductance and divided equally between the two secondary windings, with both diodes in conduction.

$$i_{D1} = i_{D2} = \frac{1}{2} \cdot i_L \tag{6.31}$$

The voltage applied to the output filter is zero because both secondary-side diodes conduct current. The duty cycle $\boldsymbol{D}$ is defined for the control of the two transistors $\boldsymbol{Sw1}$ and $\boldsymbol{Sw2}$, and it is measured for the entire period of operation. Thus, $\boldsymbol{Dmax = 0.5}$. The transfer characteristic can be calculated from the train of pulses at V_{s0}:

$$\frac{V_0}{V_d} = 2 \cdot \frac{N_2}{N_1} \cdot D \tag{6.32}$$

Differences occur in operation when using a real transformer with finite magnetising inductance; that is, $\mu =$ finite. Real transformers can be modelled with a magnetisation inductance and two leakage inductances.

For this reason, at turn-off of a transistor within the primary circuit, the load current is not divided equally through the two diodes from the secondary circuitry. When both diodes are on within the secondary circuitry, the voltage across all windings is zero. If the smaller current between the two diodes in the secondary circuitry is zero (discontinuous current), the current circulation through the output filter and the second diode from the secondary circuitry produce voltages within all windings.

An improved case, closer to ideal, is achieved when a large magnetisation inductance is used and the converter is operated with minimum ripple of the load current. Close to the ideal case means $\boldsymbol{\mu \to \infty}$, $\boldsymbol{L_m \to \infty}$. Waveforms for this case are shown in Figure 6.28.

Waveforms for weak loads, characterised with a small load current, and for a converter with $V_{in} = 300$ V DC, $N = 5{:}1$, $D = 0.26$, are shown in Figure 6.29. This is discontinuous conduction mode.

The duty cycle for the push-pull converter has a maximum value of 0.5. Due to circuit asymmetries, the values of the two currents achieved for conduction of either transistor can be different for the same duty cycle, imposing a closed-loop control of the current in order to avoid saturation of the magnetic core. The diodes in parallel with transistors have the role of limiting the currents due to the leakage inductances (Figure 6.30).

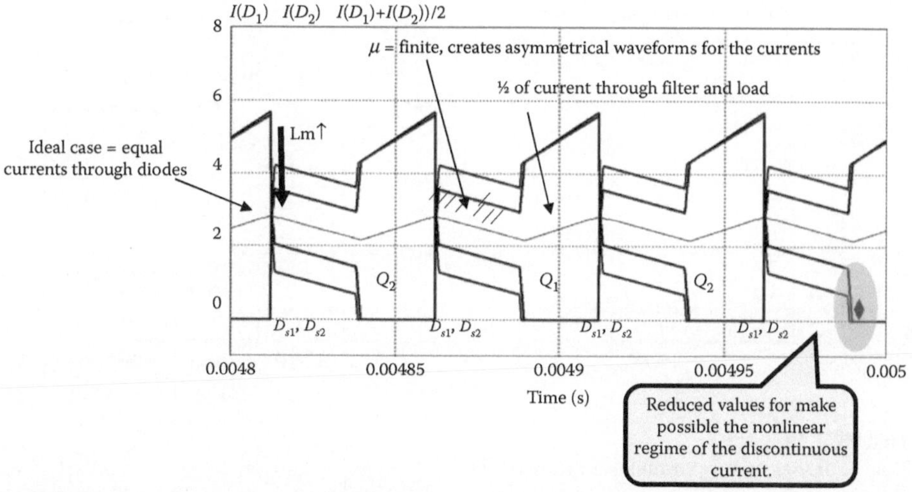

FIGURE 6.28
Waveforms for operation of the push-pull converter.

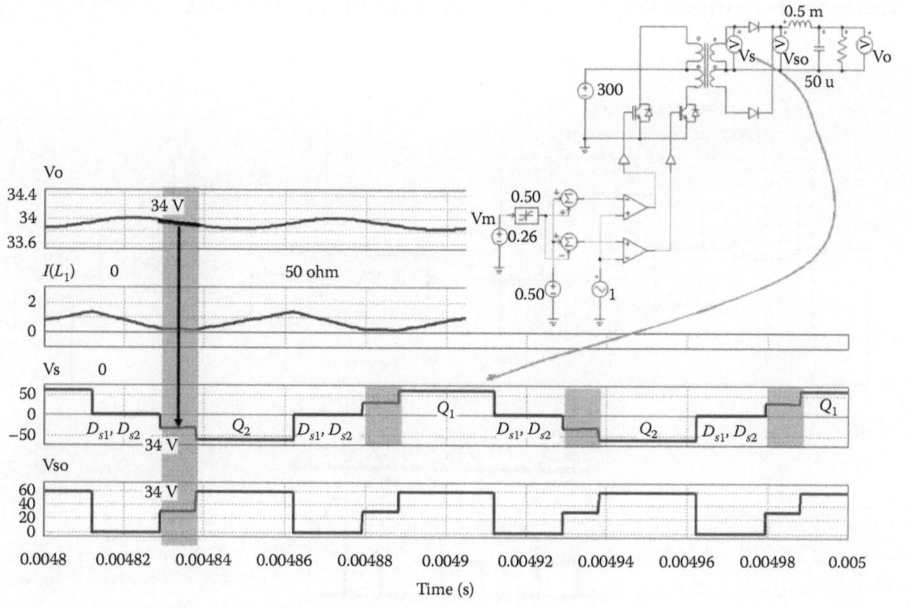

FIGURE 6.29
Push-pull converter operated in discontinuous conduction mode.

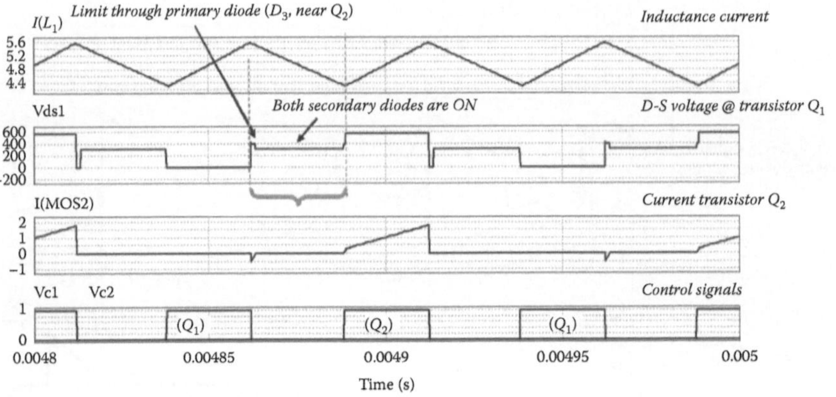

FIGURE 6.30
Role of diodes in antiparallel with transistors.

6.7 The Need for a Closed-Loop Controller

All observed converters have a nonlinear dependency of the output voltage on the duty cycle (D) and the current circulation regime. In order to compensate for the effect of this dependency, a closed-loop control for the output voltage is employed. This measure is compared with a reference voltage. The error is compensated with a control circuit (Figure 6.31).

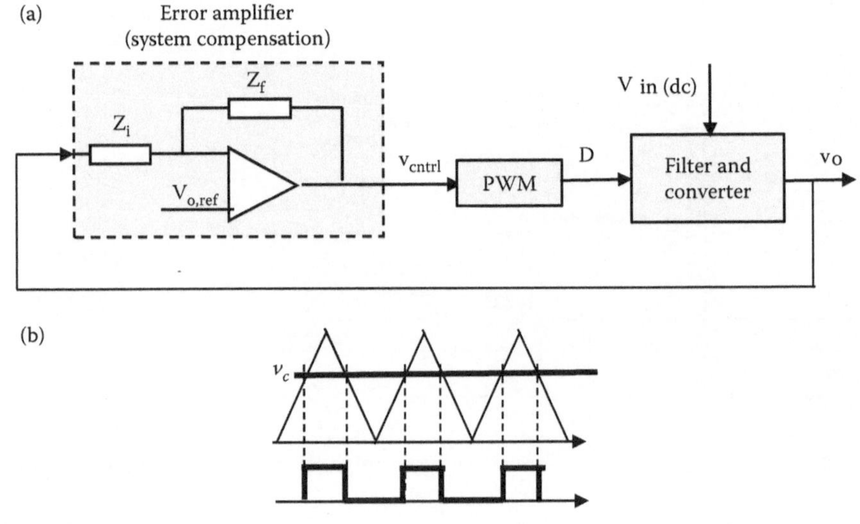

FIGURE 6.31
(a) Control circuitry, (b) waveforms for PWM generation.

Since the output filters produce a pole at low frequencies which determines a relatively slow response of the system, the dynamic performance can be improved through an internal loop for current control. The current is measured through a precision resistance or a current transformer. Usually, the control loop is constructed on the basis of the switched inductance current. The current control loop can be organised based on an average or peak current. This helps avoid core saturation.

Certain topologies like the push-pull converter require a closed-loop control of the current in order to limit the peak current and avoid saturation of the transformer.

6.8 Feedback Control of Isolated Converters: Working with TL431

The same principles of feedback control for power converters that were developed in Chapters 4 and 5 also apply to isolated converters. The difference herein consists of the requirement of measuring the DC output voltage after the isolation barrier and sending the measurement through galvanic isolation. Usually, this is accomplished with analogue-mode optocouplers: the output voltage is sensed with a resistive divider, and a compensation network is established around a TL431 circuit [9–11]. The TL431 circuit is a three-terminal adjustable shunt regulator with good thermal stability and an internal structure, as shown in Figure 6.32. The output of the TL431 circuit sets the current through an analogue optocoupler. Thus, variations of the output voltage are compensated through the optocoupler, and the output of the compensation network is further used for PWM generation on the same ground plane with the power transistor (Figure 6.33).

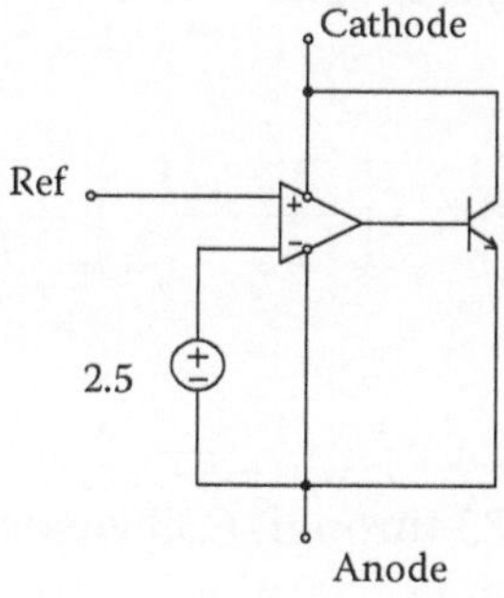

FIGURE 6.32
Internal mode for TL431.

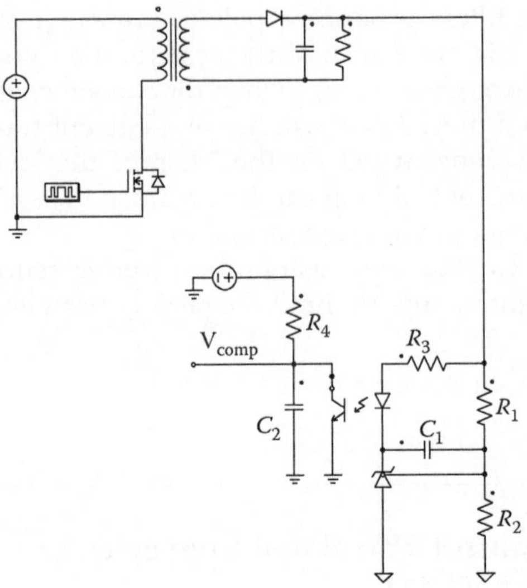

FIGURE 6.33
Setting up a feedback control circuit with TL431.

On the primary side, the output of the compensator circuit is calculated with:

$$v_c = i_{Q\_opto} \cdot (R_4 \parallel C_2) = i_{Q\_opto} \cdot \frac{\dfrac{R_4}{s \cdot C_2}}{R_4 + \dfrac{1}{s \cdot C_2}} = i_{Q\_opto} \cdot \frac{R_4}{s \cdot R_4 \cdot C_2 + 1} \tag{6.33}$$

where the current transfer through the optocoupler is defined by the device itself as:

$$i_{Q\_opto} = CTR \cdot i_{D\_opto} \tag{6.34}$$

On the load side, the optocoupler current yields from the amplifier's feedback circuit:

$$i_{D\_opto} = \frac{V_o - V_{TL431}}{R_3} = \frac{1}{R_3} \cdot \left[V_0 - \left(-\frac{V_0}{R_3} \cdot \frac{1}{s \cdot C_1} \right) \right] = \frac{1}{R_3} \cdot v_0 \cdot \left[1 + \frac{1}{s \cdot R_1 \cdot C_1} \right]$$
$$= v_0 \cdot \frac{s \cdot R_3 \cdot C_1 + 1}{s \cdot R_3 \cdot R_1 \cdot C_1} \tag{6.35}$$

Considering Equations 6.33 through 6.35 yields:

$$\frac{v_c}{v_0} = \frac{R_4}{s \cdot R_4 \cdot C_2 + 1} \cdot CTR \cdot \frac{s \cdot R_3 \cdot C_1 + 1}{s \cdot R_3 \cdot R_1 \cdot C_1} = \frac{s \cdot R_3 \cdot C_1 + 1}{s \cdot R_4 \cdot C_2 + 1} \cdot CTR \cdot \frac{R_4}{s \cdot R_3 \cdot R_1 \cdot C_1} \tag{6.36}$$

This is a typical Type II compensation (see Equation 4.20 and Figure 4.32 in Chapter 4).

Design of the compensation consists of the choice of the amplification gain, crossover frequency and phase reserve. The passive components around the TL431 circuit can be calculated from these requirements and starting from datasheet specifications like the transistor bias resistance *R*4 and the optocoupler current transfer gain CTR. The CTR is highly nonlinear and can be first estimated from the datasheet and later measured from the actual circuits.

The input resistive divider *R*1–*R*2 is easily defined from:

- The input biasing current is set around 250 μA.
- The divider gain corresponds to the ratio between the desired voltage at the converter output and the TL431 internal reference; that is, usually 2.5 V DC.

The TL431 biasing resistance *R*3 is derived from:

- From operation mode for TL431, the cathode voltage must be higher than 2.5 V DC.
- The current through the cathode and R_3 should be greater than 1 mA to get the correct regulating voltage from the internal reference.

The other components have to be calculated from Equation 6.36 and the requirements for gain, crossover frequency and phase margin. The key component C2 is actually difficult to decide on since it is composed of the parasitic capacitance of the optocoupler's transistor and some added external capacitance.

Summary

This chapter dealt with galvanically isolated converters. The analysed topologies were:

- Flyback converter
- Forward converter
- Flyback/forward converter
- Push-pull converter

An operation exists with a dependency on auxiliary phenomena like the discontinuous current through the load inductance and/or the primary-side diode conduction for discharge of reactive energy. This complicates the operation and creates nonlinear control characteristics.

The waveforms can be different from one operation point to another due to the effects of the components within the real circuitry. The operation, waveform analysis and calculation of output voltage, as well as the practical aspects, were presented.

The operation of all these converter topologies depends on magnetic devices. The magnetic materials and technologies of fabrication have evolved so much over the last decade that the magnetic device has today become a commodity product that can be bought from elite vendors, rather than a custom design for each application, made up in one's own yard.

The role of closed-loop control systems was introduced and explained. A basic design with a TL431 circuit was explained.

References

1. Picard, J., Under the Hood of Flyback SMPS Designs, 2010, Texas Instruments Power Supply Design Seminar, SEM1900, Topic 1, *Texas Instruments Literature Number: SLUP261.*
2. Fujii, M., Maruyama, H., Boku, K., 2002, "FA5553/FA5547 Series of PWM Control Power Supply IC with Multi-Functionality and Low Standby Power", *Fuji Electric Review,* 54(2), 68–72.
3. Spiazzi, G., Mattavelli, P., Costabeber, A., 2011, "High Step-Up Ratio Flyback Converter with Active Clamp and Voltage Multiplier", *IEEE Transactions on Power Electronics,* 26(11), 3205–3214.
4. Xi, Y., Bell, R., 2008, Operation & Benefits of Two-Switch Forward/Flyback Power Converter Topologies, *National Semiconductor Whitepaper,* www.embbeded.com, accessed on 21 September 2017.
5. Balogh, L., 2001, Design review: 140W Multiple Output High Density DC/DC Converter, *Texas Instruments Application Note,* pp. 6-1–6-23.
6. Neacsu, D.O., 2013, "Fault-Tolerant Isolated Converter in Low-Voltage Technology for Automotive AC Auxiliary Power", *IEEE Industrial Electronics Conference IECON,* Vienna, Austria, November 10–13, 2013, pp. 8184–8189.
7. Chandhaket, S., Konishi, Y., Ogura, K., Hiraki, E., Nakaoka, M., 2003, "A Sinusoidal Pulse Width Modulated Inverter Using Three-Winding High-Frequency Flyback Transformer for PV Power Conditioner", *IEEE 34th Annual PESC 2003, 15–19 June 2003, Acapulco, Mexico, pp.* 1197–1201.
8. Zhang, F., Yan, Y., 2009, "Novel Forward–Flyback Hybrid Bidirectional DC–DC Converter", *IEEE Transactions on Industrial Electronics,* 56(5), 1578–1584.
9. Wang, E., 2014, "Feedback Control Design of Off-Line Flyback Converter", RichTek Application Note AN017.
10. Han, M., Ye, Z., 2013, "Compensation Design With TL431 for UCC28600," Texas Instruments Application Report SLUA671.
11. Schönberger, J., 2011, "Design of a TL431-Based Controller for a Flyback Converter", Plexim Internet Paper, ver. 12–11. Also at https://www.plexim.com/files/plecs_tl431.pdf, accessed on 21 September 2017.

7

Converters with Reduced Power Loss

7.1 Context

The previous chapter explained the role of isolated DC/DC converters. At higher power levels, converter efficiency is very important since large quantities of energy are lost. For example, a 1-kW converter with an efficiency of 90% produces a 100-W loss.

Semiconductor power loss [1] can be found in:

- *Conduction state* – The loss of power equals the voltage drop across the device multiplied by the conducted current.
- *Switching loss* – Loss occurs at transition from one state to another due to finite transition time intervals.

While conduction loss depends on semiconductor technology, switching loss can be reduced with circuit techniques able to keep the current or voltage at zero during the transition from conduction to blockage or during direct transition. The operation mode determines a decrease to zero of the current through a device or of the voltage across a device before the respective transition.

Possible DC/DC converter topologies [2,3] able to achieve this goal are:

- *Resonant converters* – Where an additional L-C resonant circuit determines a certain evolution of the current or voltage.
- *Phase-shifted converters* – Where the operation determines commutation at zero voltage for all power semiconductors.

Conventional buck or boost converters can be set up with resonant converters for cancelling the voltage or current before the switching process (Figure 7.1).

Phase-shifted converters are used especially in building 1/4- or 1/2-brick converters (Figure 7.2) [4]. Technological development has allowed processing increasing levels of energy for the same dimensions of an electronic circuit. For instance, 1/4 brick converters built around the phase-shift converter principle have been developed up to 400 W when converting energy

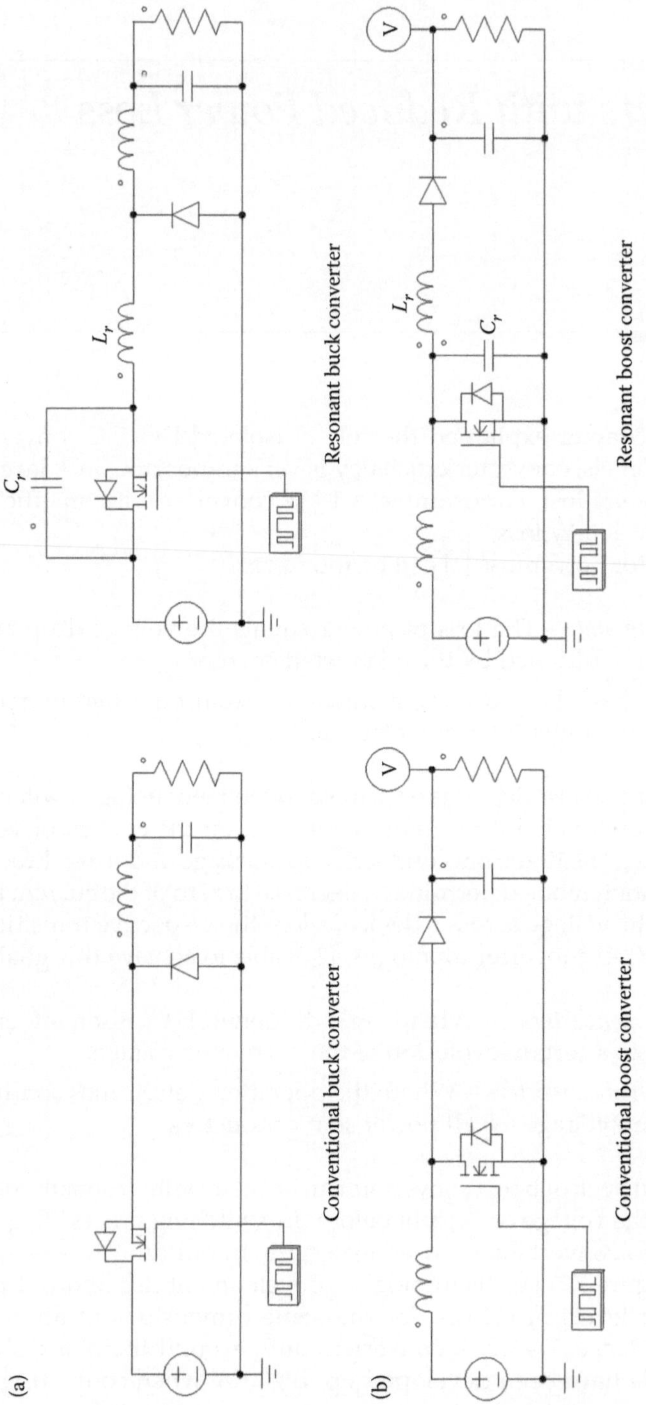

FIGURE 7.1
Resonant buck (a) and resonant boost (b) converters.

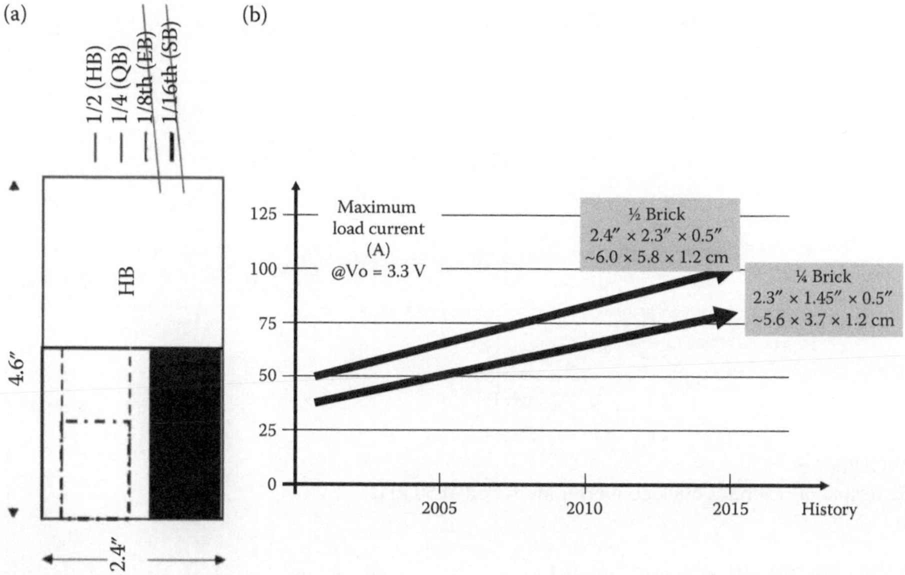

FIGURE 7.2
Phase-shifted converters are particularly used in building 1/4 or 1/2 brick converters: (a) dimensions, (b) current availability.

from 48 V DC input to an output voltage of 12 V, 5 V or 3.3 V DC. Hence, a maximum current under 100 A results in load circuit.

The use of a phase-shift converter allows a power density between 7.36 and 6.26 W/cm^3 (for example, components made by GE or Murata). Details of operation are presented within this chapter.

7.2 Resonant Converters

7.2.1 Step-Down Conversion

Figure 7.3 shows a single-switch buck converter with commutation at zero voltage, which is achieved with an additional LC circuit. The switch from Figure 7.1a is now replaced by a MOSFET transistor with an antiparallel diode. The numerical values are for educational purposes and do not reflect a real power supply case.

The power switch within this power converter uses the same control characteristics as the conventional buck converter, and the output filter can be considered nearly ideal. Hence, the load current can be considered a constant DC load current and is denoted by I_L.

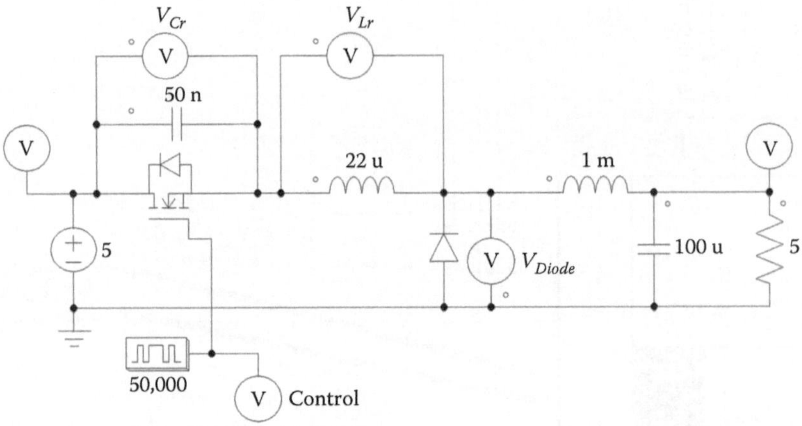

FIGURE 7.3
Example of resonant buck converter switched at 50 kHz.

Figure 7.4 illustrates possible operation waveforms, and analytical details of operation follow.

The voltage across the transistor is near zero during conduction and equals the product between the drain current and $R_{ds(on)}$. When the turn-off process starts with a turn-off command to the transistor, the load current I_L circulates through the resonant inductor L_r and the resonant capacitor C_r. This yields a linear increase of the capacitor voltage.

$$v_{C_r}(t) = \frac{I_L}{C_r} \cdot t \tag{7.1}$$

Since the load current is constant, the voltage across the resonant inductor L_r is zero. This determines a linear variation of the voltage across the buck diode D_1, from deeply reversed bias by supply voltage V_{in} to a positive bias. It yields:

$$v_D(t) = V_{in} - v_{C_r}(t) = V_{in} - \frac{I_L}{C_r} \cdot t \tag{7.2}$$

At t_1 (Figure 7.4), the diode D_1 has a positive voltage across it and turns on.

$$t_1 = \frac{V_{in}}{I_L} \cdot C_r \tag{7.3}$$

The generic $L_r - C_r$ resonant circuit can be characterised with a resonant frequency f_r and a characteristic impedance Z_r.

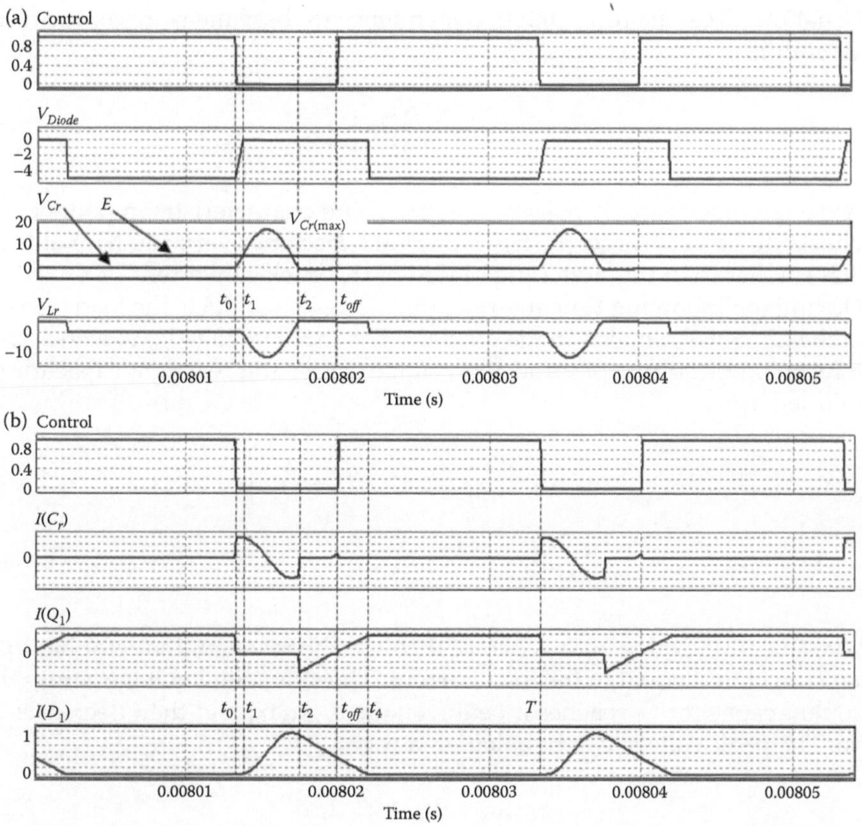

FIGURE 7.4
(a) Voltage and (b) current waveforms.

$$f_r = \frac{1}{T_r} = \frac{1}{2 \cdot \pi \cdot \sqrt{L_r \cdot C_r}} \tag{7.4}$$

$$Z_r = \sqrt{\frac{L_r}{C_r}} \tag{7.5}$$

These are also related by the following equations:

$$\omega_r \cdot Z_r = \frac{1}{C_r} \Rightarrow C_r = \frac{1}{\omega_r \cdot Z_r} \tag{7.6}$$

$$L_r = \frac{Z_r}{\omega_r} \tag{7.7}$$

Equation 7.3 can be rewritten in dependence to these measures of the resonant circuit.

$$t_1 = \frac{V_{in}}{I_L} \cdot C_r = \frac{V_{in}}{I_L} \cdot \frac{1}{\omega_r \cdot Z_r} \tag{7.8}$$

A design discussion is possible herein: if the characteristic impedance Z_r is chosen to be very large, the time interval t_1 becomes very small, or at least much smaller than the switching period of the buck converter.

During the following time interval, the diode D_1 conducts the load current and the MOSFET remains in the off state. The input voltage V_{in} is seen across the resonant circuit $L_r - C_r$, and the voltage across the resonant capacitor C_r is defined as:

$$L_r \cdot C_r \cdot \frac{d^2 v_{C_r}(t)}{dt^2} + v_{C_r}(t) = V_{in} \tag{7.9}$$

$$v_{C_r}(t) = V_{in} + Z_r \cdot I_L \cdot \sin[\omega_r \cdot (t - t_1)] \tag{7.10}$$

where the initial value of the current through the resonant inductor has also been considered equal to the load current. The sinusoidal voltage across the resonant capacitor C_r reaches a peak voltage $v_{Cr(}max_{)}$ and then decreases to zero. The peak voltage can be calculated with:

$$v_{C_r(\max)} = V_{in} + Z_r \cdot I_L \tag{7.11}$$

The moment of time when the voltage reaches zero is:

$$t_2 = t_1 + \frac{1}{\omega_r} \cdot \left[\pi + \arcsin\left(\frac{V_{in}}{Z_r \cdot I_L}\right)\right] \tag{7.12}$$

The current through the resonant circuit in this moment is given by:

$$i_{Cr}(t_2) = i_{Lr}(t_2) = -I_L \cdot \sqrt{\left[1 - \left[\frac{V_{in}}{Z_r \cdot I_L}\right]^2\right]} \tag{7.13}$$

The current through the output diode D_1 at t_2 is:

$$i_D(t_2) = I_L + I_L \cdot \sqrt{\left[1 - \left[\frac{V_{in}}{Z_r \cdot I_L}\right]^2\right]} \tag{7.14}$$

When the resonant voltage prepares for the negative swing, the antiparallel diode across the MOSFET turns on. This ensures the circulation of the current through the inductance L_r until the energy is discharged. During this interval, the voltage across L_r is maintained as constant. Hence, the current passing through L_r and the antiparallel diode follows a linear variation:

$$i_{D@MOSFET}(t) = i_{Lr}(t) = \left[-I_L\sqrt{\left[1-\left[\frac{V_{in}}{Z_r \cdot I_L}\right]^2\right]}\right] + \frac{V_{in}}{L_r}\cdot(t-t_2) \tag{7.15}$$

The current through the output diode D_1 is:

$$i_D(t) = I_L \cdot \left[1+\sqrt{\left[1-\left[\frac{V_{in}}{Z_r \cdot I_L}\right]^2\right]}\right] + \frac{V_{in}}{L_r}\cdot(t-t_2) \tag{7.16}$$

The moment of time t_3 when the current through the resonant inductor and the antiparallel diode reaches zero is found by setting Equation 7.14 equal to zero:

$$t_3 = t_2 + \frac{1}{\omega_r}\cdot\frac{Z_r \cdot I_L}{V_{in}}\cdot\sqrt{\left[1-\left[\frac{V_{in}}{Z_r \cdot I_L}\right]^2\right]} \tag{7.17}$$

When the MOSFET turns on, the current circulates through the MOSFET, the resonant inductor L_r and the output diode D_1. Because the output diode is still in the on state, the voltage across the resonant inductor equals the input voltage V_{in}. The current will continue the linear variation from zero to the load current I_L, determining the same variation of the current through the diode antiparallel to the MOSFET, L_r and output diode D_1.

$$i_{DSw}(t) = i_{Lr}(t) = \left[-I_L\sqrt{\left[1-\left[\frac{V_{in}}{Z_r \cdot I_L}\right]^2\right]}\right] + \frac{V_{in}}{L_r}\cdot(t-t_2) \tag{7.18}$$

$$i_D(t) = I_L \cdot \left[1+\sqrt{\left[1-\left[\frac{V_{in}}{Z_r \cdot I_L}\right]^2\right]}\right] + \frac{V_{in}}{L_r}\cdot(t-t_2) \tag{7.19}$$

The current through D_1 eventually turns the diode off at zero current, at a moment of time t_4 that is given by:

$$t_4 = t_3 + \frac{1}{\omega_r} \cdot \frac{Z_r \cdot I_L}{V_{in}} = t_2 + \frac{1}{\omega_r} \cdot \frac{Z_r \cdot I_L}{V_{in}} \cdot \left[1 + \sqrt{1 - \left[\frac{V_{in}}{Z_r \cdot I_L} \right]^2} \right] \tag{7.20}$$

where the results from Equation 7.6 and Equation 7.7 are considered.

After the moment t_4, the MOSFET is the only device carrying current towards the load through the resonant inductor L_r. The conduction state of this MOSFET can be changed at any time, and the entire operation cycle is repeated.

It is important to note that the moments t_2 and t_4 are very important, as they limit the possible variation of the on and off time intervals within the controller operation. Their values are given by (processed from Equations 7.12 and 7.20):

$$t_2 = \frac{V_{in}}{I_L} \cdot C_r + \frac{1}{\omega_r} \cdot \left[\pi + \arcsin\left(\frac{V_{in}}{Z_r \cdot I_L} \right) \right] \tag{7.21}$$

$$t_4 = t_2 + \frac{1}{\omega_r} \cdot \frac{Z_r \cdot I_L}{V_{in}} \cdot \left[1 + \sqrt{1 - \left[\frac{V_{in}}{Z_r \cdot I_L} \right]^2} \right] \tag{7.22}$$

It can be seen that these depend on both the passive components L_r and C_r as well as on the load current and supply voltage. While V_{in}, L_r and C_r are fixed by hardware, the load current is the only variable parameter during the operation of the power stage, and it may cause loss of resonant operation. The proper operation of the resonant circuit should respect the following constraint:

$$Z_r \cdot I_L \geq E \tag{7.23}$$

It can be seen that zero voltage transition (ZVT) can be achieved for certain current levels only, and that light loads do not lead to reduction of the switching loss. Since the switching loss for a light load is already reduced, this drawback does not really matter.

The resonant buck converter can control the output voltage only by changing the period of the entire cycle (T) or the moment when the switch is turned off (t_{off}). The actual duration of the off state is dictated by the circuit conditions. The averaged output voltage can be calculated with:

$$V_c = E \cdot \left[1 - \left(\frac{1}{2} + \frac{Z_r \cdot I_L}{2 \cdot \pi \cdot E} \right) \right] \cdot \frac{2 \cdot \pi}{\omega_r \cdot T} \tag{7.24}$$

The output voltage is therefore dependent on both the load current and the input voltage. It can be controlled only by the cycle period (T) and not the duty cycle as it was in the case of a conventional buck converter.

The voltage across the power MOSFET is increased to $V_{Cr(max)}$ from the input voltage E, which was the rating parameter for the conventional buck converter. Depending on the load current level, this can be up to twice as much as the input voltage. This obviously means overrating the power switch.

When using resonant circuits with IGBTs for higher-voltage applications, there is not much difference in the conduction losses in devices of 600 and 1200 V [5]. However, using MOSFETs implies a substantial difference between $R_{ds(on)}$ of devices rated at 20 V or 60 V, as they are the result of different technologies. The ratio of $R_{ds(on)}$ can be up to twice as much. Additionally, modern computer or telecom applications determine a very large current passing through the switch, and the conduction loss becomes important. A fair loss comparison should definitely include both the optimisation of the switching loss and the change in the conduction loss.

7.2.2 Step-Up Conversion

Many power conversion applications require a power transfer from a low-voltage source to a high-voltage load through a step-up conversion [6]. Figure 7.5 shows an example.

The voltage and current waveforms defining the operation of this resonant boost converter are shown in Figure 7.6.

The input inductance L can be modelled with a current source. The general operation of this boost converter is not modified by the presence of the resonant circuit, and the control pulses are delivered as in the case of a conventional boost converter. The analysis starts at the moment the switch is turned off. At that moment, the current through L_r and the voltage across C_r are zero.

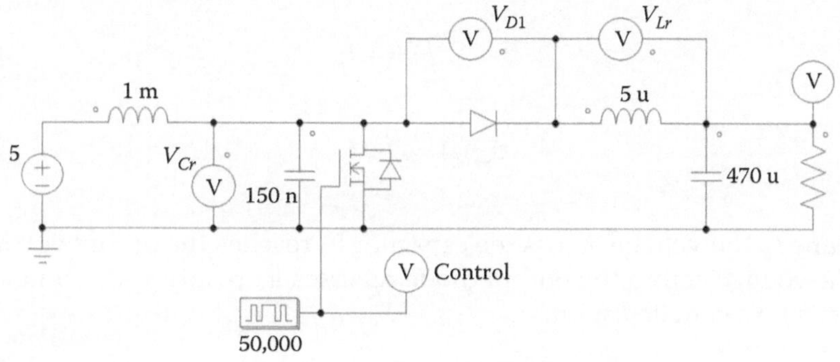

FIGURE 7.5
Example of resonant buck converter.

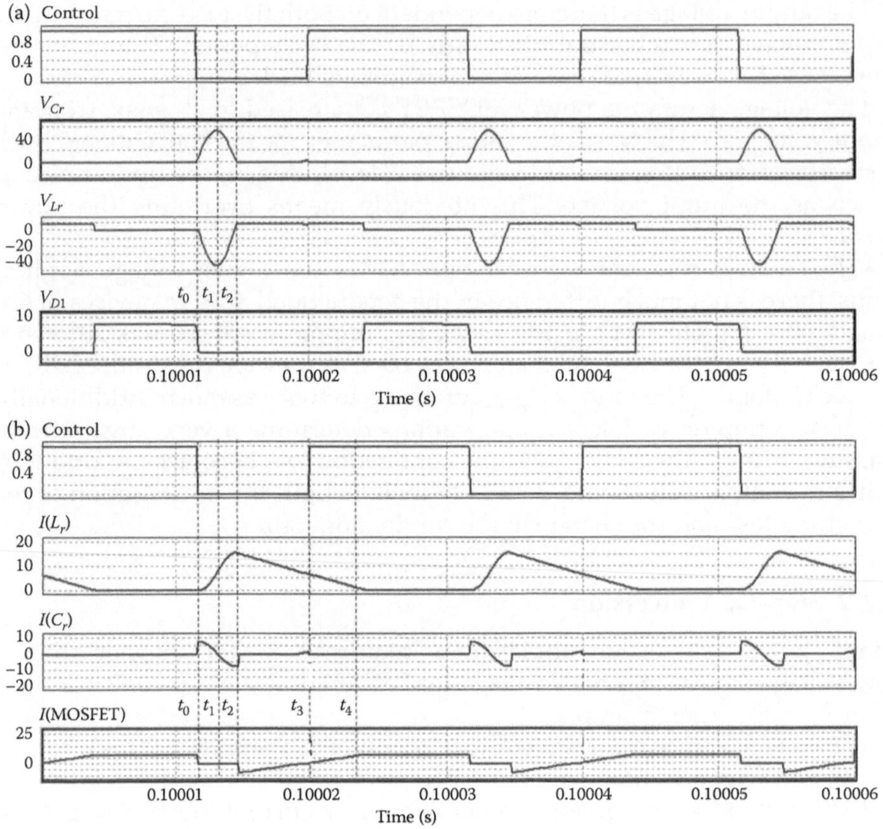

FIGURE 7.6
Voltage (a) and current (b) waveforms for a resonant boost converter.

The output diode is also in the off state. The quasiconstant input current is linearly charging C_r.

$$C_r \cdot \frac{dv_{Cr}(t)}{dt} = I_L \tag{7.25}$$

$$v_{C_r}(t) = \frac{I_L}{C_r} \cdot t \tag{7.26}$$

At time t_1, the voltage across the capacitor C_r reaches the output voltage V_0 and the voltage across the output diode reverses its polarity. The time interval t_1 can be calculated with:

$$t_1 = \frac{V_0}{I_L} \cdot C_r = \frac{1}{\omega_r} \cdot \frac{V_0}{Z_r \cdot I_L} \tag{7.27}$$

where the previous notations for ω_r and Z_r are considered.

At moment t_1, the diode turns on and the MOSFET maintains its off state. The input current is now shared between the resonant capacitor C_r, the output branch of L_r and the output diode D. The inductance L_r and the capacitor C_r form a resonant circuit supplied by the load-equivalent voltage V_0. This resonant circuit starts with the voltage across the capacitor equalling the load voltage.

$$L_r \cdot C_r \cdot \frac{d^2 v_{Cr}(t)}{dt^2} + v_{Cr}(t) = V_0 \tag{7.28}$$

The voltage across the resonant capacitor increases from the output voltage V_0 to a maximum value and then decreases to zero.

$$v_{Cr}(t) = V_0 + Z_r \cdot I_L \cdot \sin\omega_r(t - t_1) \tag{7.29}$$

At moment t_2, this voltage reaches zero. This moment can be calculated from:

$$t_2 = t_1 + \frac{1}{\omega_r} \cdot \left[\pi + \arcsin\left(\frac{V_0}{Z_r \cdot I_L}\right)\right] \tag{7.30}$$

The operation strongly depends on the load current, and the resonant swing of the voltage reaches zero only if $Z_r * I_L > V_0$. The current through C_r is:

$$i_{Cr}(t) = C_r \cdot \frac{dv_{Cr}(t)}{dt} = I_L \cdot \cos\omega_r(t - t_1) \tag{7.31}$$

and it has the final value:

$$i_{C_r}(t_2) = -I_L \cdot \sqrt{1 - \left(\frac{V_{in}}{Z_r \cdot I_L}\right)^2} \tag{7.32}$$

Analogously, the resonant current through L_r and the output diode is given by:

$$i_{L_r}(t) = I_L - i_{C_r}(t) = I_L - I_L \cdot \cos\omega_r(t - t_1) \tag{7.33}$$

After moment t_2, the resonant circuit has a tendency to swing the voltage across the capacitor to negative values, but the antiparallel diode turns on and clamps this voltage. L_r has a constant voltage across its terminals, and the current through this inductance varies linearly from its initial value at t_2.

$$i_{L_r}(t) = I_L \cdot \left[1 + \sqrt{1 - \left(\frac{V_0}{Z_r \cdot I_L}\right)^2}\right] - \frac{V_0}{L_r} \cdot (t - t_2) \tag{7.34}$$

The current through the antiparallel diode is given by:

$$i_{DSw}(t) = I_L - i_{L_r(t)} = \left[-\sqrt{1 - \left(\frac{V_0}{Z_r \cdot I_L}\right)^2}\right] + \frac{V_0}{L_r} \cdot (t - t_2) \tag{7.35}$$

The moment t_3 when the current decreases to zero is determined as:

$$t_3 = t_2 + \frac{1}{\omega_r} \cdot \frac{Z_r \cdot I_L}{V_0} \cdot \sqrt{1 - \left(\frac{V_0}{Z_r \cdot I_L}\right)^2} \tag{7.36}$$

Unfortunately, this time interval depends on both the input current (through I_L) and the output voltage (V_o) and cannot be influenced by control. The MOSFET should be controlled for the on state at any moment during the time interval (t_2, t_3), before the antiparallel diode would turn off. In this way, the MOSFET turns on after the resonant current reverses its direction at t_3.

The linear discharge of the resonant current through the output L_r continues until t_4.

$$t_4 = t_2 + \frac{1}{\omega_r} \cdot \frac{Z_r \cdot I_L}{V_0} \cdot \left(1 + \sqrt{1 - \left(\frac{V_0}{Z_r \cdot I_L}\right)^2}\right) \tag{7.37}$$

The MOSFET stays in the on state for the rest of the switching cycle, and the converter's diode remains in the off state.

7.3 Phase-Shift Converters

7.3.1 Operation of the Phase-Shift Converter

The phase-shift converter (Figure 7.7) is made of a four-transistor bridge which transforms the DC supply voltage into AC voltage with a high frequency

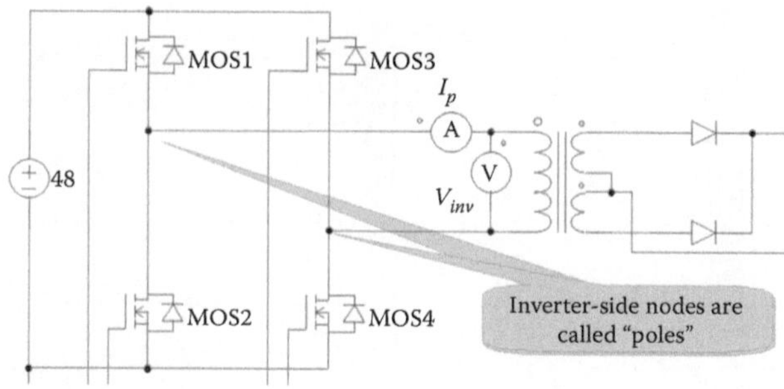

FIGURE 7.7
Phase-shift converter.

(50–150 kHz) [7]. The AC pulsed voltage is sent through a transformer. A simple rectifier transforms these pulses into DC voltage within the secondary circuitry. Finally, an L-C filter improves the output voltage quality.

Each inverter leg is controlled with a duty cycle fixed at 50% (Figure 7.8). There is a variable phase shift between the left-side leg and the right-side leg. To simplify the explanation, the right-side leg is considered to switch after (following) the left-side leg. Adjusting this phase shift regulates the output voltage.

The immediate turn-on command for a transistor pair is avoided right after the other transistor pair is turned off. Due to finite time intervals, it would be possible for two transistors to conduct simultaneously for a short time. This would produce a short-circuit of the supply voltage. A time interval is inserted to avoid simultaneous conduction, and this is called *dead time* (Figure 7.9). There are many other power converter topologies which use dead time. However, the dead time has a more extended role here. On the secondary side of the transformer, the output voltage passes through rectification and filtering (Figure 7.10).

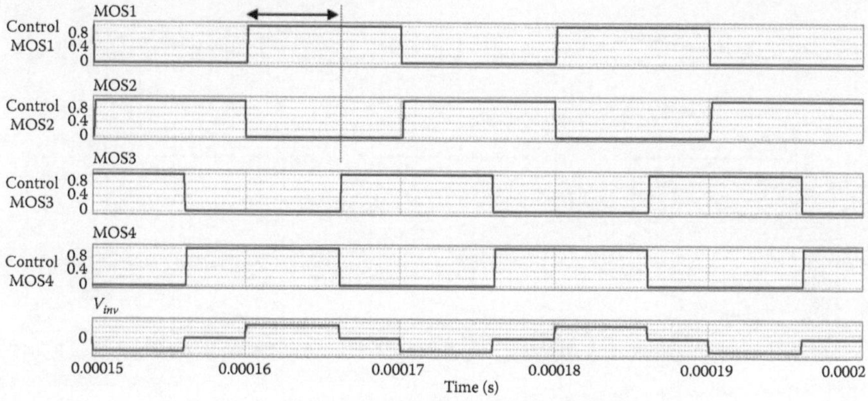

FIGURE 7.8
Waveforms for operation of the phase-shift converter.

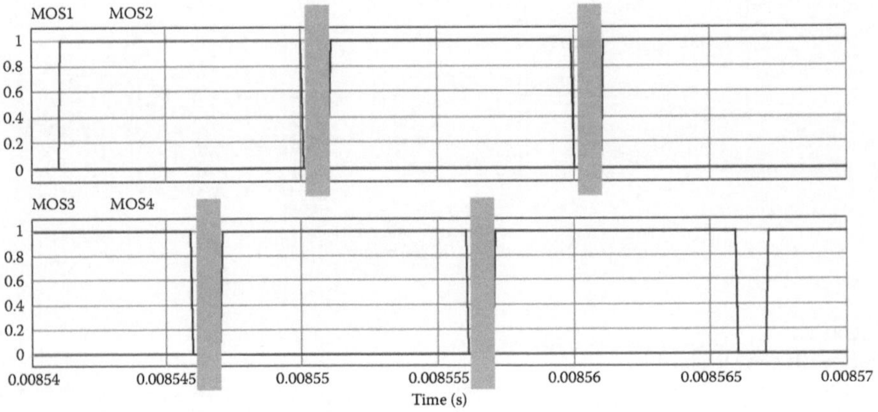

FIGURE 7.9
Dead time used with the control of a phase-shift converter.

Figures 7.11 and 7.12 show the details of operation. During the interval (t_2, t_3), transistors within the H-bridge diagonal (MOS1, MOS4) are in conduction and the entire supply voltage is applied to the primary winding of the transformer. This determines an approximately linear increase of the current through this winding due to the magnetising inductance. Depending on the magnetising inductance, load current, filter inductance and various operation modes can occur.

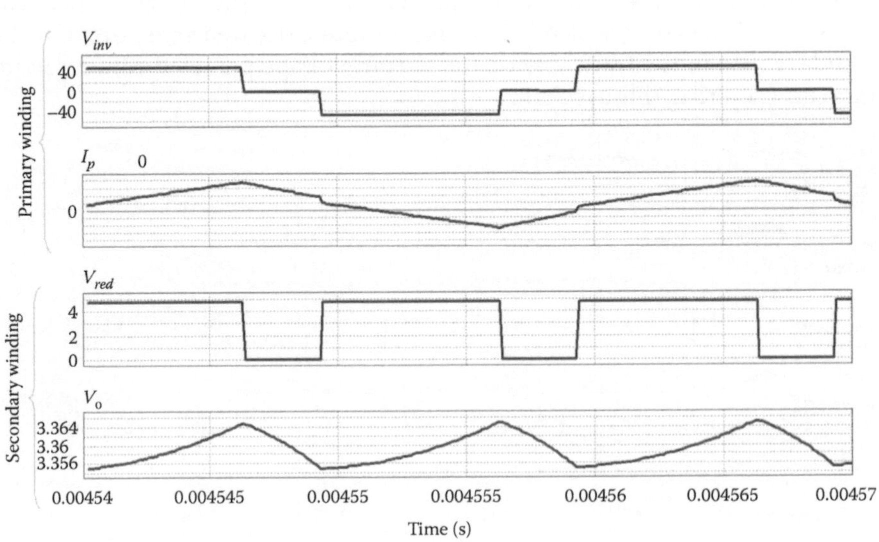

FIGURE 7.10
Transformer waveforms.

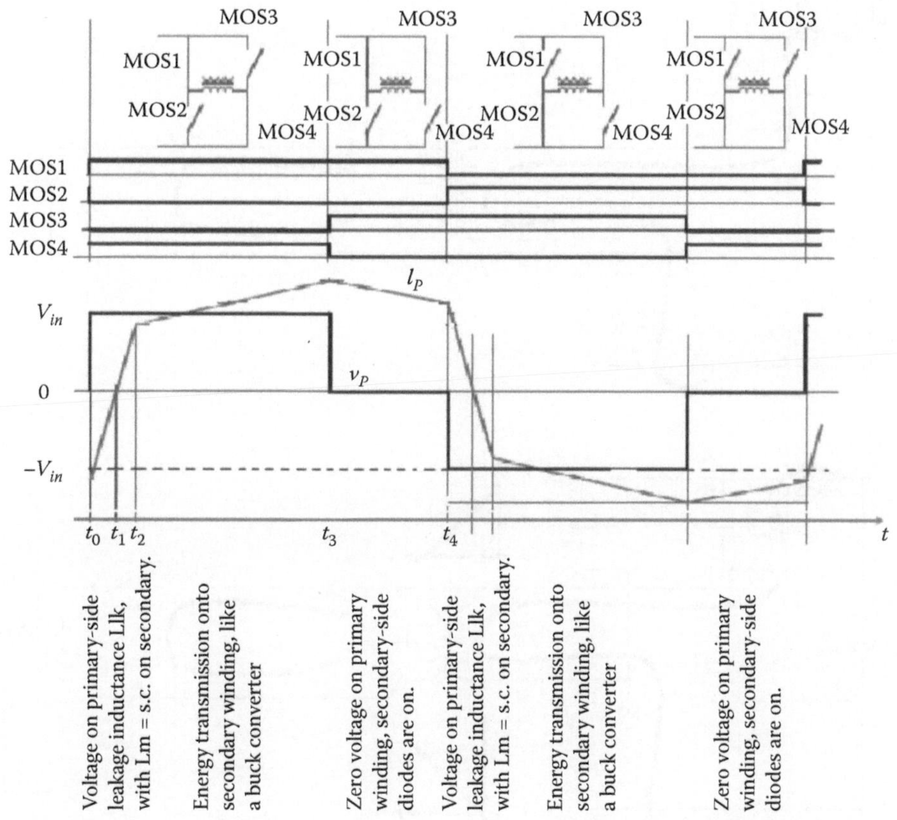

FIGURE 7.11
Details of operation.

At t_3, transistor MOS4 is turned off through a command. The C_{ds} capacitance of transistor MOS4 is hence charged with the primary-side current and its drain-source voltage increases. The capacitance of MOS3 is discharged with the primary current. The primary-side voltage converges towards zero when $V_{pri} < V_o/N$, the primary side no longer sends power into the secondary.

During the interval (t_3, t_4), the current through the transformer inductance can continue the current circulation even after the voltage across the two capacitances reaches V_{in}. Therefore, the MOS3 antiparallel diode can enter conduction (*clamped freewheeling interval*). The voltage across any of the transformer windings becomes zero. The filter inductance maintains the current circulation through the secondary winding and one or both rectifier diodes.

At t_4, transistor MOS1 is turned off. The left-side current charges the voltage across the two drain-source capacitances. When the voltage reaches V_{in}, the antiparallel diode at MOS2 turns on.

After t_4, the supply voltage is applied across the primary winding through MOS2 and MOS3, with a reversed polarity in respect to the

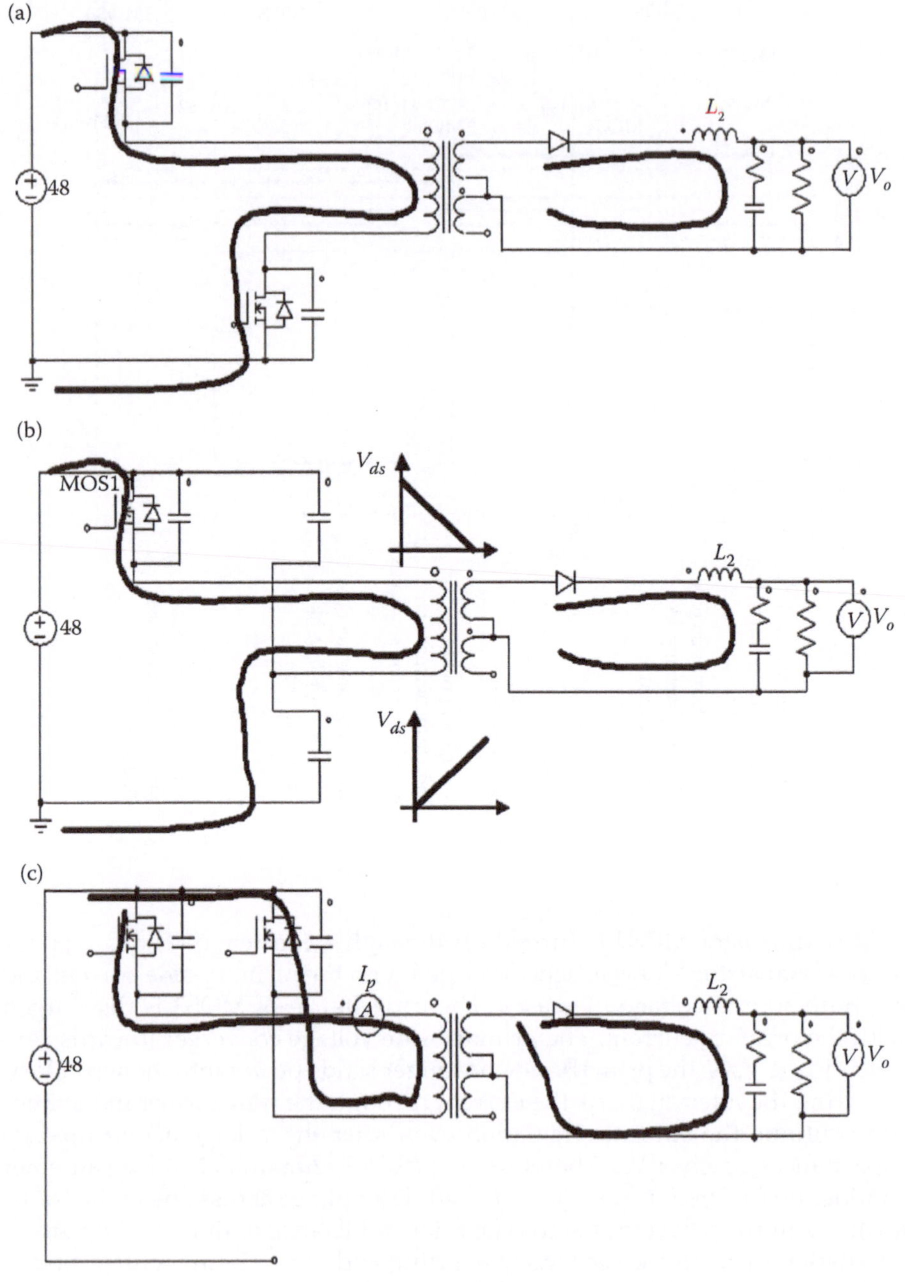

FIGURE 7.12
Circuits at various conduction states: (a) during (t_2, t_3), (b) at t_3, (c) during (t_3, t_4). (*Continued*)

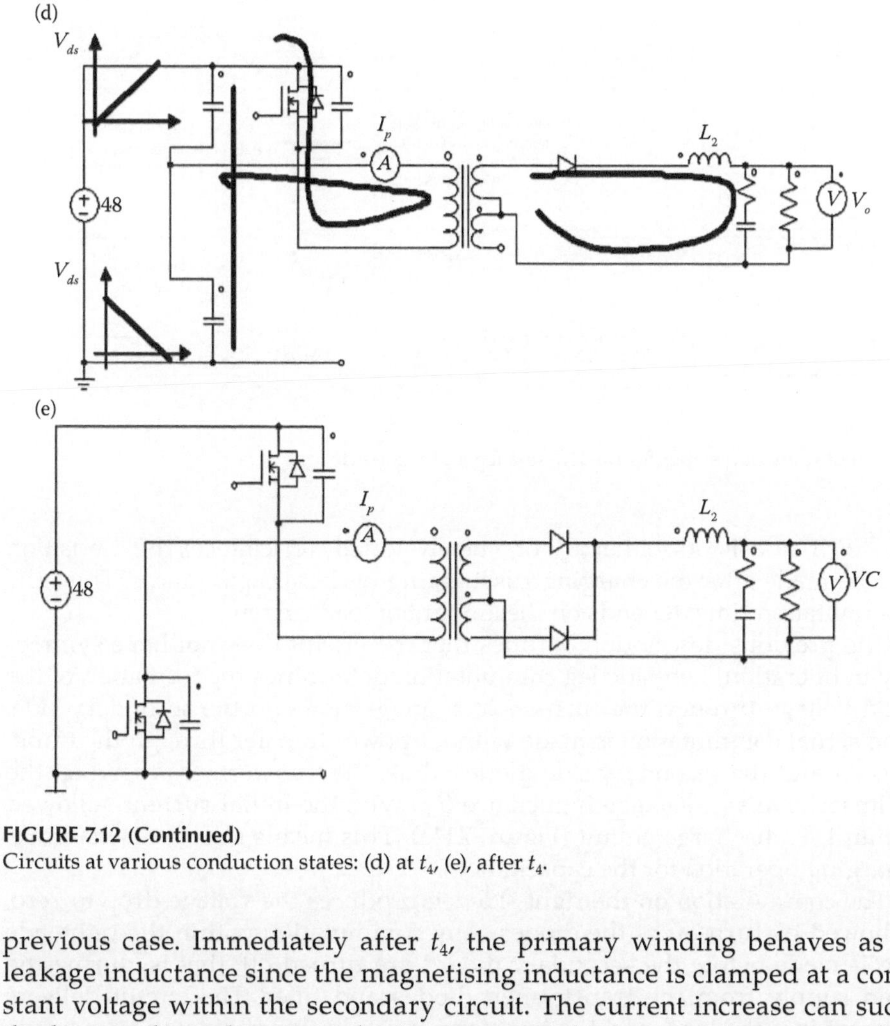

FIGURE 7.12 (Continued)
Circuits at various conduction states: (d) at t_4, (e), after t_4.

previous case. Immediately after t_4, the primary winding behaves as a leakage inductance since the magnetising inductance is clamped at a constant voltage within the secondary circuit. The current increase can suddenly pass through zero, and it increases until it becomes larger than the load current, through a transformer ratio of I_o/N. After that, the transfer is made somewhat normal with the influence of the magnetising inductance and energetic transfer to the load through a single diode within the secondary circuit.

After turning off MOS3, the operation cycles (Figure 7.13).

7.3.2 Zero-Voltage Transition

The details of commutation within any converter leg are shown in Figure 7.14. The inductance within the load circuitry produces a voltage in order to maintain the current circulation. This load current would normally turn on the diode antiparallel to MOS2. When the capacitances are connected parallel to

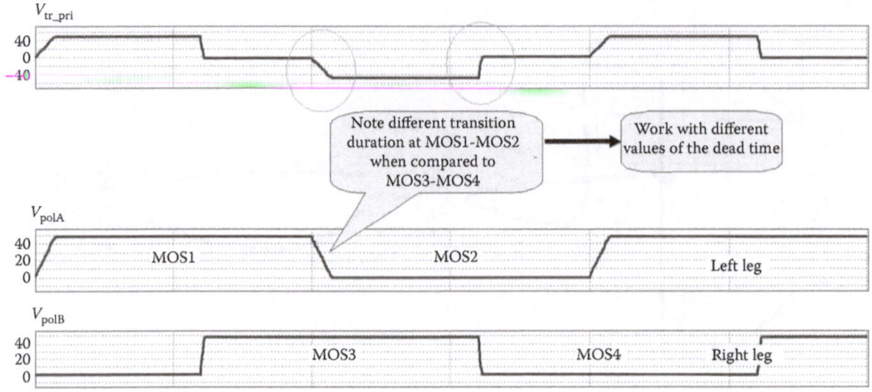

FIGURE 7.13
Different transition slopes for the left-side leg and right-side leg.

MOSFET transistors (either C_{oss} or a newly added capacitance), this transition can be made after the charging/discharging of these capacitances. Thus, the commutation time depends on the amount of load current.

The previous description outlines that the circuit does not have symmetry in operation. Left-side leg commutation determines the formation of the load voltage through the turn-on of a single diode on the secondary side. The actual commutation is made without power transfer through the transformer and the secondary-side diodes; thus, the transformer is seen on the primary side as a leakage inductance (L_{lk}) with the initial current, followed by an L-C discharge circuit (Figure 7.14a). This means slower discharging-charging operation for the capacitances.

The commutation on the right-side leg produces the voltage drop to zero, followed by turn-off of the diodes. Any commutation within the right-side leg is made before the secondary diodes are turned off; that is, during the load supply from the transformer, diodes and filter. The circuit delivers current into the load, and the transformer can be approximated with a current source (Figure 7.15b). This means faster charging/discharging of the capacitances.

The circuit is not symmetrical in operation in terms of transition times (Figure 7.16):

- Transition within the left-side leg is *slower.*
- Transition within the right-side leg is *faster.*

The dead-time duration must be defined differently for the left-side leg and the right-side leg.

Capacitances are associated with the MOSFET transistors within the time-domain simulation. At any transistor turn-off, the collector-emitter

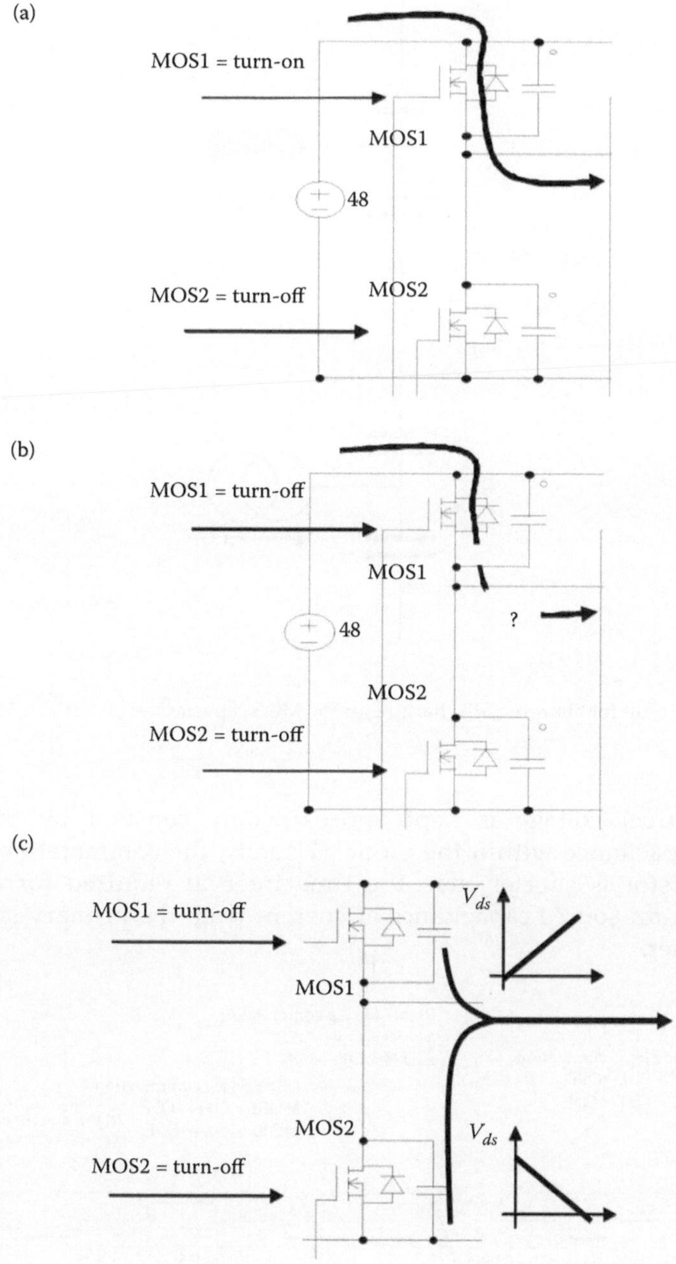

FIGURE 7.14
Explanation of transition within any converter leg (a) before commutation, (b) immediately after the turn-off command, (c) voltage evolution and transition time depend on the load current.

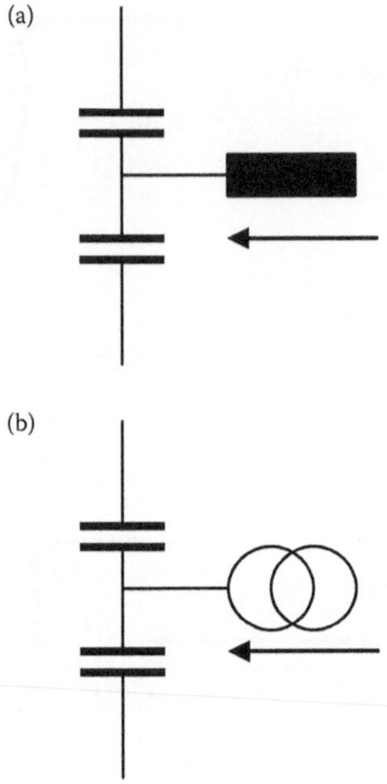

FIGURE 7.15
Equivalent circuit for charging/discharging of the MOS capacitance: (a) left-side leg, (b) right-side leg.

(drain-source) voltage is kept approximately constant by the drain-source capacitance within the model. Usually, the commutation time for the transistor is shorter than the time interval required for discharging the drain-source capacitance at current from the primary side of the transformer.

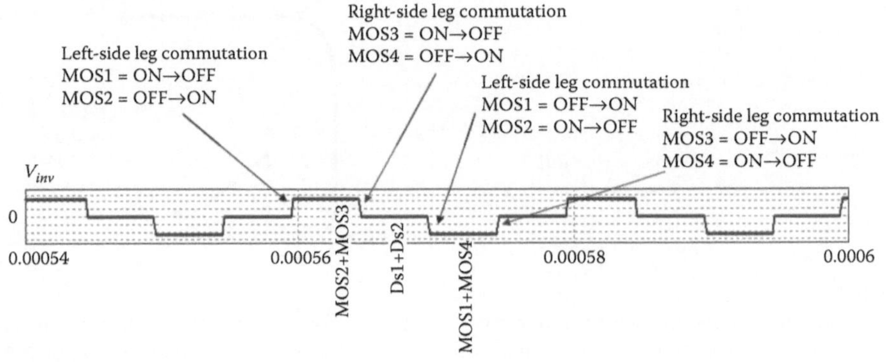

FIGURE 7.16
Different transition times for left- and right-side commutation.

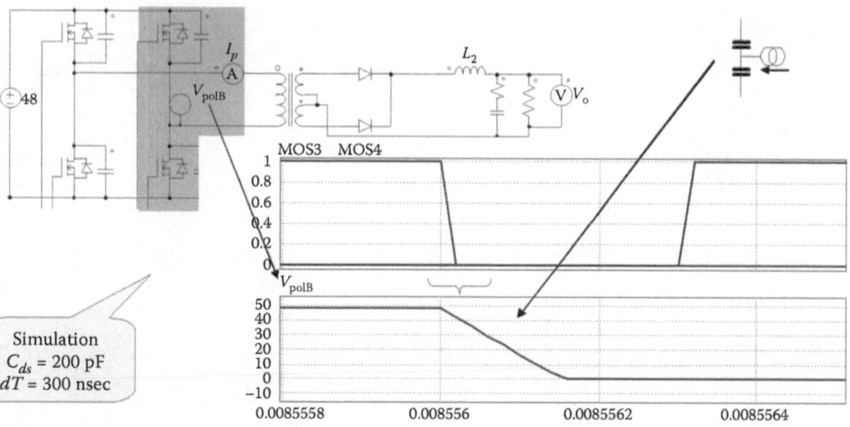

FIGURE 7.17
Commutation processes within the right-side leg.

The right-side leg commutation processes are shown in Figure 7.17. The dead-time interval is longer than the transition time. The transition time of the pole voltage is provided by the discharging/charging of the drain-source capacitances from the current source formed within the transformer. The voltage across MOS3 is maintained near zero for a longer time than the commutation time. This means the turn-off for the transistor MOS3 is made at zero voltage, with minimal loss.

If *dead time* > *discharge time* for the capacitor, then the pole voltage changes from supply voltage to zero voltage before the command for the turn-on of MOS4 (Figure 7.18). The energy of the transformer current is not lost, as it is stored within the capacitance. When the pole voltage reaches zero, the current circulation from the primary winding continues through the antiparallel diode at MOS4 (this process is called *clamping*). During this conduction interval for the antiparallel diode at MOS4, the transistor MOS4 is turned

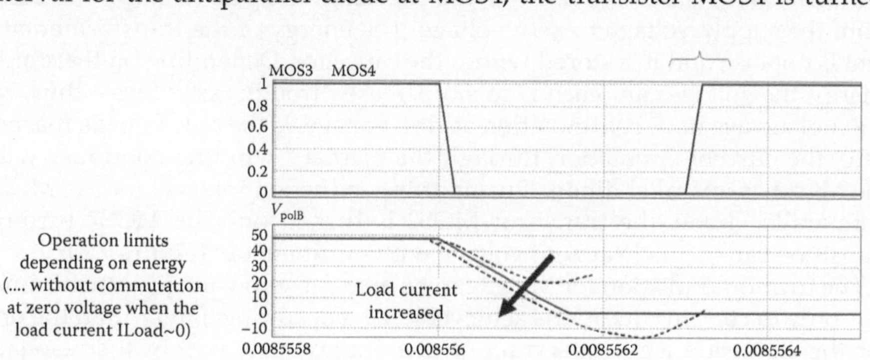

FIGURE 7.18
Pole voltage transition.

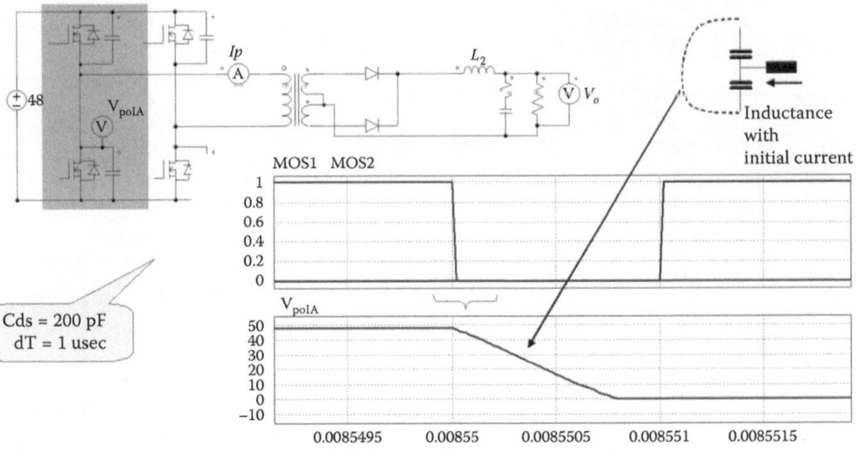

FIGURE 7.19
Commutation processes within the left-side leg.

on. Depending on the current value, the transistor MOS4 can turn on at zero voltage.

Analogously, the other commutation processes for the right-side leg can be made at zero voltage as well. That is, the turn-off transition of MOS4 is made at zero voltage, or the turn-on transition of MOS3 is made at zero voltage.

The left-side leg commutation processes are shown in Figure 7.19. At transistor turn-off, the collector-emitter (drain-source) voltage will be maintained at approximately constant by the model capacitance. Usually, the turn-off transition time for the transistor is shorter than the time interval required for the transistor discharge through the resonant circuit with the primary-side inductance. The MOS1 turn-off is made at zero voltage with minimal loss.

If *dead time > discharge time* for the capacitor, the pole voltage decreases from the supply voltage to zero voltage. The energy of the transformer current is not lost, and it is stored within the capacitor. Depending on the stored energy, the voltage can reach zero or decrease through resonance – thus, we will not always have commutation at zero voltage. If the pole voltage reaches zero, the current circulation through the primary winding continues with the MOS2 antiparallel diode. During this conduction interval for the MOS2 antiparallel diode, the transistor MOS2 is turned on. The MOS2 turn-on transition can be made at zero voltage with minimal loss (Figure 7.20).

The turn-on transition for MOS2 can be made at zero voltage with minimal loss only in certain conditions achievable for a dead time larger than the one for the right-side leg. Depending on the energy level within the resonant circuit (L-C), we can add capacitance to the circuit in order to meet the zero-voltage transition requirement (Figure 7.21).

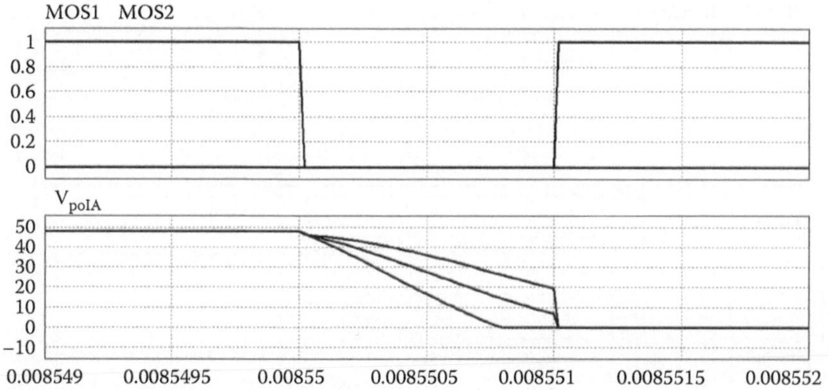

FIGURE 7.20
Transition on the left-side leg.

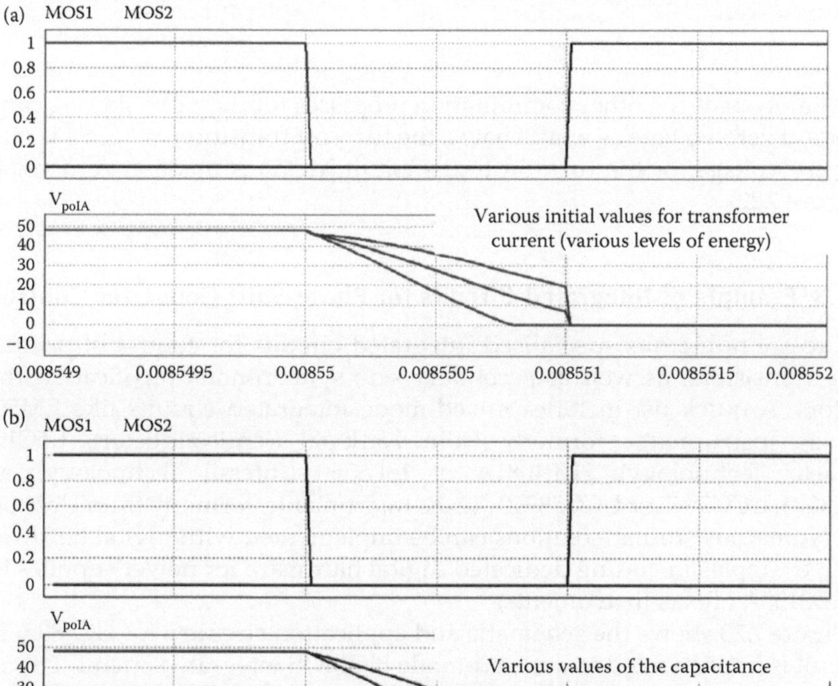

FIGURE 7.21
Operation constraints for the left-side leg: (a) behaviour at various levels of initial energy, (b) various values for the added capacitance.

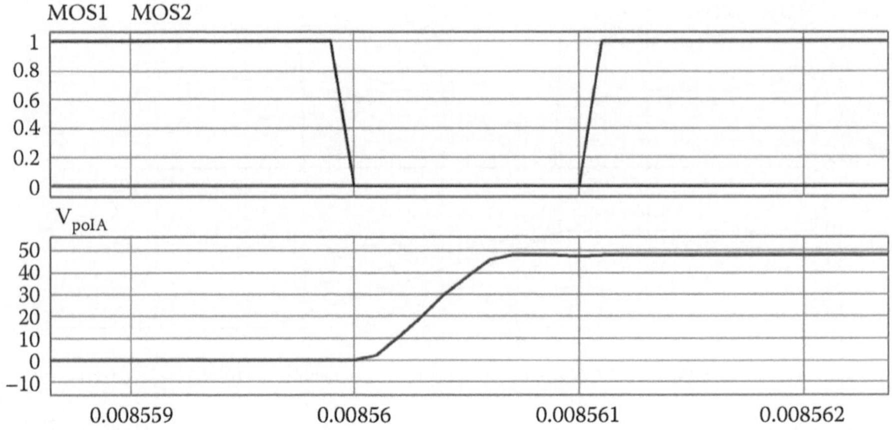

FIGURE 7.22
Turn-off transition of MOS4 is made at zero voltage, and turn-on transition of MOS3 is made at zero voltage.

Analogously, the other commutation processes for the right-side leg can be made at zero voltage as well. That is, the turn-off transition of MOS4 is made at zero voltage, or the turn-on transition of MOS3 is made at zero voltage (Figure 7.22).

7.3.3 Example of Integrated Circuits for Phase-Shift Converter Control

There are numerous specialised integrated circuits for control of primary-side transistors, as well as secondary-side synchronous rectification transistors. A quick list includes mixed-mode integrated circuits like LM5046 (Texas Instruments, formerly from National Semiconductor); LTC1922 (Linear Technology); HIP4081A or ISL6550 (Intersil Technology); and UC3875, UCC3895 or UCC28950 (Texas Instruments, formerly from Unitrode) [9]. Numerous similar solutions can be implemented with digital hardware, with examples including dedicated digital hardware for power supplies like UCD3138A (Texas Instruments).

Figure 7.23 shows the schematic and application circuitry for UC3895. The circuit is implemented as a mixed-mode BiCMOS integrated circuit. This circuit provides control signals for four H-bridge MOS devices.

Special functions implemented within this integrated circuit can be understood by following the pinout and external connections:

- Adaptive-delay-set pin ADS sets the ratio between the maximum and minimum programmed output delay dead time. The maximum delay modulation occurs when ADS is grounded. The ADS pin

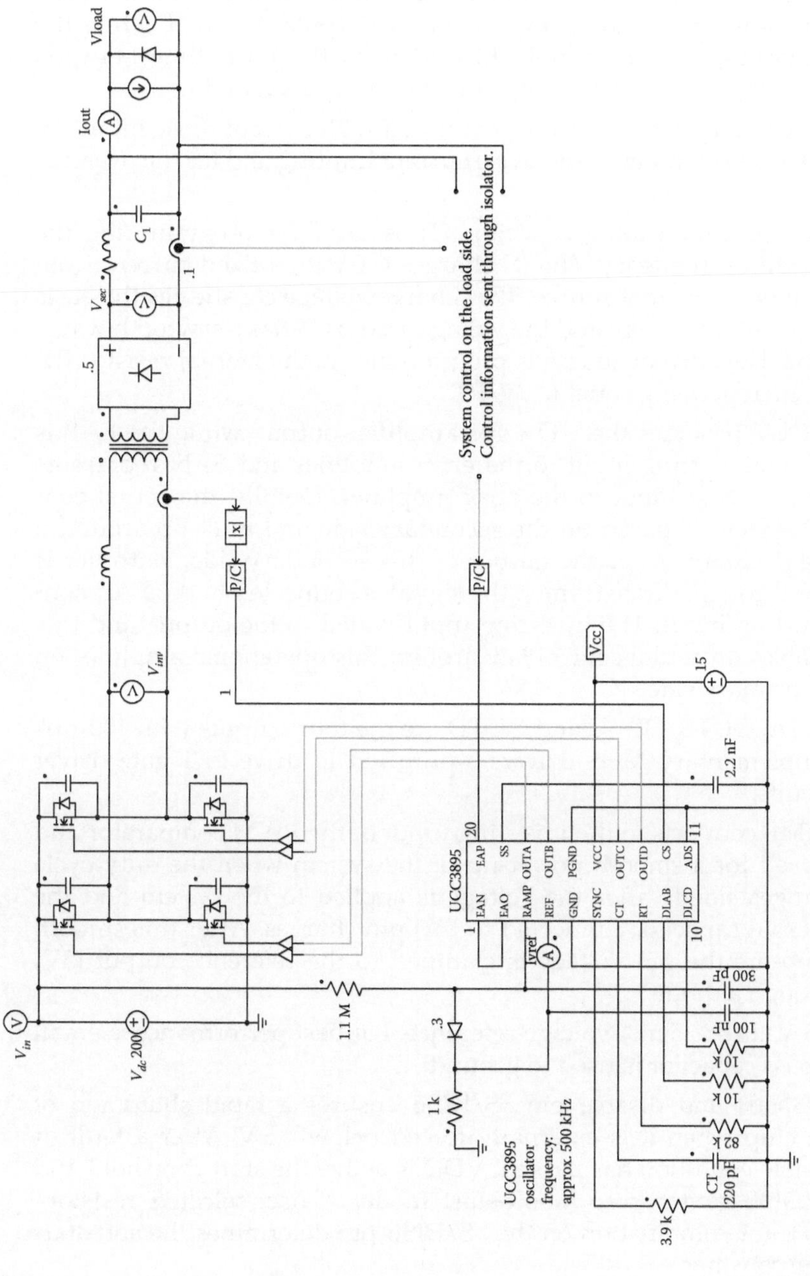

FIGURE 7.23
Schematic and application circuitry for UC3895. [Anon, 2013, BiCMOS Advanced Phase-Shift PWM Controller, Texas Instruments, SLUS157P – Revised June 2013.]

changes the output voltage on the delay pins DELAB and DELCD according to a setup equation.

- The delay-programming pin DELAB defines the dead time between switching of left-side complementary outputs A and B, while the delay-programming pin DELCD defines the dead time between switching of right-side complementary outputs C and D.
- The primary-side current is monitored with current sense input CS, and it is used for cycle-by-cycle current limiting and for the overcurrent comparator.
- The oscillator timing capacitor CT is used for programming the switching frequency. The IC charges CT with a fixed current from an internal current source. This charge voltage constitutes the basis of PWM generation, and the voltage across CT has a sawtooth waveform. The current source is programmed with external resistor RT, the current being equal to 3V/RT.
- EAOUT provides the I/O error amplifier output, while the EAP is the noninverting input to the error amplifier and EAN represents the inverting input to the error amplifier. Usually, the actual control system is paced on the secondary side and built up around a TL431 circuit. After the output of this secondary-side controller is sent through a transformer, the signal is connected to the EAP noninverting input. The inverting input is tied to the output, and this disables an actual control law around this operational amplifier on the primary side.
- OUTA, OUTB, OUTC and OUTD are the four outputs with 100-mA complementary MOS drivers, optimised to drive FET gate driver circuits.
- RAMP connects to the inverting input of the PWM comparator and is used for a smooth power-up of the system when the duty cycle changes slowly after the voltage is applied to the system and the external capacitor connected to this pin charges. After this smooth ramp-up, the pin voltage is clamped to the reference output (5V) through a diode.
- A 5 V, ±1.2% 5 mA voltage reference. For best performance, a small bypass capacitor is used to ground.
- Soft-start and disable pin SS/DISB ensures a rapid shutdown of the chip when it is externally forced below 0.5 V. After a fault or disable condition has passed, VDD is above the start threshold and SS/DISB switches to a soft-start mode. A user-selected resistor/capacitor combination on the SS/DISB pin determines the soft-start time constant.
- The SYNC pin forces a synchronisation of the oscillator.

7.3.4 Brick Converters Using Phase-Shift Converters

The main application for phase-shift converters is half- or quarter-brick power converters used as intermediate bus converters within a telecom power system [8]. Multiple solutions are reported for either digital or analogue control with power levels of hundreds of W.

The digital solutions allow for extended improvement of efficiency, up to 99% [10], for a conversion from 400 V DC to 48 V DC, and an operation at 50% of the rated power of 5 kW and a power density of $\rho = 2.3$ kW/litter (38 W/in^3). This result recommends phase-shift converter technology for multiple applications in the future of telecom power systems as an intermediate bus converter. Other solutions easily run at mid-90% efficiency for conversion from 400 V DC to 48 V DC or from 48 V DC to 12 V DC, on power levels appropriate for quarter, half or even full brick.

7.4 LLC Converters

7.4.1 Operation

The LLC converter [11–13] represents a topology similar to that used as an isolated DC/DC converter, and it is operated from a high-voltage bus to the traditional low-voltage telecom DC bus of 48 V or directly to the 12 V DC distribution DC bus required by modern architecture. The input voltage is provided with a telecom rectifier. For instance, the European 220 V rms AC grid can be rectified with a diode bridge rectifier into a voltage of 324 V DC.

The circuitry of the LLC resonant converter together with a numerical example is shown in Figure 7.24. It includes an inverter branch operated with a 50% duty cycle and a small dead time, a series inductance L_r and a series capacitor C_r, followed by a transformer with a magnetising inductance L_m. Sometimes, the primary-side power stage is actually made up of a bridge inverter, such as in the case of a phase-shift converter. In either case, a square wave voltage is applied to the passive network.

The two inductors together with the series capacitor form a resonant tank that will create a quasisinusoidal current through the power transistors, hence the name LLC resonant converter. The secondary side is completed with a rectifier and an output filtering capacitor.

The power transistors are operated with zero voltage switching, which yields a high efficiency. The high efficiency of the LLC resonant converter at large DC bus voltages recommends this topology for telecom DC bus controller applications. The power range is usually within hundreds of watts.

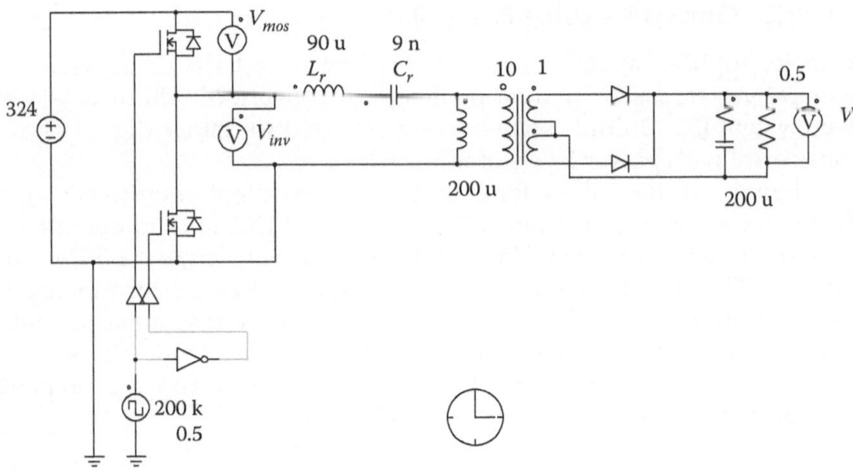

FIGURE 7.24
Example of circuitry for a LLC converter used as a DC bus converter, from a high voltage of 324 V DC to a low voltage of 12 V DC, after transformer isolation.

The LLC converter avoids the multiple operation modes of the phase-shift converter and generally ensures a resonant operation over a wider range. Additionally, the MOSFET body diode recovery mechanism in a phase-shifted ZVS full bridge DC/DC converter can produce losses and stress the application. This may not be present with an LLC converter except during startup, when system reliability issues caused by an incomplete body diode reverse recovery of the power MOSFET need to be considered.

On the downside, a series capacitor is needed on the primary side of an LLC converter to carry the entire load current. Such a series capacitor may pose cooling problems in certain applications.

Since both phase-shift and LLC converters are operated with a 50% duty cycle, there are some similarities in control.

The production of the output DC voltage can be understood with Figure 7.2. The pulsed voltage from the inverter branch is applied to the passive network from Figure 7.2. The voltage transfer function yields:

$$H(s) = \frac{V_o}{V_{in}} = \frac{1}{n} \cdot \frac{(n^2 \cdot R_{load}) \| (s \cdot L_m)}{Z_{total}} = \frac{1}{n} \cdot \frac{(n^2 \cdot R_{load}) \| (s \cdot L_m)}{s \cdot L_r + \dfrac{1}{s \cdot C_r} + (n^2 \cdot R_{load}) \| (s \cdot L_m)} \tag{7.38}$$

Along with the actual design of the compensation law, it makes sense to assess the steady-state transfer characteristic, also known as DC gain. In this respect, we have to consider energy transfer on the fundamental

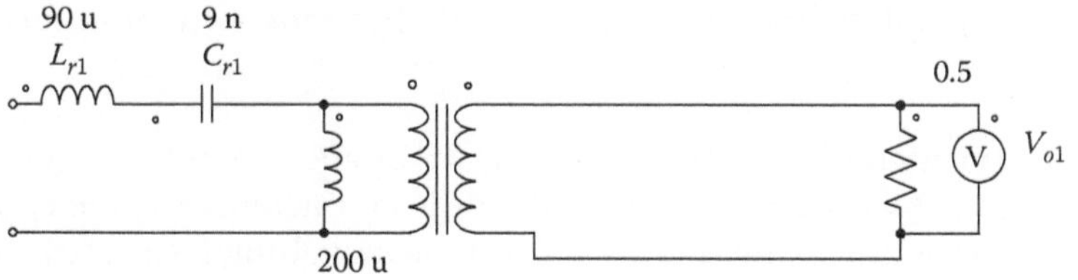

FIGURE 7.25
Passive component network for transfer function depiction.

component of the square wave produced by the inverter branch. The steady-state gain has to be calculated at this fundamental frequency since this is the only lever of control. The square wave is obtained with a constant 50% duty cycle, and the DC component of the output voltage can be adjusted with the switching frequency only. The switching frequency equals the square-wave frequency and therefore its fundamental frequency (Figure 7.25).

After some calculation, the steady-state transfer characteristic yields:

$$Gain(f_{sw})=|H(s)|=\frac{1}{n}\frac{1}{\sqrt{\left[\frac{L_r}{L_m}+1-\frac{L_r}{L_m}\cdot\left(\frac{f_{res}}{f_{sw}}\right)^2\right]^2+\left(\frac{\pi^2}{8\cdot n^2}\cdot\frac{P_{out}}{V_{out}^2}\cdot\sqrt{\frac{L_r}{C_r}}\right)^2\cdot\left(\frac{f_{sw}}{f_{res}}-\frac{f_{res}}{f_{sw}}\right)^2}} \tag{7.39}$$

where the resonance frequency is considered:

$$f_{res}=\frac{1}{2\cdot\pi\cdot\sqrt{L_r\cdot C_r}} \tag{7.40}$$

Using numerical data from Figure 7.24, the dependency of the output voltage on the switching frequency is calculated in MATLAB and plotted in Figure 7.26.

When a square wave is produced by a single inverter branch, the output voltage is achieved by multiplying the gain by half the DC input voltage. Later on, Figures 7.27 and 7.28 show simulation results for the ZCS and ZVS cases. A gain of 0.07 is read from characteristics in Figure 7.26 and multiplied by 162 V DC to yield an output voltage of about 12 V DC. Conversely, if a single-phase bridge inverter operated with a 50% duty cycle is used on the primary side, the full input voltage has to be considered when assessing the transfer with the characteristics of Figure 7.26.

Different operation modes occur within different switching frequency ranges:

- At a considerably smaller switching frequency than the resonance frequency (considered at the peak of the characteristics), the characteristic monotonically increases with the switching frequency.
- At a considerably larger switching frequency than the resonance frequency, the characteristic monotonically decreases with switching frequency.

It can be seen that monotony of the characteristic in Figure 7.26 changes with the switching frequency.

7.4.1.1 Operation with Switching Frequency above the Resonance Frequency

If the variation range of the switching frequency is considered far from the resonance frequency, a quasilinear dependency can be achieved in either case: a change in switching frequency can therefore change the output voltage.

The possible perturbation factors acting against the output voltage are load resistance (or load current) and input voltage. The transfer characteristic has been calculated with a series with simplified assumptions, particularly neglecting loss in the circuit. Any loss component depends on the current level. Hence, a real-time test is required to assess operation at different points of the characteristic. Figure 7.27 illustrates the changing of the operation point when the load changes. At a higher load current ($R = 0.5$ Ohm),

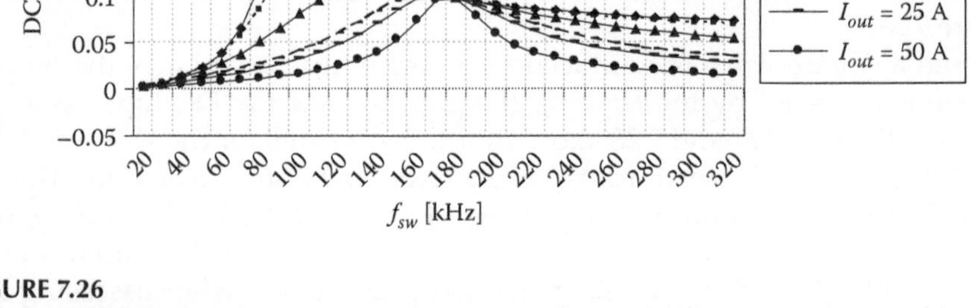

FIGURE 7.26
Steady-state (DC transfer gain) characteristic expressed as dependency to the switching frequency. Results depicted from MS-EXCEL calculation.

voltage loss occurs and the transfer characteristic needs to move at lower switching frequencies where the gain is larger. More voltage is expected towards the load, and this should compensate for the voltage loss.

The average values of the output voltage for the two operating points are similar, while the output ripple depends on the load current: a larger load current commands a larger voltage ripple on the output capacitor. This test has featured a load of maximum 300 W. As mentioned, various applications of the LLC circuitry are also within the kW range.

Figure 7.27 also demonstrates operation with ZVS. The transistors are commanded to turn on when the antiparallel diodes conduct the resonant current. The current needs first to go to zero, wherefrom the MOSFET actually turns on. At that moment, the voltage drop is zero due to previous diode conduction.

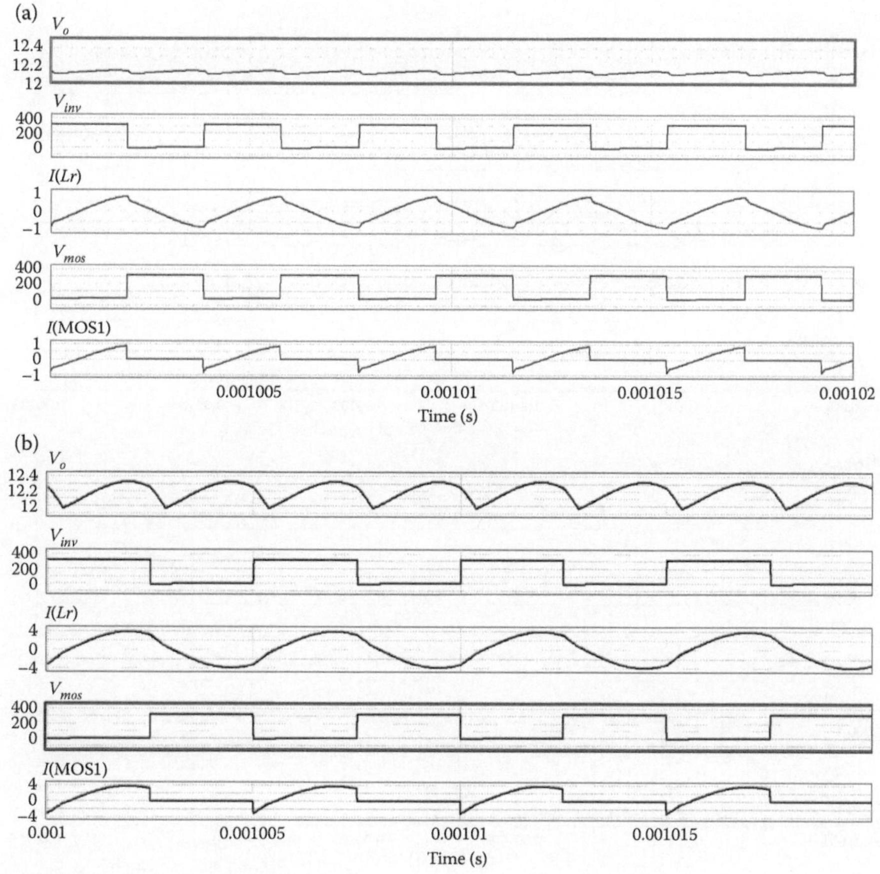

FIGURE 7.27
Main waveforms during operation with ZVS at different points: (a) $R_{Load} = 5$ Ohm and $f_{sw} = 260$ kHz, (b) $R_{load} = 0.5$ Ohm and $f_{sw} = 200$ kHz.

7.4.1.2 Operation with Switching Frequency below the Resonance Frequency

A second linear portion of the gain dependency on switching frequency occurs for switching frequencies considerably lower than the resonant frequency. This is considered in the results in Figure 7.28, where a switching frequency of 137 kHz is used for a large load current, while just 68 kHz are enough to provide a similar load voltage when the load resistance is 10 times larger. These results follow the DC transfer gain characteristic plotted in Figure 7.26 for the considered numerical data.

This operation mode demonstrates the turn-off of the power MOSFET at zero current, followed by the conduction of the antiparallel diode, while the turn-on process takes over the complementary diode's current.

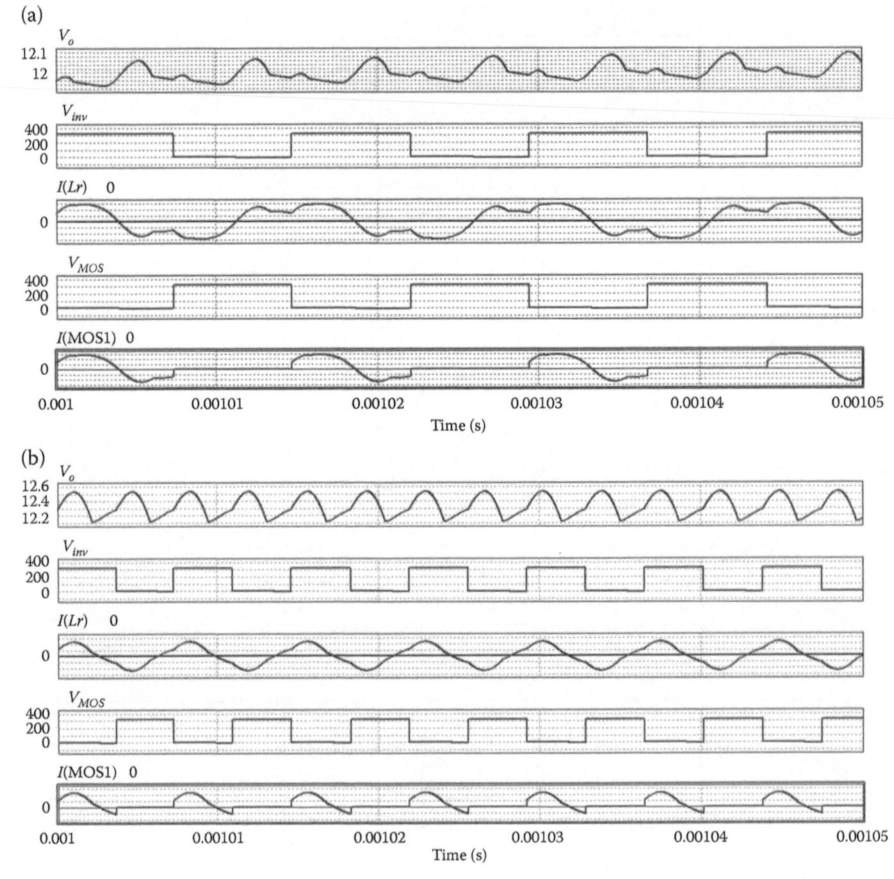

FIGURE 7.28
Main waveforms during operation with ZCS (a) $R_{load} = 5$ Ohm and $f_{sw} = 68$ kHz, (b) $R_{load} = 0.5$ Ohm and $f_{sw} = 137$ kHz.

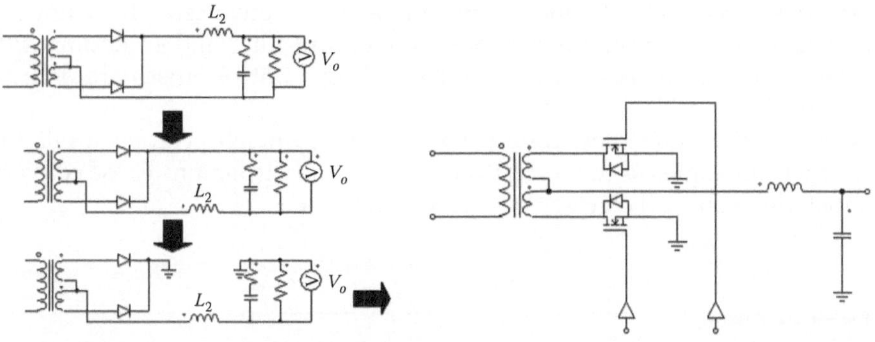

FIGURE 7.29
Various circuits for synchronous rectification.

7.5 Synchronous Rectification on Secondary Side

Analogously to the operation of the synchronous converter presented in Chapter 3, the rectifier diodes can be replaced with MOSFET transistors on the secondary side in order to allow reduction of conduction loss and elimination of the recovery currents' effect. The circuit is redrawn for the same ground circuit within the control scheme (Figure 7.29).

Control of the MOSFET transistors within the synchronous rectifier is defined with the operation intervals of primary side MOSFETs. This is done

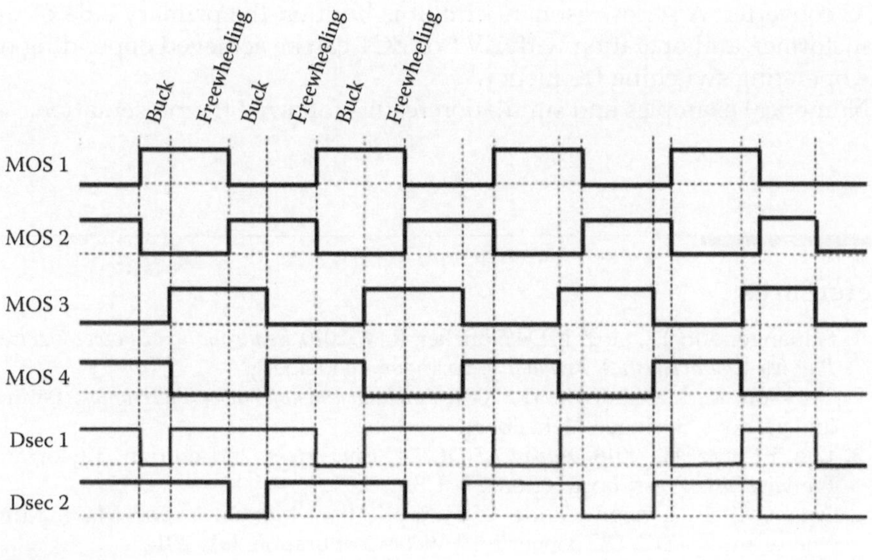

FIGURE 7.30
Waveforms for operation and control for synchronous rectification.

analogously with the demonstration for a buck converter. This allows a free circulation of the current through the output filter and secondary-side diodes, and this process is called freewheeling. Control pulses are defined in Figure 7.30.

Note that the same synchronous rectification principles can be applied to any other topologies of converters with transformer isolation. These converters include push-pull converters or LLC converters.

Summary

In order to decrease switching loss, power semiconductor devices have to switch at zero current or zero voltage. Resonant circuits equipped with passive components were introduced near the power switch to achieve either zero current transition (ZCT) or zero voltage transition. Equations and limits of operation were produced.

The operation of a H-bridge converter with galvanic isolation through a transformer was defined with a fixed 50% duty cycle and phase-shift control between the left-side leg and right-side leg. The converter operational states were explained separately, and it was shown that, in certain conditions, zero-voltage switching can be achieved. There are specialised integrated circuits for transistor control on the primary side, as well as on the secondary side. This is the phase-shift power converter.

An alternative to the phase-shift power converter is offered with the LLC converter. A series resonant circuit is built on the primary side of the transformer, and operation with ZVT or ZCT can be achieved depending on the operating switching frequency.

Numerical examples and simulation results followed the presentation.

References

1. Schlangenotto, H., Lutz, J., De Doncker, R.W., 2010, *Semiconductor Power Devices: Physics, Characteristics, Reliability*, Springer, Heidelberg.
2. Erickson, R., Maksimovic, D., 2001. *Fundamentals of Power Electronics*, 1st and 2nd editions, Springer, Heidelberg.
3. Luo, F.L., Ye, H., 2016, *Advanced DC/DC Converters*, 2nd edition, Taylor and Francis, CRC Press, Boca Raton, FL, USA.
4. Anon., 2014, "Design Guide and Application Manual – For Maxi, Mini, MicroFamilies DC/DC converters", Vicor Corporation, July 2014.
5. Neacsu, D.O., 2013, *Switching Power Converters – Medium and High Power*, CRC Press, Taylor and Francis, Boca Raton, FL, USA.

6. Mohan, N., Undeland, T., Robbins, W., 1995, *Power Electronics*, John Wiley & Sons, Hoboken, NJ, USA.
7. Aigner, H., Dierberger, K., Grafham, D., 1998, "Improving the Full-Bridge Phase-Shift ZVT Converter for Failure-Free Operation under Extreme Conditions in Welding and Similar Applications", Advanced Power Technology, Application Note APT9803, also presented at IEEE IAS.
8. Kankanala, R., 2010, "Quarter Brick PSFB DCDC Converter Reference Design Using the dsPIC® DSC", Microchip Technology Application Note 1337.
9. Anon, 2013, BiCMOS Advanced Phase-Shift PWM Controller, Texas Instruments, SLUS157P – Revised June 2013.
10. Badstuebner, U., Biela, J., Kolar, J.W., 2010, "An Optimized, 99% Efficient, 5kW, Phase-Shift PWM DC-DC Converter for Data Centers and Telecom Applications, 2010", *IEEE International Power Electronics Conference*, Singapore, pp. 626–634.
11. Florescu, A., Oprea, S., 2013, "High Efficiency LLC Resonant Converter with Digital Control", *Revue Roumaine du Science Techniques – Électrotechnique et Energétique*, 58(2), 183–192.
12. Liu, J., Zhang, J., Zheng, T.Q., Yang, J., 2017, "A Modified Gain Model and the Corresponding Design Method for an LLC Resonant Converter", *IEEE Transactions on Power Electronics*, 32(9), 6716–6727.
13. Yu, R., Ho, G.K.Y., Pong, B.M.H., Ling, B.W.K., Lam, J., 2012, "Computer-Aided Design and Optimization of High-Efficiency LLC Series Resonant Converter", *IEEE Transactions on Power Electronics*, 27(7), 3243–3256.

8

Power Management of Embedded Systems

8.1 Requirements for Digital Control

Modernisation of telecommunication systems brings new standards and new requirements for power supplies in addition to efficiency and size [1–4]. Examples include:

- Uninterrupted operation when the grid fails
- Sleep mode with power shutdown during the time when the entire system or a subsystem of the equipment is not in use
- Reduction of the DC component within the AC grid network
- Connection or disconnection of certain modules during operation (hot swap)

Most of the new requirements for power supplies can be implemented on a digital platform with digital (intelligent) control taking advantage of a communication channel between the power supplies. The usage of a digital control platform on power supplies has been avoided for a long time due to inherent power loss within digital circuits and additional cost. The improvement of CMOS technologies for digital integrated circuits allowed options with very low loss, and it has favoured the digital control of power supplies.

Digital control and digital management of power supplies are used mostly within applications related to telecom, data or computing systems, where multiple operating nodes impose sequential control, with multiple operating states or modes. Conversely, digital platforms are very rarely used in standalone power supplies.

8.2 Power Control versus Digital Management of Power Supplies

The rapid evolution of digital platforms has favoured an unfortunate confusion of terms between digital control and digital management.

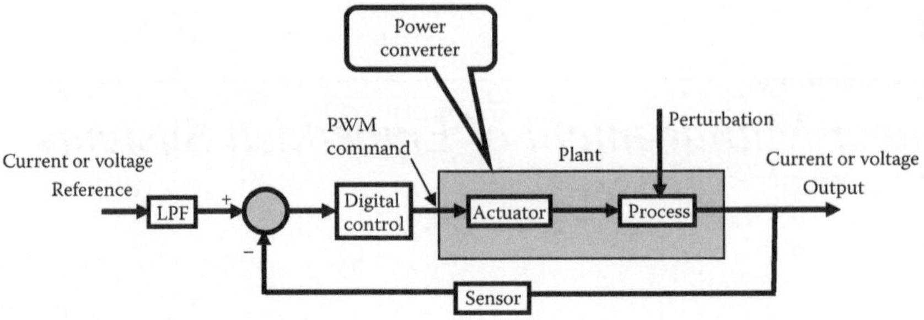

FIGURE 8.1
Principal structure of a digital control loop.

Digital control of the power transferred through power supplies ensures the adjustment of the converter's operation mode in order to satisfy requirements. The output variables such as the voltage or current are controlled through digital control methods, also known as frequency compensation circuits. This is usually done through feedback control loops which implement a frequency compensation law (Figure 8.1). The duration of the PWM pulses is therefore controlled, period after period, to ensure the desired technical performance.

Usually, control or frequency compensation laws are implemented in analogue mode using *mixed-mode integrated circuits*. However, solutions for digital implementation can be considered with microcontrollers or field programmable gate array (FPGA). The analogue and digital options for platform selection are explained in Figure 8.2.

Digital management of power supplies ensures the selection of the most appropriate operation mode at each moment.

Examples include:

- Selection of certain operating modes in order to reduce power loss, such as *sleep mode* or *standby*
- Operation with variable output voltage level to adjust for improved efficiency based on load current or load power
- Usage of various communication modules with the upper hierarchical level for up-to-date reference and operational requirements
- Failure mode management, such as management of special events like short-circuit, overvoltage or overtemperature
- Usage of optimal operation algorithms, with a parameter change because of various operation constraints (for example, dead-time variation along the PWM control)
- Automatic selection of the number of branches within a multiphase converter depending on the load current
- Correlation of operation from multiple power sources, with communication between them

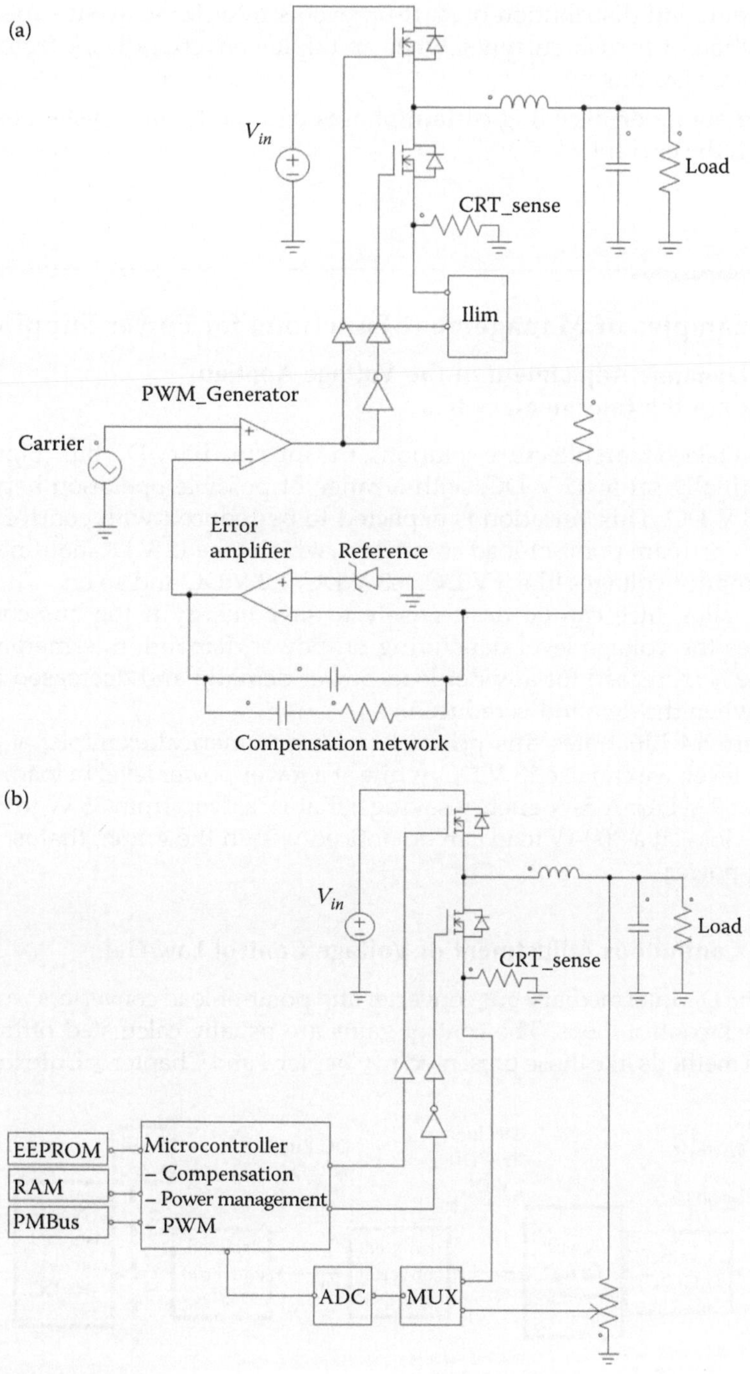

FIGURE 8.2
(a) Analogue and (b) digital options for platform selection.

- Temporal distribution of start-up events in order to avoid superposition of inrush currents, when multiple converters work from the same DC bus
- Enabling or disabling certain phases of a multiphase system based on the current

8.3 Examples of Management Functions for Power Supplies

8.3.1 Dynamic Adjustment of the Voltage Applied onto the Intermediary Bus

In most telecom architecture solutions, the intermediary DC bus (Figure 8.3) is nominally set at 12 V DC, with a range of possible operation between 9 and 13 V DC. This variation is expected to be reduced with control within the downstream point-of-load converters, which take 12 V DC nominal input and produce voltages like 5 V DC, 3.3 V DC, 1.2 V DC and so on.

This allowance can be used wisely to save energy if the bus converter changes the voltage level depending on power demand: intermediary bus voltage is increased for a wider load power demand and decreased accordingly when the demand is reduced.

Figure 8.4 illustrates this principle with a numerical example: at a large power level, we can use 13 V DC, while at a lower power level in load, we can use just 9 V DC. A 5-W energy savings; that is, a drop from 35 W to 30 W in energy loss, at a 300-W load can be noticed within the graph; that is, 1.6% of useful power.

8.3.2 Continuous Adjustment of Voltage Control Law Gains

Both the DC intermediary bus converter and point-of-load converters work with feedback control loops. The control gains are usually calculated offline with design methods like those presented in Chapter 4 and Chapter 5. Unfortunately,

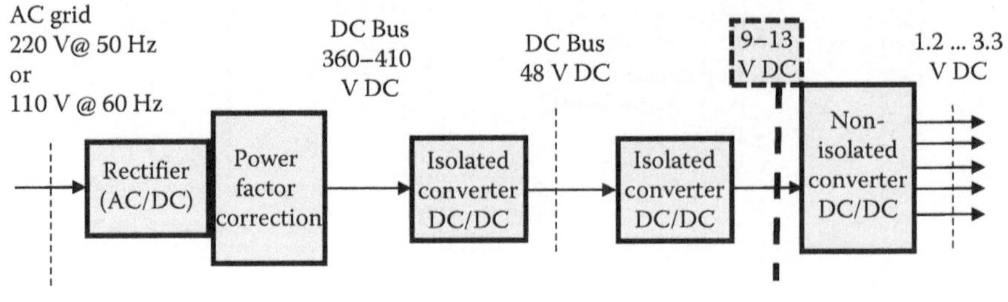

FIGURE 8.3
A 12-V DC intermediary bus within the telecom power system architecture.

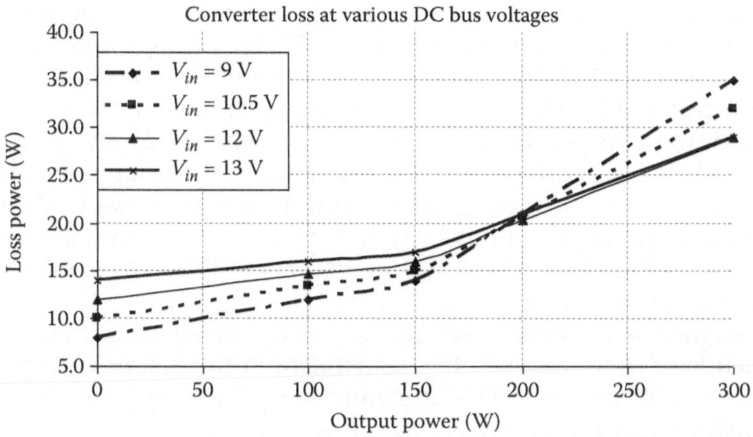

FIGURE 8.4
Numerical example, with measurements for power loss at various load demand power levels.

all these design methods provide gains dependent on the peculiar operation bias point defined with duty cycle, input voltage and load current.

This is an optimisation opportunity: instead of using fixed gains for the control law in any operating mode, the digital platform can change the gains during operation based on a set of gains precalculated and optimised depending on the operation point. This is specifically possible with digital platforms (Figure 8.2b) and nearly impossible with analogue-mode controllers (Figure 8.2a).

Due to stability concerns, it is impossible to change compensation gains truly continuously. For this reason, a dwell time is chosen to assess a need for gain change and gains are changed somewhat in steps. This was presented in Chapter 5 when studying the state space control.

An alternative solution is proposed within a series of power supply–integrated circuits with the PMBus interface. For instance, Linear Technology's LTC3880 designates the information from *bit*[7] of MFR_PWM_MODE_LTC3880 (command code 0xD4) to adjust gains in two steps for low and high current. Changing this bit value changes the PWM loop gain and compensation: 0 – low current range, 1 – high current range. However, changing this bit value whenever an output is active may have detrimental system results.

8.3.3 Adjustment of the Output Voltage at Point-of-Load Converters

A change of the output voltage reference is possible for energy savings and lower operating temperature. Generally, this change is made after a communication with the processor using the power supply, and it is governed by a dynamic code for voltage identification (D-VID). Usually, the VID codes are defined in accordance with a well-known *Intel standard*. This standard is continuously updated, with the most recent version from 5 December 2013 [5].

For illustration, here are some excerpts from the *Voltage Regulator-Down (VRD) 11.1, Processor Power Delivery Design Guidelines,* September 2009, Intel document number 322172-001. The voltage adjustment is made sequentially, either in steps of 12.5 mV or 6.25 mV, depending on the number of bits used in discretisation. As per the standard, the VID steps of 6.25 mV can be delivered at each 1.25 μs, while the VID steps of 12.5 mV should be delivered at each 2.50 μs. The output voltage of the power converter; that is, the supply voltage for the processor, should stay stable within 5 mV from the reference value within a time interval of 15 μs for an event D-VID at over 50 mV and at 5 μs for events with less than 50 mV change in magnitude.

This method is the base need for the power management bus, which is explained later in Section 8.6. The adjustment of the output voltage at PoL converters is a component of the dynamic voltage and frequency scaling set of techniques, which is described later in Section 8.7.

8.3.4 Adaptive Control of Dead Time

Dead time is used to prevent a short-circuit within either an inverter branch or a synchronous converter (Figure 8.5). Usually, the gate driver is set with a fixed dead time able to satisfy the worst situation; that is, operation with very long transition time intervals, calculated for one device's transition from conduction to the off state and the other device's state from the off state to the conduction state.

Unfortunately, the duration of the commutation time varies with the current. At the operation points where the preset dead-time interval remains larger than necessary, a loss in voltage is produced. There are several

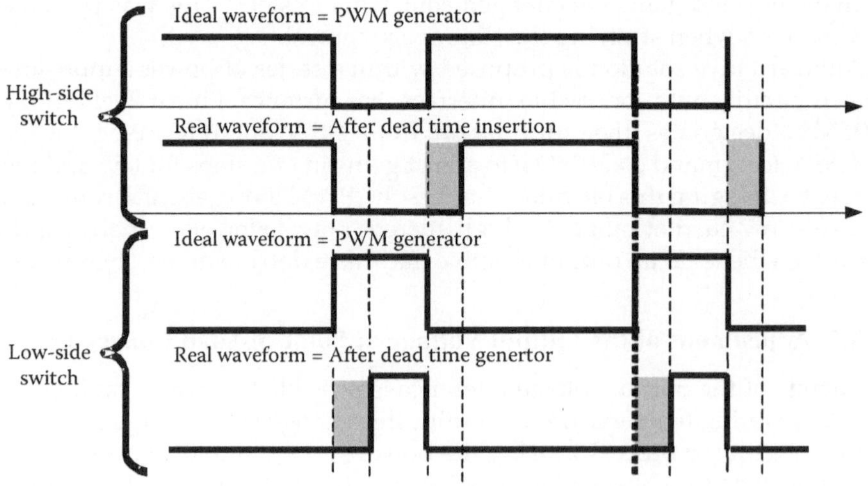

FIGURE 8.5
Time diagrams for adaptive dead time generator.

algorithms used in the industry aimed at compensating this voltage based on the latest current measurement while maintaining a constant dead-time interval [6]. In this respect, the duration of the pulses is increased with the amount of loss so that the voltage desired from that particular pulse stays the same in the load. Reversing this statement, the voltage loss can be reduced with the proper choice of dead time in correspondence with the actual current through the device. Hence, the dead time is adaptively changed during operation based on the load current.

8.3.5 Selection of the Number of Branches within a Multiphase Converter

The multiphase buck converter is used within point-of-load converters (Figure 3.24). The efficiency of this setup can be estimated for operation with a variable number of phases in operation (Figure 8.6). The system efficiency can be optimised if the operation of certain converter branches is inhibited depending on the load current. The overall system efficiency stays on the envelope of the four curves.

8.4 Connectivity

All examples presented in the previous section emphasise the importance of digital intelligent control of power supplies. Such optimisation measures are very difficult or limited in analogue implementation. Moreover, since these methods depend on the actual operation of the converter, most of the decisions about digital management of power supplies require communication (connectivity) with various power supplies or between these power sources and the higher hierarchical level; that is, the processor system being supplied with energy.

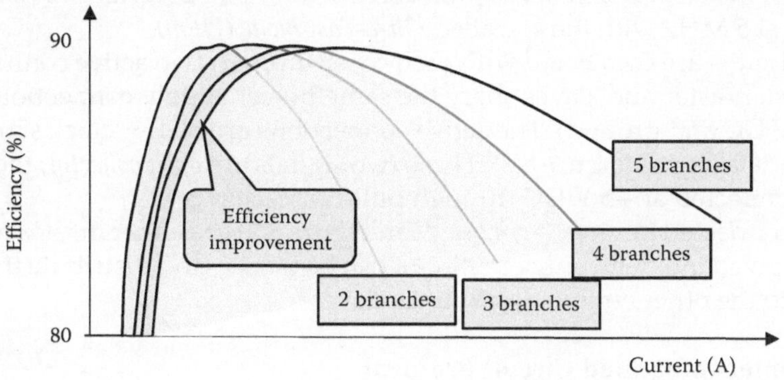

FIGURE 8.6
Efficiency calculated when using various converter configurations, with 2, 3, 4 or 5 branches.

Digital microcontrollers offer a wide range of serial communication interfaces, such as

- Communication protocols RS232, RS422 and RS485 on hardware support for the serial communication like universal asynchronous receiver–transmitter (UART)
- Protocol and communication area network (CAN) hardware
- Communication through 1, 2 and 4 bits, on serial peripheral interface (SPI)
- Universal serial bus (USB)

Despite numerous software and hardware resources available for these serial interfaces, a novel protocol has been developed to simplify the hardware structure. This new simplified protocol is derived from the hardware structure for interintegrated circuit (I^2C) connectivity, previously developed for serial communication between integrated circuits, usually on the same power ground circuitry.

8.4.1 Inter-integrated Circuit Hardware

Interintegrated circuit communication represents a master–slave communication that was invented by Phillips in 1982 as a serial communication at 100 kHz. The original protocol has not been used for a long time. Since 1990, other corporations have adopted the I^2C standard for integrated circuits. These include Siemens AG, Infineon Technologies AG, Intel Mobile Communications, NEC, Texas Instruments, STMicroelectronics (previously SGS-Thomson), Motorola (now Freescale Semiconductor), Intersil and so on. As of October 2006, there are no license fees for the implementation of I^2C. Many digital integrated circuits today offer this interface for communication with other integrated circuits, with a reduced number of connections on the printed circuit board. The original protocol has been upgraded with technology development. For instance, the last version of the protocol (version 4.0 of 2012) allows communications at 5 MHz with the so-called *Ultra-Fast mode (UFm).*

I^2C circuits are connected with each other through two active connections, and both master and slaves share the same power supply connections (usually, 5 V DC and ground). The active connections are SCL = clock signal and SDA = data signal (Figure 8.7). These two signals are *open collector*; they need to be connected at +5 V DC through pull-up resistances.

The connected devices (#1, #2 or #3 in Figure 8.7) are either master or slave. At any given moment, only one circuit can be master and distribute the clock signal to the other circuits working as slaves.

8.4.2 Inter-integrated Circuit Protocol

The protocol associated with the hardware from Figure 8.7 is a start/stop communication protocol with dedicated START/STOP sequences (Figure 8.8). For

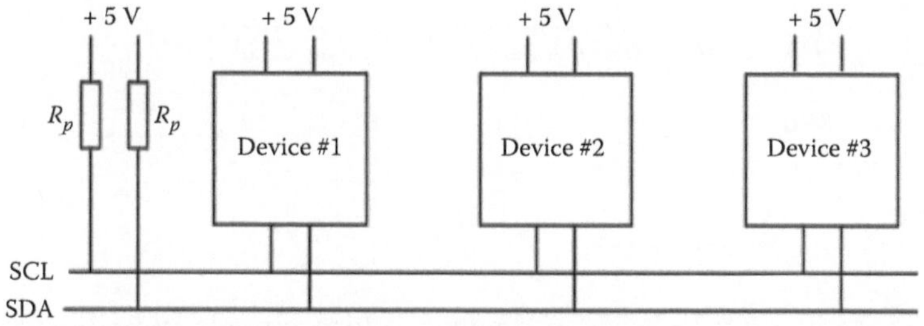

FIGURE 8.7
Schematic illustration of the I²C connections.

the START condition to occur, the data line goes LOW when the clock signal line is HIGH. For the STOP condition to occur, the data line goes HIGH when the clock signal line is HIGH.

The CLOCK and DATA buses operate under the rules outlined in Figure 8.9.

Data is transmitted in 8-bit sequences from most significant bit (MSB) to least significant bit (LSB) (Figure 8.10). The device receiving data pulls the data line LOW during the ninth clock pulse to acknowledge receipt (ACK) or stays HIGH to not acknowledge (NACK) (Figure 8.10).

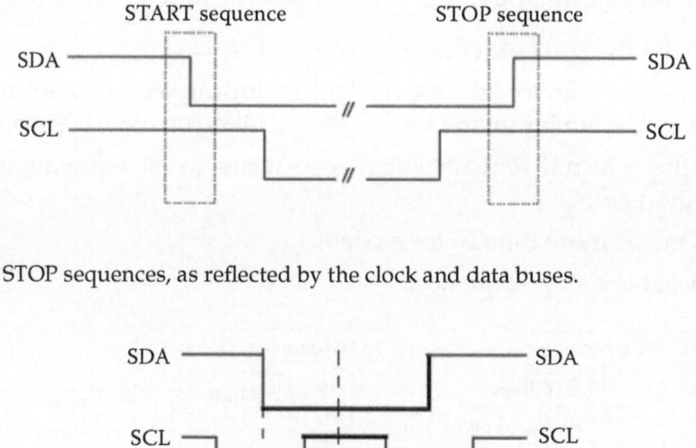

FIGURE 8.8
START and STOP sequences, as reflected by the clock and data buses.

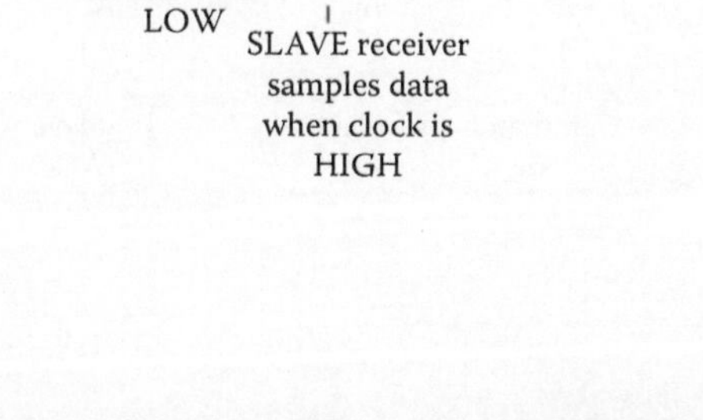

FIGURE 8.9
Clock and data buses.

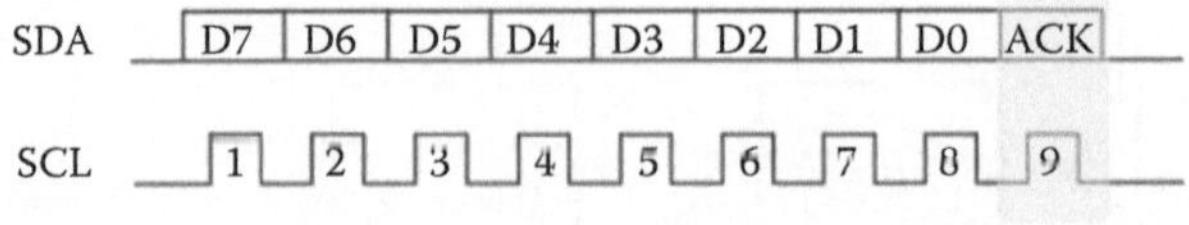

FIGURE 8.10
Slave acknowledgment on the 9th bit.

Since there are multiple slave devices, addressing is usually done on 7 bits or rarely on 10 bits. The more popular 7-bit address is placed towards the MSB of the 8-bit word, and the 8th bit is used to indicate if the current instruction is a READ or WRITE operation. The device address is sent immediately after the start sequence and followed by an address for the internal register within the device to which the communication is targeted.

This particular addressing mode requires designer attention since some manufacturers report addresses on 7-bit, while others report addresses already aligned on 8-bit.

8.4.3 Inter-integrated Circuit Software Sequence

There are two major sequences used during communication: WRITE and READ. Details for each one are herein provided.

Writing to an external device is done in several steps (Figure 8.11):

1. Broadcast the start sequence.
2. Send the slave device address on 7-bit, including where to write to, followed by the 8th bit set to LOW ($R/W = LOW$) for a WRITE operation.
3. Send the internal register address to write to or, equivalently, the command code.
4. Send one or more data bytes as data.
5. Broadcast the stop sequence.

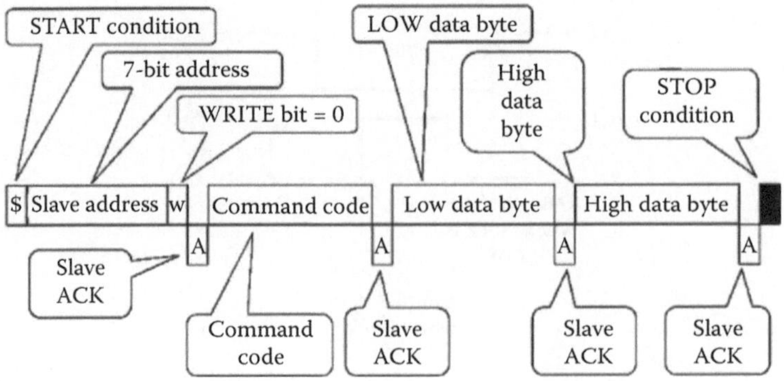

FIGURE 8.11
WRITE format.

Reading from an external device is done in several steps:

1. Broadcast the start sequence.
2. Send the slave device address on 7-bit, with the 8th bit set as LOW for a WRITE operation (*R/W* = *LOW*).
3. Send the internal register address to be read from or, equivalently, the command code.
4. Repeat the start sequence.
5. Send the slave device address on 7-bit, followed by the 8th bit set as HIGH logic level for a READ operation (*R/W* = *HIGH* voltage level).
6. Read one or more bytes of data.
7. Broadcast the stop sequence.

If a slave device is not ready to process the communication, it has the ability to keep the clock line = *LOW* logic level after the 9th clock bit, thus stopping any communication. This operation is called *clock stretching* and is not allowed during the first 8 bits of transmission.

8.4.4 Examples of Integrated Circuits with Inter-integrated Circuit Interface

A series of integrated circuits have the I^{2}C interface, and they are mostly dedicated to:

- Specific sensors for temperature, voltage or humidity
- Digitally controlled potentiometers, for instance the Microchip Technology 7-bit MCP401x I^{2}C digital potentiometer

More recently, integrated circuits with power management bus aspirations include hardware for I^{2}C and/or the power management bus protocol. For instance, Infineon – International Rectifier IR3564 is a digital control–integrated circuit used for a double power supply, working as a multiphase asynchronous buck converter. Similarly, Texas Instruments UCD9240 was released after a previous UCD9112.

Fully PMBus-compliant integrated circuits can be classified into two classes:

- Power supply–integrated circuits with one or multiple channels, able to directly deliver high power as a source
- Power management–integrated circuits, able to control and collect data from multiple conventional power supplies, mostly with analogue operation and without their own PMBus interfaces

8.5 Historical Perspective on System Management Bus

The system management bus (SMBus) was developed for motherboards (processor-based circuits), specifically for communication between processors or power supplies. It was developed in 1995 by Intel and maintained within an industrial consortium, with a last version (v.2.0) from August 2000. It has never required licensing. Usually, the SMBus is not user configurable.

Despite being built on the I²C infrastructure, there are differences at both the hardware and protocol levels. However, the I²C and SMBus devices are compatible with each other under 100 kHz and are completely identical for READ/WRITE of a single byte. Differences occur with the transfer of data packets, also called multibyte communication. That is where the new *packet error checking* and *host notify* protocols are introduced. Since the SMBus is not widely used today, the hardware details are briefly presented in Table 8.1, while protocol details will be replaced with a later presentation of the PMBus protocol.

8.6 Introduction of Power Management Bus

8.6.1 Hardware

The power management bus [7] represents a standard protocol for power supply management, with a well-defined set of commands, which allows communication with power converters as well as other devices within the power system. The PMBus protocol is implemented over the previous serial interface, SMBus, with hardware similar to I²C. The PMBus can be used with multiple compatible slave devices, built with either analogue or digital technologies.

This is a relatively new protocol, established after 2007, with many products under development. The standard is maintained by the PM-Bus

TABLE 8.1

Electrical Parameters of I²C and SMBus Hardware Circuits

Parameter		I²C		SMBus	
		min	**max**	**min**	**max**
V_{IL}	Fixed level	−0.5	1.5 V	n/a	0.8 V
Input LOW	~ from $V_{DD} = 5$ V		$0.3 \times V_{DD}$	n/a	n/a
V_{IH}	Fixed level	3 V	$V_{DDmax} + 0.5$ V	2.1 V	n/a
Input HIGH	~ from $V_{DD} = 5$ V	$0.7 \times V_{DD}$		n/a	n/a
V_{OL}	V_{OL} at 3 mA	0	0.4 V	n/a	n/a
Output LOW	V_{OL} at 350 µA	n/a	n/a	0	0.4 V
I_{PULLUP}		n/a	n/a	100 µA	350 µA
I_{LEAK}		−10 µA	10 µA	−5 µA	5 µA

Implementers Forum. This forum allows a simultaneous and impressive development of all tools for using these devices, including libraries, simulators, emulators and so on.

As a major difference from SMBus, the PMBus introduces two new signals with optional usage (Figure 8.12):

- CONTROL signal = For ON/OFF control
- ALERT signal = As an interrupting signal due to an emergency

The CONTROL signal acts as a chip enabler for slave devices to define either a start moment or a stop due to various imminent hardware constraints. The name itself has been adopted differently by various manufacturers. For instance, Linear Technology calls this pin RUN within its products, and Texas Instruments calls it CNTL.

For this reason, integrated circuit CONTROL pins are bidirectional: a logic HIGH applied on these pins enables the controller, while the open-drain output holds the pin LOW until the device is out of reset. This means a pull-up resistor to 3.3 V DC is required in the application of the PMBus protocol.

All start-off, soft-start, soft-off and ramp times are defined with respect to this hardware-enabled pin.

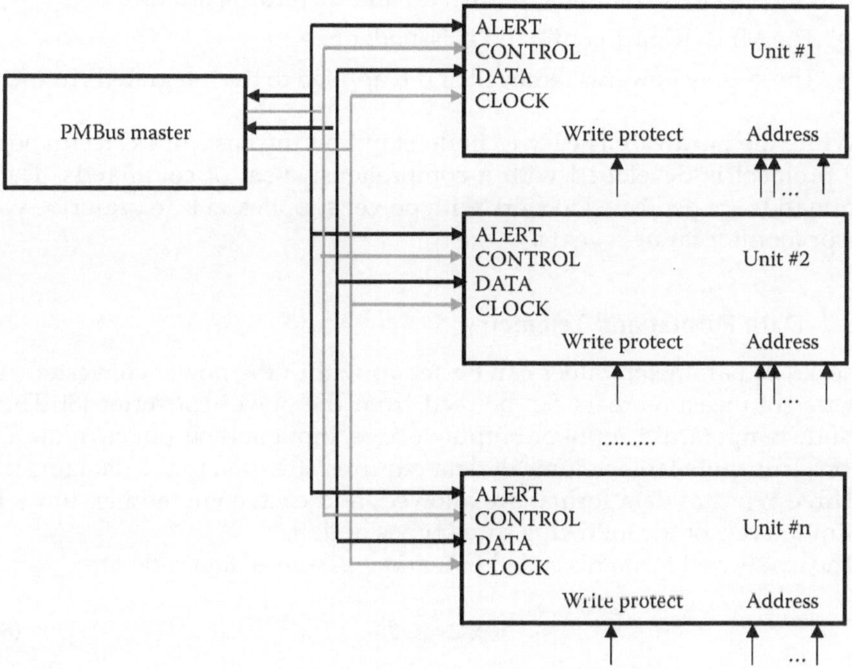

FIGURE 8.12
Hardware for PMBus, including two new signals, CONTROL and ALERT.

The CONTROL signal has its roots in the optional /SMBSUS from the system management bus. That signal in a LOW state shows when the system has entered suspend mode. Suspend mode refers to a low-power mode where most devices are stalled or powered down. Upon resumption, the /MBSUS returns HIGH and all devices resume their operational state. The /SMBSUS signal is optional and was included since many of the functions served by the SMBus relate to suspension and resumption of the entire system's operation.

The ALERT signal from the PMBus is active in LOW voltage level, and it is sometimes called /SMALERT from the ancillary system management bus. This pin also requires a pull-up resistor since it can work bidirectionally as input for an externally generated alert or as output when the alert is generated by the integrated circuit itself. Any fault or warning event will set the ALERT pin to LOW. The pin will remain LOW until the CLEAR_FAULTS command is issued by the communication master.

The CLEAR_FAULTS command does not cause a power supply that has turned off for a fault condition to restart. Power supplies that have shut down for a fault condition are restarted when:

- The output is enabled through the CONTROL signal, the OPERATION command or the combined action of the CONTROL signal and OPERATION command to turn off and then turn back on;
- The MFR_RESET command is issued; or
- The supply power is removed and reapplied to the integrated circuit.

While the hardware is easy to understand on the basis of I^2C technology, the protocol is developed with a comprehensive set of commands. These commands are dedicated to work with power supplies, able to prescribe voltage or monitor failure events.

8.6.2 Data Format and Telemetry

A series of parameter values can be set up within the power converter, and a series of measurements can be read from the power converter [8]. These include temperature, input or output voltage, input and output currents and so on. This quite large volume of data requires attention to the data format.

Three types of data format are allowed, and each command assumes by definition one of the following three types of data.

The most-used format is called L11, and it assumes data coded as:

$$X = Y \cdot 2^N \tag{8.1}$$

where N is coded on 5 bits (white $4 \rightarrow 0$, in Figure 8.13) and Y is coded on 11 bits (grey tone $10 \rightarrow 0$, in Figure 8.13). All voltage, temperature and current

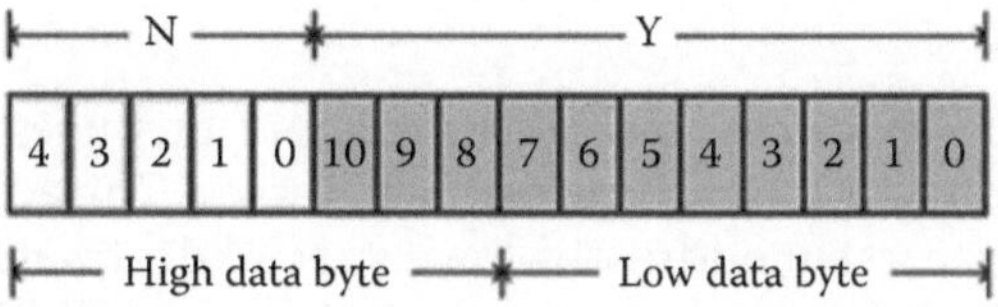

FIGURE 8.13
Data format L11.

measurements are reported in L11, while all time settings are also sent to devices in L11.

In order to get the highest resolution from the setup, the mantissa *Y* has to be as large as possible, while the exponent *N* has to be as small as possible. From all options for selection of *N* and *Y*, the best choice is the one which allows the largest *Y* under 1024; that is, 2^{10}.

For instance, when setting up a time interval in L11 format: in hardware, the L11 format is implemented at reception with a clock at a fixed frequency equal to *N* times the base local clock frequency. A counter counts clock pulses up to a value equal to the mantissa *Y*.

When a time interval of 1 second is required, this could be achieved with 0 × 5001 (0 × 0101-0000-0000-0001b), which means counting one single pulse with a very large clock period. This means an error of half a counting clock; that is, a 0.5-second error. On the contrary, the best resolution is achieved with 0 × 83FF (0 × 1000-0011-1111-1111b) from $1023{\cdot}2^0$. This has a maximum error of 1/1024 of a second.

Finally, the zero value is preferably coded with 0 × 8000.

Other formats include direct data in 2's complement, or L16, format. This format is used for most output voltage–related settings, like reference value, margins, protection threshold, power good threshold and so on. It actually works very similarly to L11 and Equation 8.1, with the exponent *N* set by the VOUT_MODE command during the preliminary or offline programming. The 16-bit mantissa is sent by the VOUT_COMMAND command during operation whenever the output voltage needs to change. The same format applies to all settings concerning the output voltage, like margins, faults, warnings or measurements.

Finally, the third format is a direct data transfer, coded on 7-bit. This is a rare case caused by certain commands which need further decoding within slave-side electronics.

Telemetry allows continuous monitoring of the converter's operating parameters, like input voltage, output voltage, inductor current, temperature, operating frequency, duty cycle, current settings of the power supply's parameters, fault vectors and so on. The number of measurements depends on the integrated circuit used for power supply management. Each designer writes his or her own software to loop through these measurements for all slave devices at a low frequency rate of several Hz.

8.6.3 Addressing

Similarly to the SMBus, the power management bus uses 7-bit addressing, with rare extension to a 10-bit format for addresses. The address space is not fully useable since certain addresses have special destinations, according to Table 8.2. The only difference between the SMBus and PMBus consists of the addition of two dedicated addresses, 0 × 28 and 0 × 37, that were added for PMBus 3.1 zone operations. The address space uses 90 addresses among the 128 possible, and there is no central authority in further assigning the useable addresses; it all comes down to the system designer.

Starting from this useable space, different implementation solutions for PMBus software and protocol further dedicated some of them to certain global tasks. For instance, the PMBus solution offered by Linear Technology uses 0 × 5A and 0 × 5B to allow the system master to talk to more than one device at a time. Hence, the address 0 × 5A is a global address used with the LTC388X family, while address 0 × 5B is a paged global address for the LTC388X family and a paged global address for the LTC297X family of integrated circuits. A third global address is 0 × 7C, and it is used to signal if LTC388X family devices have CRC errors in their EEPROM.

Since most manufacturers use some addresses for their own arrangements, it is suggested that each designer create an address map picking up

TABLE 8.2

Address space for PMBus and SMBus

Address	Destination
0 × 00	START
0 × 01 to 0 × 0B	Battery
0 × 1C	Alert response address
0 × 1D to 0 × 27	Useable
0 × 28	ZONE READ
0 × 29 to 0 × 2B	Useable
0 × 2C to 0 × 2D	RESERVED
0 × 2E to 0 × 36	Useable
0 × 37	ZONE WRITE
0 × 38 to 0 × 3F	Useable
0 × 40 to 0 × 44	RESERVED
0 × 45 to 0 × 47	Useable
0 × 48 to 0 × 4B	PROTOTYPE
0 × 4C to 0 × 60	Useable
0 × 61	DEFAULT ADDRESS
0 × 62 to 0 × 77	Useable
0 × 78 to 0 × 7B	Mark for 10-bit
0 × 7C to 0 × 7F	RESERVED

addresses from various devices on the bus, yet paying attention to the 7-bit format, as many circuits report this as an 8-bit word with the last bit null. For instance, 0 × 7F is reported as 0xFE.

8.6.4 Set of Commands

The structure of each command can be seen in Figure 8.14. Each instruction starts with the address of the destination, followed by the command code and the data operand on one or more bytes. In some cases, the *parity error check* (PEC) byte is used at the end of the transmission.

The serial interface supports the following protocols defined in the PMBus specifications: (1) send command, (2) write byte, (3) write word, (4) group, (5) read byte, (6) read word and (7) read block. Depending on the actual hardware – processor, computer or microcontroller – low-level software routines are written for each operation. On the other hand, dedicated integrated circuits with the PMBus interface allow interpretation of each such transmission string.

The instruction set is well defined with an extensive number of commands, which can be grouped in several categories:

- Control commands
- Input commands
- Output commands
- Fault limit commands
- Fault response commands
- Time setting commands
- Status commands
- Monitor commands
- Identification commands
- Other configuration commands
- Supervisory commands

While most commands are generic, 57 command codes are reserved for manufacturer-specific functionality. From the maximum of 127 commands possible when coded on 7-bit, just a few codes are available for further development. The others define everything about power supply operation. Obviously, not all systems use all features.

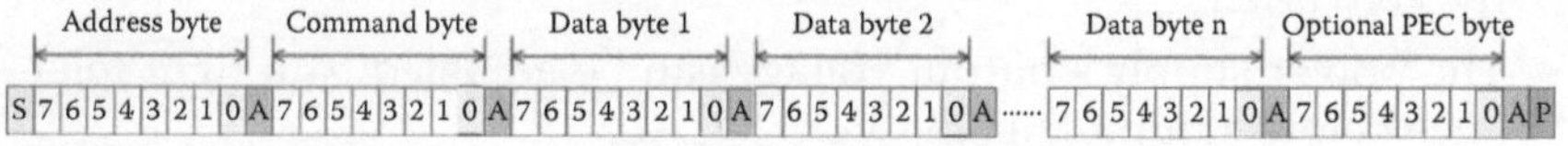

FIGURE 8.14
Possible transmission structure, similar to Figure 8.11.

Integrity of communication is checked with the parity error check. First, the PEC_REQUIRED bit is set to active. Any PEC write-cycle errors and any attempts to access unsupported commands or write invalid data to previously mentioned commands will result in a command fault. This sets the CML bit in the STATUS_BYTE and STATUS_WORD, and the ALERT signal line is pulled low. This allows an action faster than the usual communication transaction.

8.6.5 Voltage Setting

The most important feature of the PMBus is the ability to set and modify the level of the output DC voltage from a power supply. The desired voltage is set with VOUT_COMMAND. Once this is established, the system manager has to send to the power supply a series of parameters also related to the output voltage. These are:

- Margin voltages (VOUT_MARGIN_HIGH and VOUT_MARGIN_LOW) are used for system testing and correspond to good operating voltages larger and smaller than the actual command. A good practice would recommend setting them within ±5% of the command voltage.
- Warning voltages (VOUT_OV_WARN_LIMIT and VOUT_UV_WARN_LIMIT) are set to allow the system manager to assess potential problems without shutting down the entire operation. A good practice would recommend setting them within ±10% of the command voltage.
- Fault voltages (VOUT_OV_FAULT_LIMIT and VOUT_UV_FAULT_LIMIT) are set to shut down operation when overvoltage or undervoltage occurs during operation. This way, components within the power supply are protected. A good practice would recommend setting them within ±15% of command voltage. A vector is defined with a special command word to define what operation is performed when this fault occurs: ignore fault, immediately off or deglitch. Furthermore, a variable delay can be inserted before an action.
- Absolute maximum output voltage is set with VOUT_MAX, and this is an unconditional shutdown and protection.
- The slew of the output voltage can also be protected with VOUT_TRANSITION_RATE. For example, the rate of the output voltage variation can be set at 250 V/sec.
- Trimming voltage (VOUT_TRIM) allows the fabrication process to perform a last check on the actual voltage achieved at the output of the converter.
- The power supply's output voltage can be adjusted to 'a tight tolerance before shipping' using VOUT_CAL.
- When passive current sharing is used, the VOUT_DROOP command is used to set the output to 'lossless resistance'.

Similar settings can be imposed on the converter's current. Each parameter is saved in the nonvolatile memory of the power supply's manager. Section 8.6.7 provides details on operation with a nonvolatile memory associated with each power supply manager.

The output voltage for each power supply can also be programmed in terms of startup procedure. This is especially necessary when the system comprises multiple converters supplied from the same voltage bus. In order to avoid stressing the bus with a large inrush current, each local power supply is controlled to start with some delay. Both the time delay at startup (TON_DELAY) and the slope of the startup voltage climb (TON_RISE) are programmable with the PMBus. They follow any command for turn-on.

Analogously, each power supply with management features can have a programmable time delay at stop by command (TOFF_DELAY) and a slope for the voltage decline (TOFF_FALL).

When the same integrated circuit that is a SLAVE to the PMBus controls multiple channels of power supplies, the designer can use the command PAGE to switch between them and to program each channel more or less individually.

8.6.6 Fault Management

Various FAULT signals from hardware are collected within the integrated circuit used as the system administrator. If FAULT signals are coming from external hardware, they are tied into an open drain output/input pin and all FAULT signals are active low. The power supply IC or system administrator IC process individual FAULT signals and generate a PMBus ALERT signal. This is a fast action signal, global for all devices and able to interrupt the processor on a higher hierarchical level. The host has to decide on stop or continue the entire operation. This process is the preferred action, while it is possible to set the entire fault management based on polling different slaves on the bus.

There are two types of actions in fault conditions:

- Any individual power supply has to respond to another incoming fault from another power supply, aiming to shut down when another power supply faults.
- Any individual power supply propagates its fault to other channels.

The response to any type of hardware fault is generally programmable for each power supply individually. By default, the integrated circuits are configured to shut down operation if other devices fault and to propagate their own fault via the FAULT pins.

A complete fault management system includes input voltage undervoltage (UV) and overvoltage (OV), input current overcurrent, high input power, output voltage undervoltage and overvoltage, output current overcurrent, high output power, improper operation (POWER_GOOD), overtemperature and overlapping the maximum on-time.

Responses to FAULT can be programmed to:

- Immediate stop, with latch and restart
- Immediate stop and no restart
- Inhibit operation as long as the fault is active
- Delay turn-off
- Limit parameters according to program
- Continue operation

All of these are coded within a fault response data byte.

Other faults can come from impossible read operations, unsupported command codes or possibly invalid or out-of-range data.

8.6.7 Using Nonvolatile Memory within Integrated Circuits

The initial use of the power management bus was dedicated to setting the output voltage with the aim of changing this voltage depending on the operation mode. In this respect, the first generation of power supply–integrated circuits offered a digital reference input. This is the case with MC33470, a synchronous buck controlled with a 5-bit VIN reference input set as a parallel bus.

As ideas about the power management bus settled in and a complete picture of the advantages was discovered, new integrated circuits emerged. They also came with a paradigm shift from allowing changes during operation to an opportunity for trimming or debugging the application circuit. Moreover, a backcompatibility problem with conventional analogue circuits was suggested.

The outcome was an internal architecture based on a sizable nonvolatile memory able to store all operation parameters either from a factory setting or in the field. This architecture allows operation of the power supply with or without connection to a power management bus. Conversely, programming the memory can be done independently of the operation of the power train.

All parameters influencing any detail of operation are preset in nonvolatile memory (NVM, of the EEPROM type). When the PMBus is present, the parameters installed by the factory in NVM can be written and changed by an upper hierarchical level through the PMBus into memory (Figure 8.15).

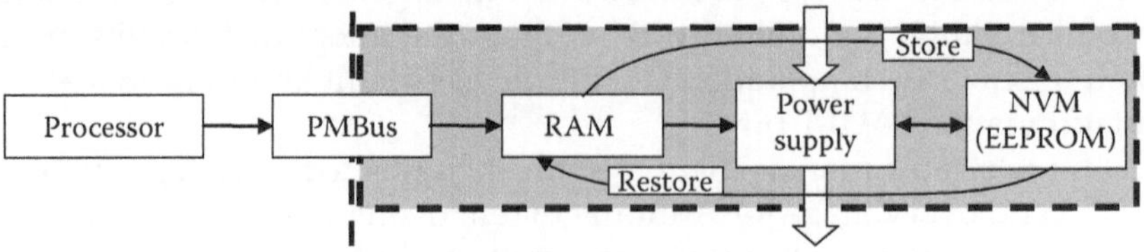

FIGURE 8.15
Using nonvolatile memory for saving parameters.

Nonvolatile memory also includes information from vendors to users. This includes:

- Revision number for the PMBus software
- Manufacturing model number
- Manufacturing ID serial number
- Manufacturing revision of the product
- Manufacturer's location, for tax and standards compliance
- Date of fabrication
- Nominal voltages and extreme values
- Manufacturer's maximum input and output currents
- Manufacturer's specification for power
- Manufacturer's information about pinout and pin accuracy
- Manufacturer's information about temperature ratings
- Manufacturer's information about efficiency

The user can also store data with 16 direct commands (USER_DATA_00 … USER_DATA_15).

Despite the impressive development of the power management bus (over the last decade from 2007 to 2017), not all industrial equipment has digital power supplies with a PMBus interface. To secure backcompatibility, a series of integrated circuits have been released to manage and supervise ancillary power supplies. This class of circuits is different from conventional power supply–integrated circuits since they behave as system monitoring devices. The industry often calls them 'digital wrappers' (Figure 8.16).

8.7 Methods for Power Savings in Computing Systems

8.7.1 Power Loss Components

While the previous section addressed the interfacing of a power supply with an upper hierarchical level for adequate power management, this section discusses strategies applied to computing systems in one's own software or that can benefit from the power management bus.

The development of embedded systems has prompted concerns about energy consumption within these systems. Embedded systems range from applications with a power budget of several mW in wearable electronics to systems with a power budget of kW, like powerful server computers. Due to the digital nature of all systems within these ranges, most practices for energy savings are similar. Energy management and energy savings plans for

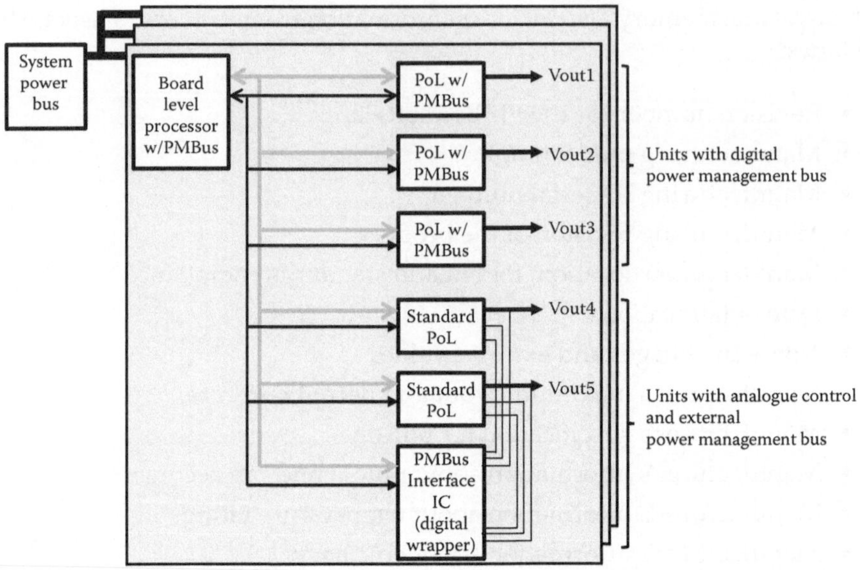

FIGURE 8.16
PMBus within a very complex system, composed of modern digital power supplies and conventional analogue power supplies.

embedded systems are at the early stages, with some techniques well adopted by the industry and others just at the stage of being studied by researchers.

The power management bus was introduced and perfected by power supply engineers to meet these demands. Initially, a change in voltage was suggested, with levels defined with 5 bits, then 7 bits, and many ICs with the PMBus feature allow operation with voltages defined on full-byte words. The power management bus is a tool, an enabling technology allowing the computing system operator to decide on the operation of the power supply depending on the architecture and operation mode of the computing system.

Numerous studies from embedded systems operators have tried to quantify the benefits of the usage of the PMBus. For instance, the study in [9] analysed the effect of having multiple levels of output voltage and concluded that systems with three or four levels of supply voltage produce nearly the same energy savings as systems with an ideally infinite number of levels.

The power loss through any electronic equipment has two components: a dynamic component and a static component.

- The *dynamic component* is due to charging and discharging the capacitances in the circuit, mostly due to MOSFET and CMOS transistor structures. Hence, this power loss component yields:

$$P_{dyn} = C \cdot v(t) \cdot \frac{dv(t)}{dt} \approx k \cdot C \cdot V^2 f \tag{8.2}$$

where C is the dynamic capacitance in operation, V is the operating rail voltage and f is the operation frequency. Since the dynamic power component depends on squared voltage, the power can be reduced drastically with a decrease in voltage. This is possible if the operation frequency is also slightly decreased.

- The *static component* of power loss comes from leakage currents within the embedded circuits, and it can be expressed as:

$$P_{st} = I_{leak} \cdot V \tag{8.3}$$

Different strategies were first developed to address dynamic and static losses separately. More recent studies consider both under the same optimisation structure.

8.7.2 Deployment of Dynamic Voltage and Frequency Scaling Methods

Techniques for reduction of dynamic power loss consist of dynamic voltage and frequency scaling. Dynamic voltage and frequency scaling is the adjustment of voltage and frequency settings within an embedded system in order to optimise resource allotment for tasks and maximise power savings when various resources are not needed.

The deployment of DVFS methods has been imposed by the realities of modern embedded systems:

- Either in mobile battery-powered systems or in grid-supplied computing centres, energy supply is a crucial limitation. Size or space is at premium, and heating during the processing of large sets of data is possible due to limited heat removal capability. Reducing power loss saves on the volume of embedded system equipment.
- Operation with reduced loss enables a longer lifetime and improves the reliability of the system. Conversely, a system stressed with operation at higher temperatures is subject to more faults and a reduction in lifetime.
- A computer centre equipped with 10 servers is usually designed to work with a power supply system able to take the 10 servers at full load. In practice, this never happens. This means the conservative principles of system design prevent one from connecting new servers within the same centre due to limited grid availability. On the contrary, a good energy management system can plan for the right usage of resources with runtime adaptation and relaxation of schedules. The same grid connection can now allow more servers to operate simultaneously, and overprovisioning is thus avoided.

- DVFS methods align the design of the power system within the computing centre with modern processor architectures based on multicore, cache, scratchpad memory and multilevel memory plans. Such architectures impose a special sharing of resources and time-domain planning of energy consumption.
- The internal architecture of modern processors and peripheral-integrated circuits is based mostly on CMOS technology. This allows an increase in on-chip transistor density. However, a fraction of the integrated circuit having all transistors operating at the maximum frequency allowed by technology would overheat that fraction of the die and produce a fault of the integrated circuit. Hence, strategies for spreading energy consumption within the integrated circuit are required to avoid meeting the utilisation wall.

DVFS is a general name for techniques able to adjust the voltage and frequency of an embedded system based on operational requirements. When the embedded system is highly used with all internal resources working, dynamic power loss is decreased with a decrease in operating frequency and rail voltage level. To enable both frequency and supply voltage changes, the system requires programmable clock frequency and a programmable DC/DC converter with response time on the order of microseconds. Examples of commercial microprocessors which fully support DVFS technology for saving power are AMD's PowerNow and Intel's SpeedStep.

8.7.3 Strategies Used with Dynamic Voltage and Frequency Scaling

Slowing down the execution of code may produce missed deadlines. This is generally accepted in multimedia applications and not that much in scientific computing. Missing some task deadlines may therefore not be observed by human visual and hearing systems. The DVFS strategies are based on optimisation, and various searching algorithms are proposed.

Some examples of DVFS strategies follow.

- Define the best strategy for achieving the highest completion ratio with the lowest possible energy consumption [10].
- Drop some tasks completely to create slack, which is subsequently shared between multiple processors in an effort to save energy instead of running simultaneously at full capability [2].
- Schedule based on the earliest deadline to ensure meeting all deadlines while saving energy [11].
- Precisely monitor the performance–energy tradeoff by using run-time information about the external memory access statistics. The optimal CPU clock frequency and the corresponding minimum

voltage level are chosen based on the ratio of the on-chip computation time to the off-chip access time so that the system lowers the CPU frequency in the memory-bound region of a program to keep the performance degradation to a low value.

8.7.4 Diminished Results for Dynamic Voltage and Frequency Scaling with Multicore Processors

On the negative side of DVFS techniques, the trend with modern processors is towards multicore architecture rather than the increase of operating frequency for a single-core operation. This means the power savings with DVFS techniques are diminishing over time. The processor generations in the early 2000s succeeded in using DVFS techniques since the trade-off between runtime performance and energy consumption allowed optimisation. Figure 8.17 illustrates this aspect [12].

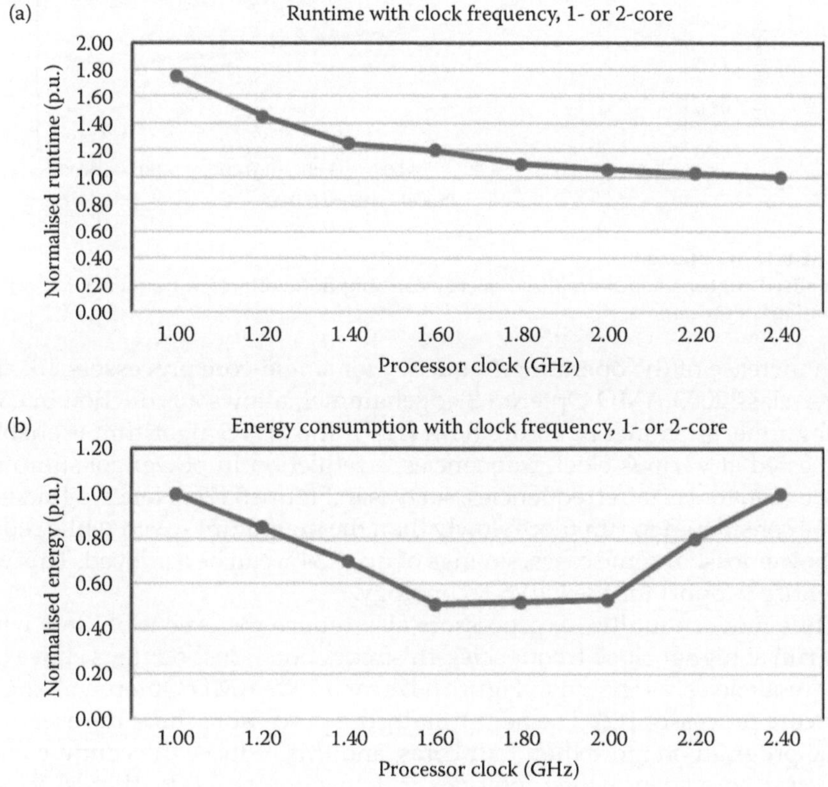

FIGURE 8.17
Normalised runtime and normalised energy consumption with clock frequency for various generations of processors.

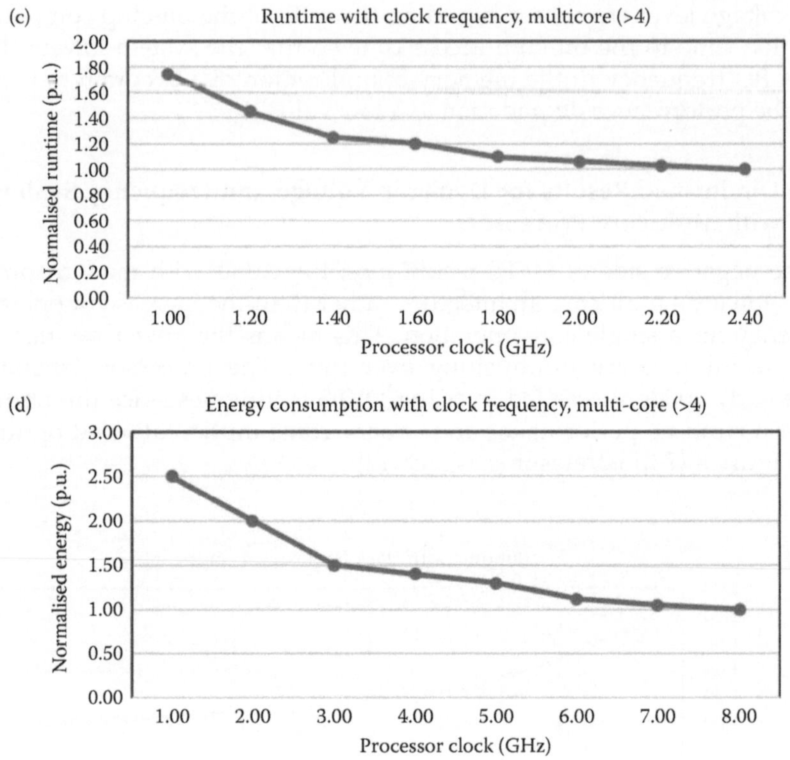

FIGURE 8.17 (Continued)
Normalised runtime and normalised energy consumption with clock frequency for various generations of processors.

An increase of the operating frequency for single-core processors, like the server-class 2003 AMD Opteron Sledgehammer, allows a reduction in computing time, as shown in Figure 8.17a [12]. If the DVFS algorithm is enabled and tested at various clock frequencies, a reduction in power consumption can be seen for certain frequencies, such as in Figure 8.17b. Hence, a trade-off can be considered to run more slowly than the maximum speed while reducing power loss. In some cases, savings of up to 34% can be achieved. This was the entire support for the DFVS technology.

While modern multicore processors also reduce the execution time when they run at higher clock frequencies, this reduction is less obvious above certain frequencies, as shown in Figure 8.17c for a 2009 AMD Opteron Shanghai four-core processor [12]. The newer multicore processors share the execution of the program on more than four cores, and this reduces the runtime. Each processor core spends long intervals in a low-power mode; that is, an idle mode or similar. Enabling the DVFS algorithm does not help in this case since the four cores do not run continuously, and it becomes more important to manage their low-power mode time and what they do during this mode.

Furthermore, in most cases, the four or more cores are on the same integrated circuit and benefit from the same power supply. This means a DVFS optimisation per core is not possible. This can be seen in Figure 8.17d. There is no energy saving possible with a clock frequency reduction. This means multicore processors lose somewhat the advantage of DFVS technologies.

Results in Figure 8.17 are for a benchmark derived from a program written in C and used for single-depot vehicle scheduling in public mass transportation. This test, called 181.mcf, was used as a worst-case memory-bound benchmark where DVFS has the best chance of being effective at reducing energy consumption. This test best suits server-class platforms, while DVFS may still be effective on other platforms, such as smartphones.

8.7.5 Strategies for Static Power Reduction

While the previous conclusion regarding DVFS strategies may be subject to the nature of the benchmark tests used in evaluation or the processor itself, it may be considered a valid idea: modern processors slowly moved the possibility of reducing energy loss from dynamic components towards static (leakage) component reduction. The leakage component is more important in multicore processors due to the increased number of transistors on chip.

Furthermore, the technology of transistors within processors had a tremendously fast development, with a design rule decrease from 10 μm in 1971 down to 5 nm recently (Figure 8.18). Using smaller transistors within the integrated circuits allows their operation with lower threshold voltage while producing increased subthreshold leakage current. Hence, the static power is once more increased.

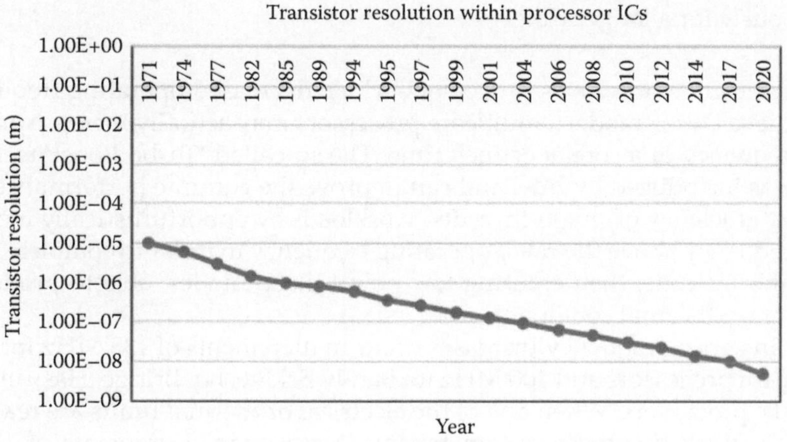

FIGURE 8.18
Evolution of design rule for integrated circuit technology.

Given the shrinking potential for saving energy with conventional DVFS, it seems only a matter of time until manufacturers abandon DVFS in favour of playing with ultralow power sleep modes. The name of the new technology is *power mode management* (PMM).

All strategies for reduction of static power loss speculate on the various operation modes and stress the planning of a suitable sequence of operation modes. Different operation modes of the same digital hardware consume different amounts of power and take different times to transition from one mode to another. Operation modes with lower energy consumption; that is, low-power modes like idle or sleep modes, also take the longest time to return to normal mode to service the next task. For saving energy while keeping the runtime performance competitive, these modes should be carefully used. This selection is considered power mode management.

There are many optimisation strategies considered in the literature. Some examples follow:

- A weight-based optimisation was proposed to consider both runtime performance and power loss.
- A statistical history of tasks helped optimisation in another study [13].
- A combination of DVFS and PMM was considered in another approach [14]. In the case when a processor is active, a processor speed is selected to balance dynamic and static power consumption. When the processor is idle, the coming tasks are delayed as much as possible, such that their deadlines are not missed and scattered, and short intertask idle intervals are merged into a few large idle intervals. Large idle intervals lead to reduced mode transition overhead, since the processor can stay in either an idle or active state continuously for a longer time.

Conversely to the 2000s processors, which decreased operating frequency to achieve DVFS, modern multicore processors may actually increase operating frequency of a core at crunch time. The so-called 'Turbo Boost' technology was introduced by Intel and can improve the runtime performance and energy efficiency of single-threaded workloads by opportunistically increasing frequency above the rated operating frequency in order to complete work in a shorter time, then entering low-power sleep modes, able to produce a temporary thermal cooldown.

For instance, frequency increases occur in increments of 133 MHz for Intel Nehalem processors and 100 MHz for Sandy Bridge, Ivy Bridge, Haswell and Skylake processors. When any of the electrical or thermal limits are reached, the operating frequency automatically decreases in decrements of 133 or 100 MHz until the processor is again operating within its designed electrical and thermal limits. Turbo Boost 2.0 was introduced in 2011 with the Sandy

Bridge microarchitecture, while Intel Turbo Boost Max 3.0 was introduced in 2016 with the Broadwell-E architecture.

Summary

This chapter described modern requirements for a digital platform used for management of power supplies. Examples of management functions applied to power supplies were detailed. In order to meet these expectations, the power management bus with I^{2}C hardware support is currently used for point-of-load converters and intermediary bus converters. The principles of using the PMBus interface were illustrated with various examples.

Some examples of dynamic voltage and frequency scaling and power mode management were included to illustrate the importance of the power management bus within a modern computing architecture.

References

1. Anon, 2007, "MPC7448 RISC Microprocessor Hardware Specifications", Freescale Semiconductor Doc Number MPC7448EC, Rev. 4, March 2007.
2. Anon, 2006, "Voltage Regulator-Down (VRD) 11.0 – Processor Power Delivery Design Guidelines – For Desktop LGA775 Socket", Intel Corporation, November 2006.
3. Jovanovic, M.M., 2011, "Power Conversion Technologies for Computer, Networking, and Telecom Power Systems – Past, Present, and Future", *IEEE First International Power Conversion & Drive Conference,* San Petersburg, Russia, pp. 1–6.
4. Anon, 2002, "Power Supplies for Telecom Systems", Maxim Integrated Application Note no. 280, 17 July 2002.
5. Anon, "PMBus Specifications – Version 1.3.1", http://PMBus.org. Accessed on 21 September 2017.
6. Neacsu, D., 2013, *Power-Switching Converters,* 2nd edition, CRC/Taylor and Francis.
7. White, R.V. 2016, "PMBus™: Review And New Capabilities", *Embedded Power Labs, Seminar APEC 2016.*
8. Anon, 2011, "3880 – Dual Output PolyPhase Step-Down DC/DC Controller with Digital Power System Management", Linear Technology Datasheet, 2011.
9. Hua, S., Qu, G., 2003, "Approaching the Maximum Energy Saving on Embedded Systems with Multiple Voltages", *IEEE/ACM International Conference on Computer-Aided Design,* San Jose, CA, USA, 9–13 November 2003, p. 26.
10. Hua, S., Qu, G., Bhattacharyya, S.S., 2003, "An Energy Reduction Technique for Multimedia Application with Tolerance to Deadline Misses", *IEEE Design Automation Conference,* Anaheim, CA, USA, 2–6 June 2003, pp. 131–136.

11. Xian, C., Lu, Y., Li, Z., 2007, "Energy-Aware Scheduling for Real-Time Multiprocessor Systems with Uncertain Task Execution Time", *IEEE Design Automation Conference*, San Diego, CA, USA, 4–7 June 2007, pp. 664–669.
12. Le Sueur, E., Heiser, G., 2010, "Dynamic Voltage and Frequency Scaling: The Laws of Diminishing Returns", *Proceedings of the 2010 International Conference on Power Aware Computing and Systems*, Vancouver, Canada, Article No. 1-8.
13. Huang, L., Santinelli, L., Chen, J.J., Thiele, L., Buttazzo, G.C., 2010, "Adaptive Power Management for Realtime Event Streams", in *Asia and South Pacific Design Automation Conference*, Taipei, Taiwan, 18–21 January 2010, pp. 7–12.
14. Niu, L., Quan, G., 2004, "Reducing Both Dynamic and Leakage Energy Consumption for Hard Realtime Systems", in *2004 International Conference on Compilers, Architecture, and Synthesis for Embedded Systems*, Chicago, IL, USA, 6–10 November 2004. pp. 140–148, ACM.

9

Diode Rectifiers

9.1 Power Rectifier Usage

Usually, energy is distributed as 50 or 60 Hz AC voltages, while most electronic circuits need to be supplied with DC voltages [1,2]. This implies the need for a conversion from AC to DC voltages (Figure 9.1). This conversion unit is also called an AC/DC converter.

9.2 Operation of a Single-Diode Rectifier

Operation of a single-phase single-diode rectifier with a resistive load is observed in the circuit from Figure 9.2 and waveforms from Figure 9.3. The operation of the same rectifier with a resistive-inductive load is demonstrated with waveforms shown in Figure 9.4.

The inductive component in the load stores energy during the diode conduction interval under a positive grid voltage, and this determines an extension of the conduction at the beginning of the following (negative) polarity of the supply voltage. The diode's conduction angle is defined within Figure 9.5 as the angular interval when the diode is in conduction. It is herein greater than 180° and can be calculated analytically based on circuit components.

Both the load voltage and current have an average value different from zero, which represents a DC component. This justifies the classification of this converter as an AC/DC converter. It is important to calculate the average value of the output voltage. Generally, the average value is calculated over a fundamental cycle:

$$V_{dc} = \frac{1}{T} \cdot \int_{0}^{T} v(t)d(t) \_\_\_\_ or \_\_\_\_ V_{dc} = \frac{1}{\omega T} \cdot \int_{0}^{\omega T} v(\omega t)d(\omega t) \tag{9.1}$$

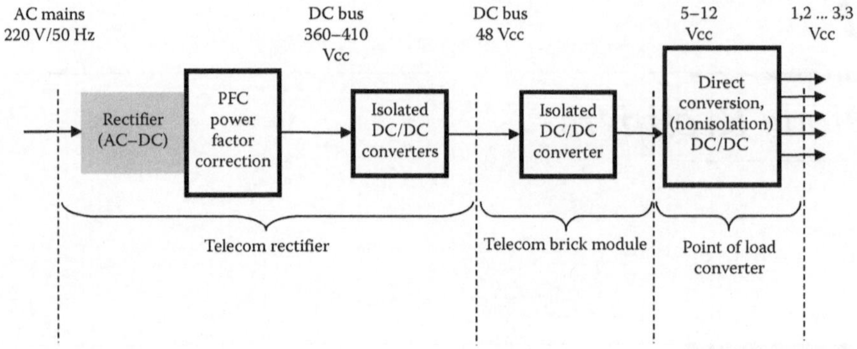

FIGURE 9.1
Rectifier location within the power supply architecture.

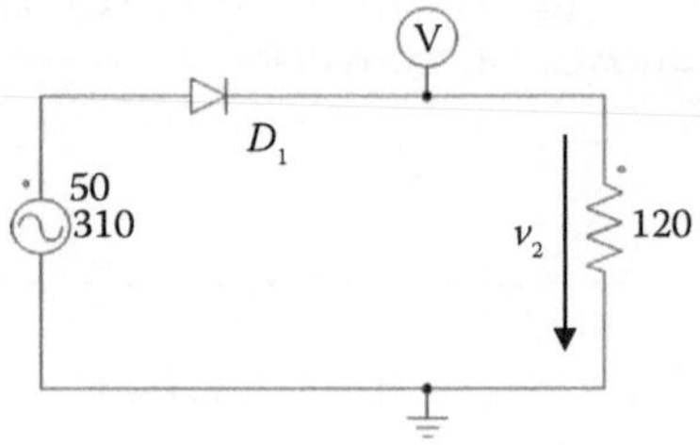

FIGURE 9.2
Circuit drawing for a single-phase single-diode rectifier.

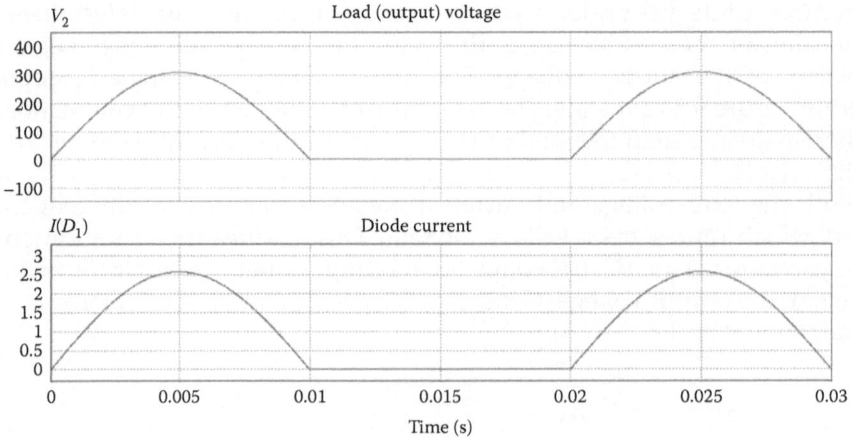

FIGURE 9.3
Waveforms characteristic of a single-phase single-diode rectifier.

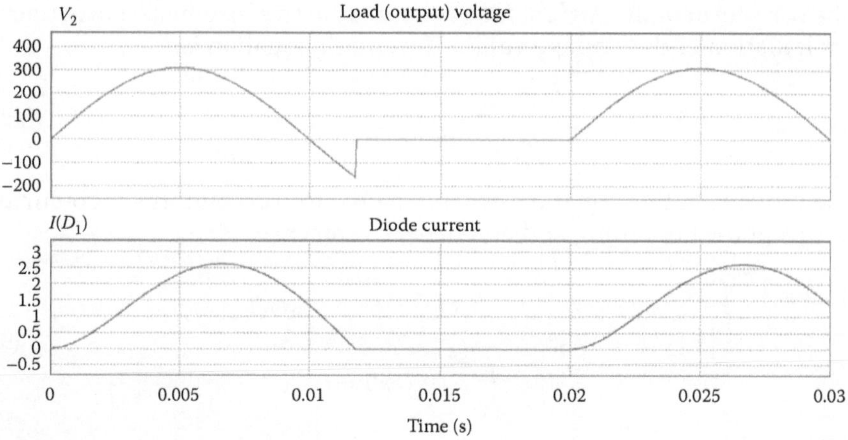

FIGURE 9.4
Waveforms for operation with a resistive-inductive load.

This has to be particular to the waveform from the previous operation example of the single-diode rectifier with resistive load.

$$V_{dc} = \frac{1}{2 \cdot \pi} \cdot \int_{0}^{\pi} V \cdot \sin(\omega t) d(\omega t) = \frac{1}{2 \cdot \pi} \cdot V \cdot [-\cos(\omega t)]_0^{\pi} = \frac{2 \cdot V}{2 \cdot \pi} = \frac{\sqrt{2} \cdot E}{\pi} \quad (9.2)$$

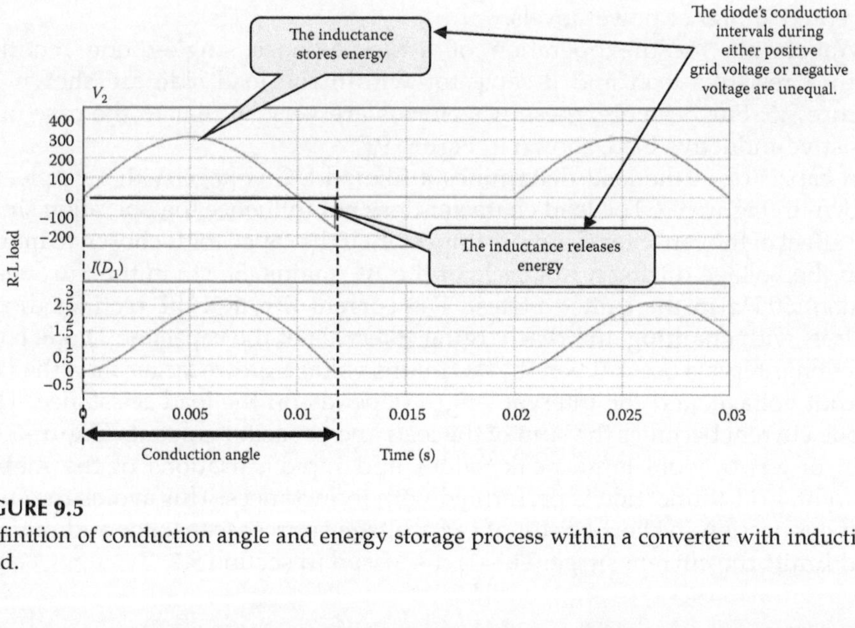

FIGURE 9.5
Definition of conduction angle and energy storage process within a converter with inductive load.

The waveform of the current for a purely inductive load helps particularise the calculation of the average value. The circuit equation is:

$$L\cdot\frac{di}{dt}=v_L \tag{9.3}$$

The integral of this equation (when we take into account the zero-current intervals at the beginning and end of each interval):

$$\begin{aligned}\frac{1}{L}\cdot\int_0^{t_{final}} v_L(t)dt &= \int_0^{t_{final}} di \Leftrightarrow \frac{1}{L}\cdot\int_0^{t_{final}} v_L(t)dt = i(t_{final})-i(0)\\ &=0\Leftrightarrow\int_0^{t_{final}} v_L(t)dt=0\end{aligned} \tag{9.4}$$

This confirms that the average value of the voltage across an inductance is zero. The energy stored during the positive polarity of the AC mains equals the energy delivered when the voltage drop is negative across the inductance.

Most applications require an output voltage without ripple (essentially, a pure DC voltage). This is also the case within the generic architecture of the telecom power system shown in Figure 9.1. A capacitor is required within the load circuit of the rectifier to filter the output voltage. This configuration is also called a *peak detector.* The capacitor can charge from the AC mains when the mains voltage is greater than the capacitor voltage; therefore, for a short interval. During the time duration of this charging interval, the diode current follows the sum of the capacitor current and load current and can be very large. This results in operation with voltage and current pulses, which distort the AC grid system. For this reason, this configuration cannot be accepted at higher power levels.

Waveforms for the operation of a single-phase single-diode rectifier with a resistive load and a capacitor within the load side are shown in Figure 9.6. Furthermore, these waveforms are very similar to the case of a resistive-inductive load, shown in Figure 9.7.

A capacitor in the load determines a filtered DC voltage with a ripple, as shown in Figure 9.7. The load character does not influence the operation since the effect of the capacitor C is more important if the capacitor is chosen to maintain the voltage during a full cycle of the AC mains (50 Hz in the European Union, 60 Hz in the United States). The current through the rectifier diode pulses, with charging and discharging intervals of the capacitor. Diode conduction intervals occur when the AC mains voltage grows larger than the DC output voltage, and the interval length depends on the load resistance. The diode current becomes the sum of the load and capacitor currents (Figure 9.8).

In order to avoid large peak values and rapid variations of the mains current, grid connection is performed with inductances. This avoids the sudden connection in short-circuit of two voltage sources (capacitor and mains) and limits the current surge. This is discussed in section 9.4.

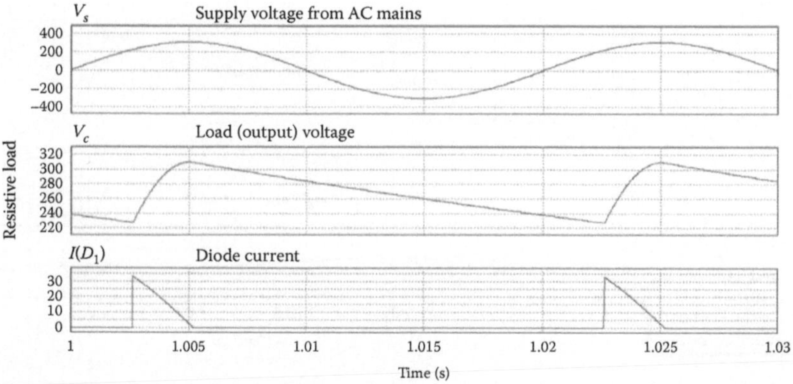

FIGURE 9.6
Resistive load for single-phase rectifier with capacitor on the load side.

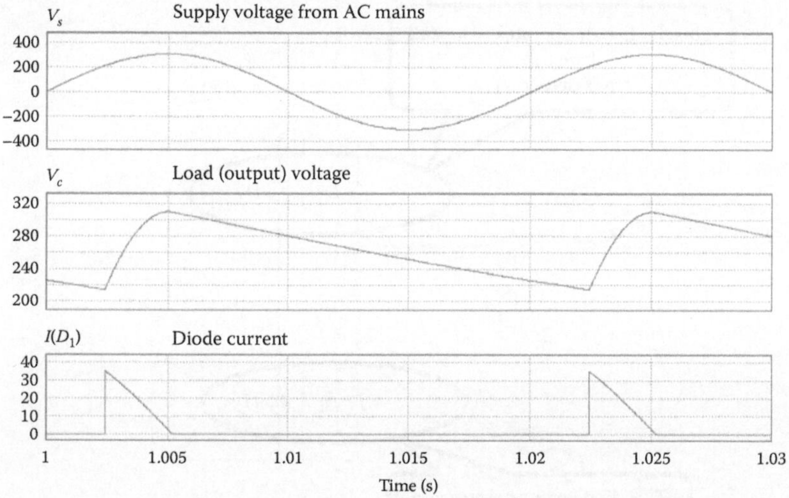

FIGURE 9.7
Resistive-inductive load for single-phase rectifier with capacitor on the load side.

9.3 Operation of a Bridge Rectifier

The circuit of a single-phase bridge rectifier is shown in Figure 9.9. It is composed of four diodes. The diodes connected with a common cathode work as a maximum (peak) detector, while the diodes connected with a common anode work as a minimum detector.

The waveforms corresponding to a resistive load are shown in Figure 9.10, and the case of a resistive-inductive load is shown in Figure 9.11. Despite the presence of an inductive component in the load, the conduction angle

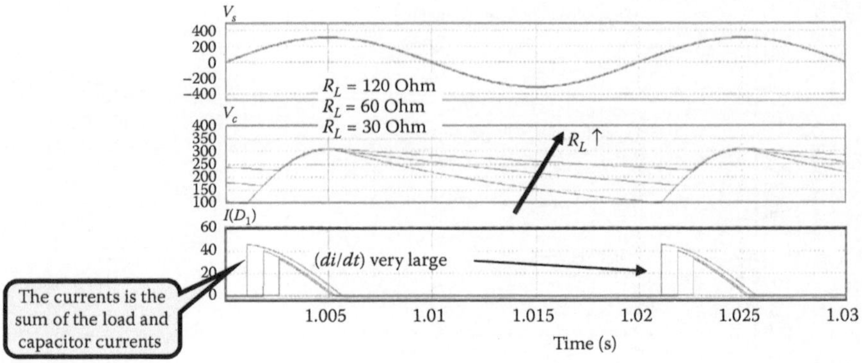

FIGURE 9.8
Results of variation of the resistive load.

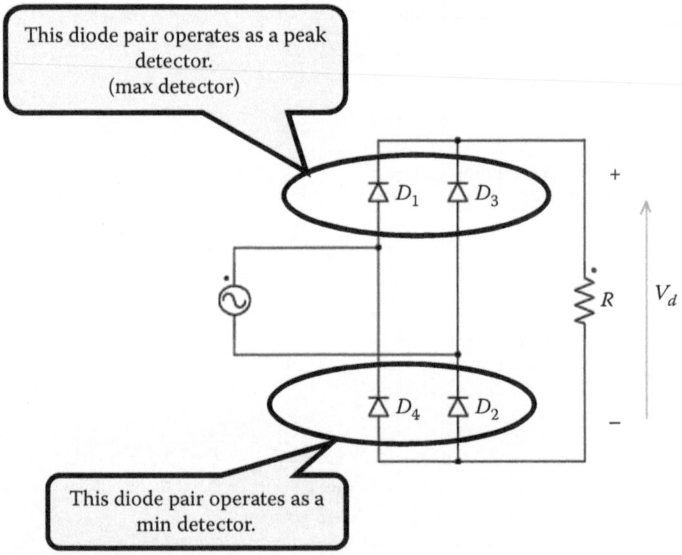

FIGURE 9.9
Single-phase bridge rectifier.

for each diode is 180^0. This is a major difference from the case of a single-phase single-diode rectifier since the bridge rectifier ensures the load current circulation from one pair of diodes to the other. Furthermore, the diodes turn off with forced commutation at nonzero current when a negative voltage applies across them.

Since the output (load) side has a voltage with an important DC component, rectifiers are AC/DC converters. The average value of the output voltage can be calculated and considered as an expression of the DC voltage.

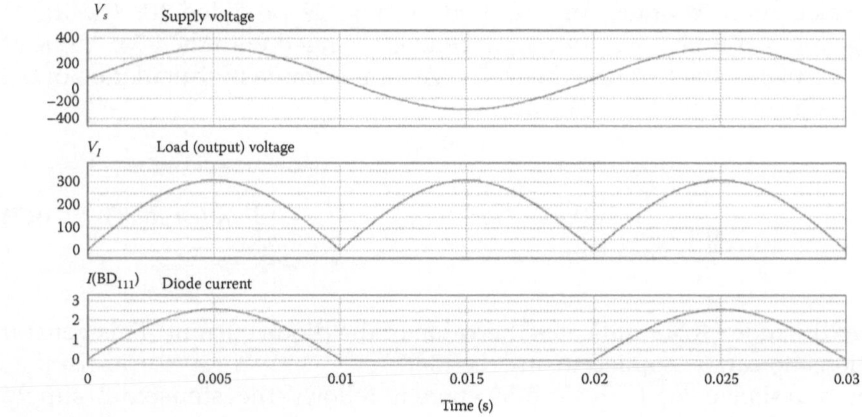

FIGURE 9.10
Waveforms for a single-phase bridge rectifier.

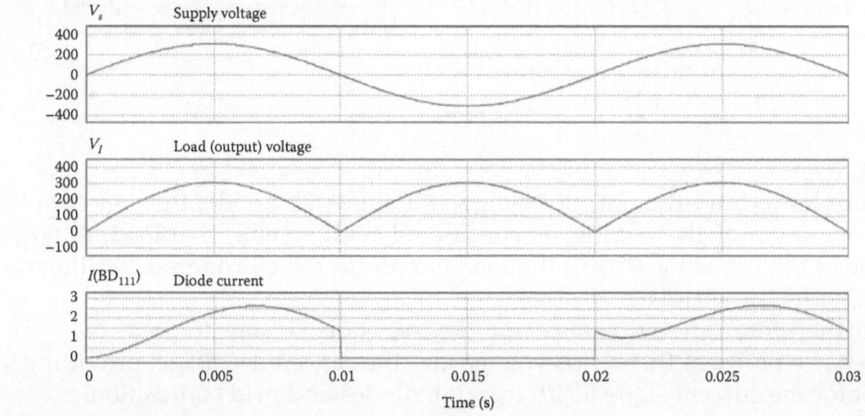

FIGURE 9.11
Operation with inductive-resistive load.

The average value of a measure is provided with the formula:

$$V_{dc} = \frac{1}{\omega T} \cdot \int_{0}^{\omega T} v(\omega t) d(\omega t) \tag{9.5}$$

where $E = RMS$ value of the supply voltage.

For the single-phase bridge rectifier, it gives:

$$V_{dc} = \frac{1}{\pi} \cdot \int_{0}^{\pi} V \cdot \sin(\omega t) d(\omega t) = \frac{2 \cdot V}{\pi} = \frac{2 \cdot \sqrt{2} \cdot E}{\pi} \tag{9.6}$$

A calculation example for the diode ratings is provided for the diode bridge rectifier with a resistive load. Diodes are rated with the RMS value of the current passing through them. The general formula for calculation of the RMS value of a signal is provided by:

$$I_{D,rms} = \sqrt{\frac{1}{T} \cdot \int_0^T i_D{}^2(t)d(t)} \_\_or\_\_ I_{D,rms} = \sqrt{\frac{1}{\omega T} \cdot \int_0^{\omega T} i_D{}^2(\omega t)d(\omega t)} \quad (9.7)$$

where the index *RMS* indicates the value of the diode current. This general relationship can be applied to any rectifier.

For a resistive load, the diode current follows the sinusoidal supply voltage, and the calculation yields:

$$\begin{aligned} I_{D,rms} &= \sqrt{\frac{1}{2\cdot\pi} \cdot \int_0^{\pi} (I\cdot\sin(\omega t))^2 d(\omega t)} = I\cdot\sqrt{\frac{1}{2\cdot\pi} \cdot \int_0^{\pi} \left(\frac{1+\cos(2\omega t)}{2}\right) d(\omega t)} \\ &= I\cdot\sqrt{\frac{1}{2\cdot\pi} \cdot \left(\frac{1}{2}\cdot\pi + \frac{1}{4}\cdot\sin(2\omega t)\Big|_0^{\pi}\right)} = I\cdot\frac{1}{2} \end{aligned} \quad (9.8)$$

In order to filter the output voltage and approach an ideal power supply, the operation of the rectifier is considered with a capacitive load. A large value of the capacitor within the load circuit is next considered for illustration, with $C = 470\ \mu F$.

Furthermore, an inductance on the AC side is able to take over the difference between the mains voltage and the capacitor voltage, providing a limit for the current slope (*di/dt*) through diodes and grid connection.

The circuit for this case is shown in Figure 9.12 and the waveforms in Figure 9.13. The effect of the grid inductance is seen on the slopes of the current through diodes.

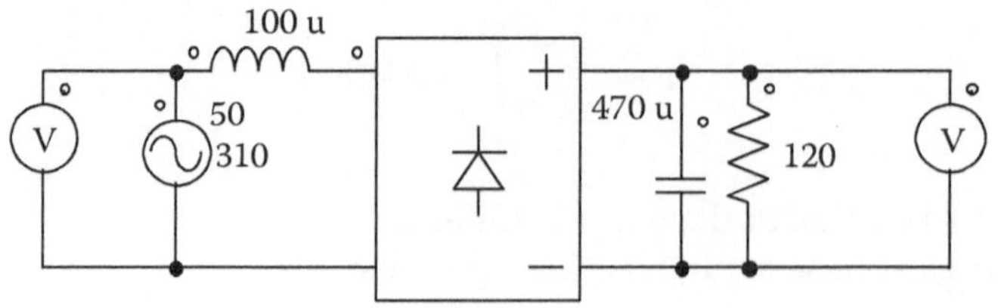

FIGURE 9.12
Circuit including a large capacitor on the DC side.

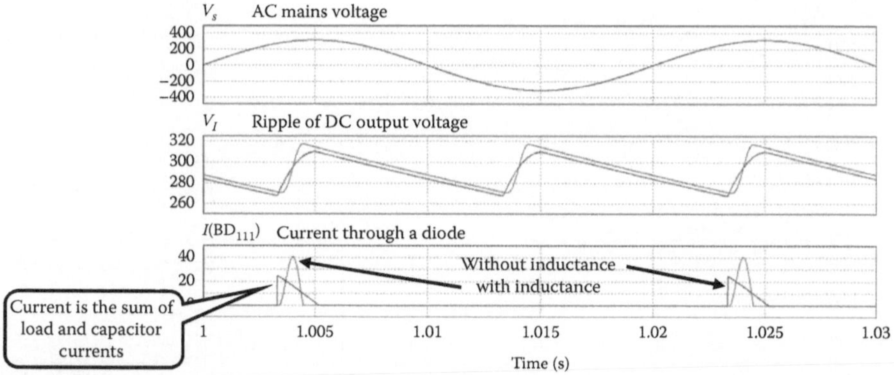

FIGURE 9.13
Waveforms for bridge rectifier with large capacitive load.

9.4 Connection Inductance Influence on the Commutation Process

The commutation processes within a rectifier with a single-diode rectifier and a diode for circulation of the inductive load current is shown in Figure 9.14. The effect of a large load inductance is modelled with a current source able to circulate a constant DC current. At the end of each polarity of the supply voltage, the current transfers from one diode to another. Due to the connecting circuit inductance, the current transition is not instantaneous; rather, it has a delay also called *commutation interval* (*commutation angle*). The commutation phenomena at current transition from one diode to another are illustrated with Figure 9.15.

At diode D1 turn-on, the intermediary states are shown in Figure 9.16. The effect of the value of the line (grid) inductance is illustrated with Figure 9.17.

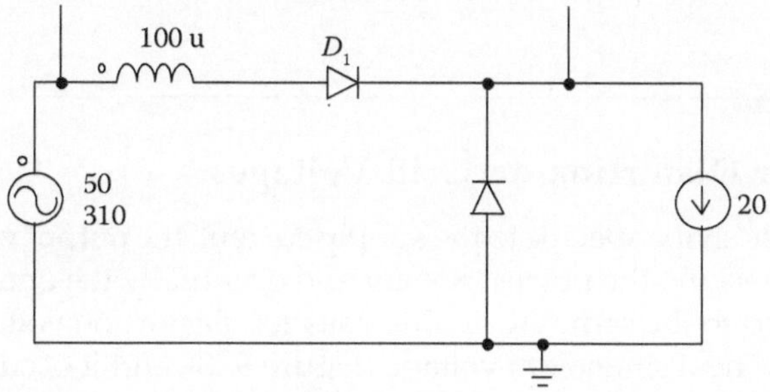

FIGURE 9.14
Circuit for study of the diode commutation process.

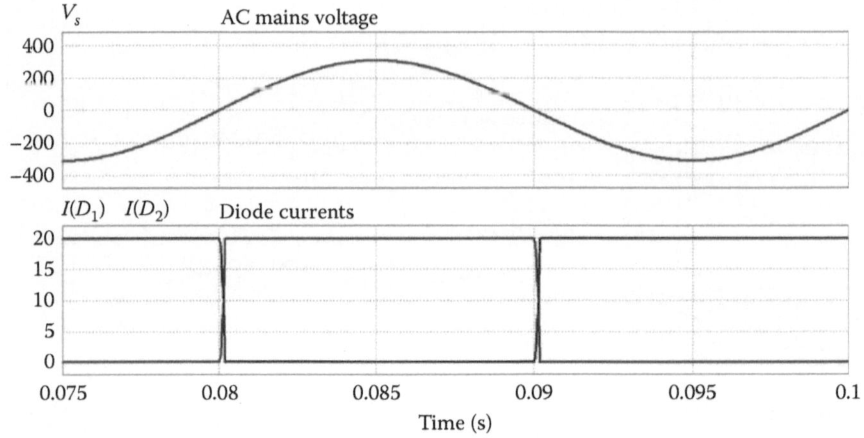

FIGURE 9.15
Current transition from one diode to another.

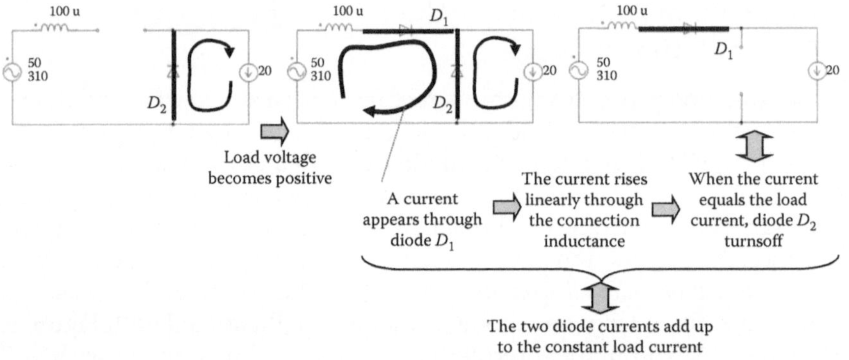

FIGURE 9.16
Intermediary states at current commutation.

9.5 Rectifier Distortion on Grid Voltages

Due to the pulse shape specific to the supply current, the output voltage is easily distorted, influencing the power transfer and eventually the operation of other loads connected to the same AC mains voltages. Distortion voltage appears in addition to the ideal sinusoidal voltage (Figure 9.18), and it occurs on the wire inductance and connection inductance ($v = L \ni di/dt$). If the current slope rises, the distortion becomes more important. If the parasitic inductance increases, the distortion becomes more important. The distortion shown in Figure 9.18 in time-domain waveforms is also illustrated in Figure 9.19 as harmonic spectra.

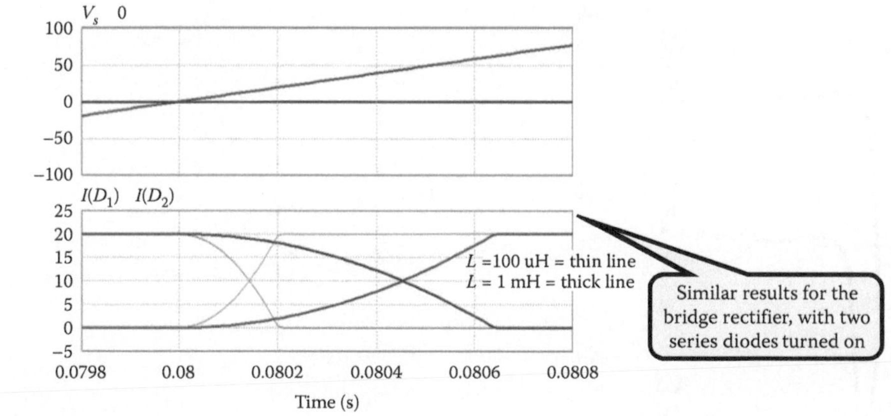

FIGURE 9.17
Currents for various values of the grid inductance.

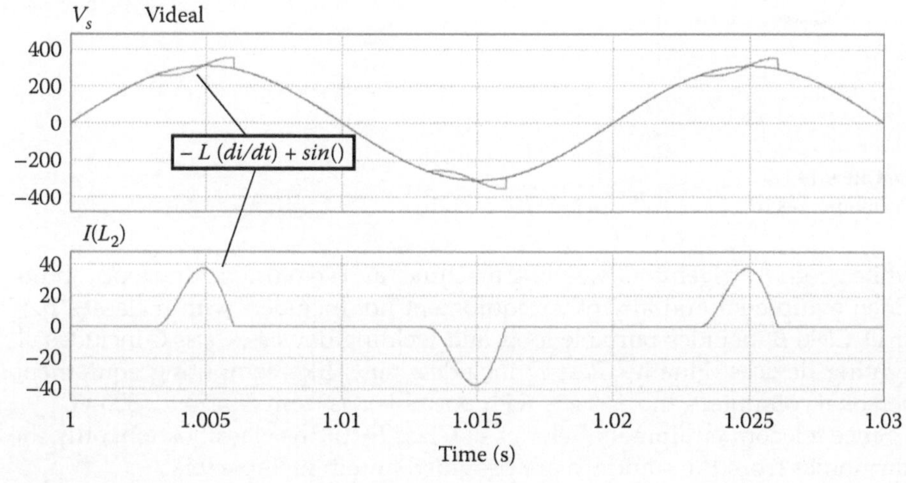

FIGURE 9.18
Waveforms including the distortion of the grid voltage.

Due to the importance of the quality of AC mains voltages, a series of professional standards have been proposed and adopted by the industry. In 2001, the European Union elaborated on standard EN61000-3-2 in order to define the admissible limits for supply currents up to the 40th harmonic. The standard has been amended yearly since. This standard builds upon previous standards created by various professional organisations like EN 60555-2, IEEE 519, IEC 61000-3-2, IEC 61000-3-6 and IEC 61000-4-7.

Several application classes are defined within EN61000-3-2, each with a different set of requirements. Thus, *Class A* corresponds to three-phase equipment,

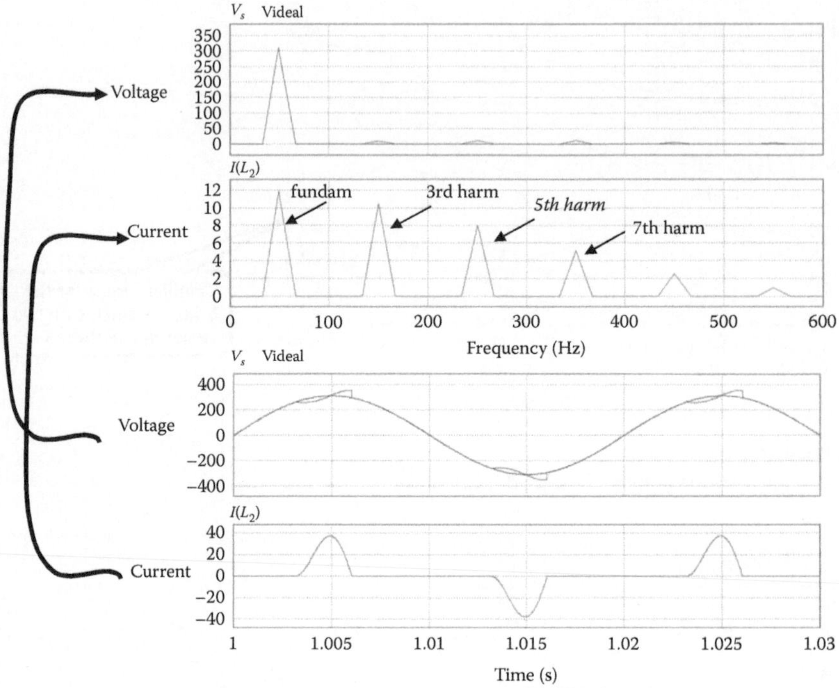

FIGURE 9.19
Harmonic spectra.

white goods (refrigerators, washing machines and so on), machine tools, audio-video equipment and any other equipment not included within classes B, C or D. *Class B* includes portable tools and welding devices. *Class C* includes all lighting devices. Finally, *Class D* includes items like computing equipment, personal computers and TV sets with power levels from (50) 75 …. 600 W.

Since telecom equipment belongs to Class D in this classification, only the harmonics from the standard are provided herein in Table 9.1.

TABLE 9.1
Harmonics for Class D Equipment

Order n	Max. Allowed Harmonic Current mA/W	Max. Allowed Harmonic Current A
3	3.4	2.30
5	1.9	1.14
7	1.0	0.77
9	0.5	0.40
11	0.35	0.33
$13 \leq n \leq 39$	$3.85/n$	

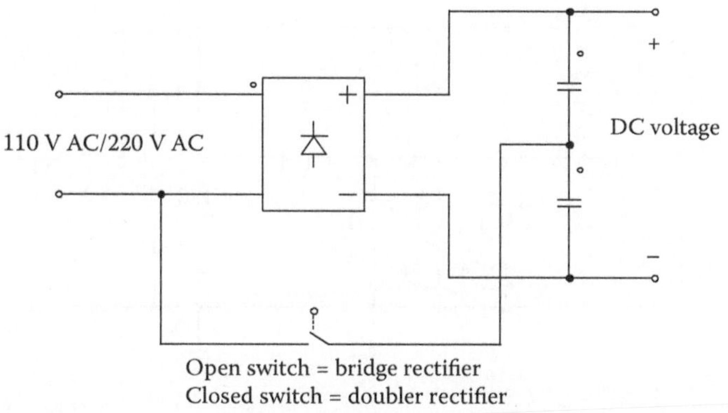

FIGURE 9.20
Circuit able to work with both 110 V AC and 220 V AC standards through a commutator function.

9.6 Other Single-Phase Rectifiers: Alternating Current Doubler Rectifier

A converter topology as shown in Figure 9.20 is used to supply the same power equipment without control from AC voltages like 110 V AC @ 60 Hz or 220 V AC @ 50 Hz. When used to supply from 220 V AC mains, the circuit works with a switch in open circuit as a bridge rectifier. When used to supply from 110 V AC mains, it works with a switch in closed circuit as a voltage doubler. It yields a single diode rectification on each polarity, alternatively charging C1 and C2.

The circuit was very often used in the early 2000s for creating multi-input power supplies. The advent of power factor correction has forced the usage of a power switch within the rectifier stage and the appropriate control of this switch. Hence, the opportunity to use the same power stage with a wider range of input voltages has arisen. Modern dual-system power supplies currently use wide input range power factor correction operation. Solutions for power factor correction are provided within Chapter 10.

9.7 Three-Phase Rectifiers

At higher power levels, three-phase rectifiers are used. The most-known three-phase diode topologies are the *midpoint rectifier* and *three-phase bridge rectifier.*

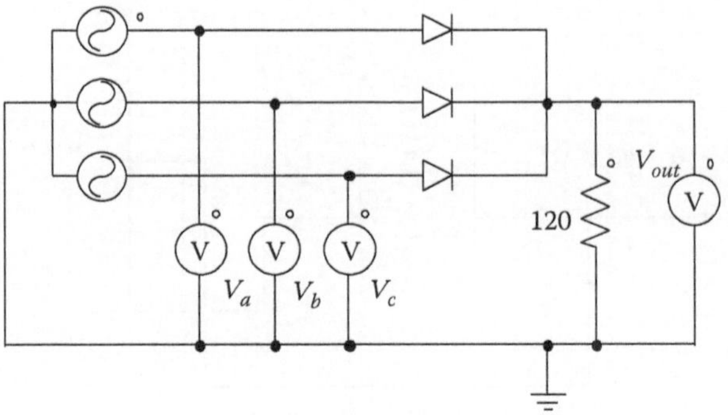

FIGURE 9.21
Midpoint three-phase rectifier.

9.7.1 Midpoint Rectifier

The three-phase diode midpoint rectifier has the circuit shown in Figure 9.21. Its operation with a resistive load can be understood from the waveforms shown in Figure 9.22, while operation with a resistive-inductive load is shown in Figure 9.23. A value of 180^0 is achieved for the conduction angle, as the three diodes share the conduction time intervals, with one in conduction at a time.

A continuous circulation of the load current is achieved when using a large load inductance, although a single diode conducts at any given moment. The inductance within the load circuit determines filtering of the load current. Through extrapolation, if the inductance has an infinite value, waveforms are quasiconstant for the load current as well as for the diodes' currents. This means $L \rightarrow \infty$ *&* $L\ di/dt = V = >i = constant$ (Figure 9.24). There is no ripple

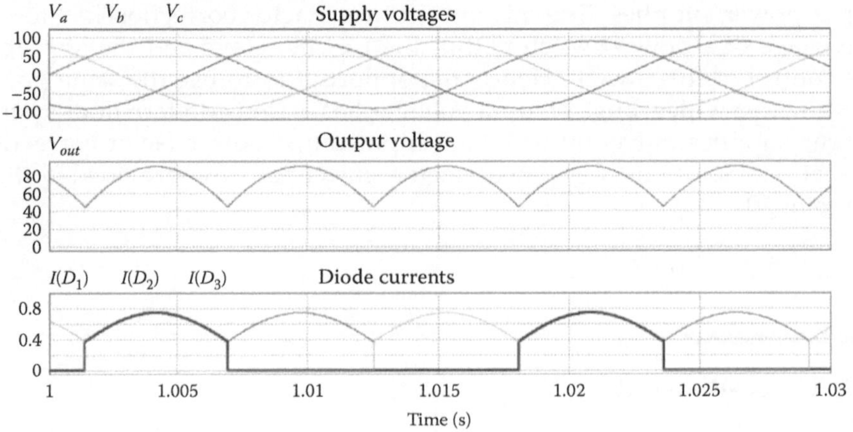

FIGURE 9.22
Waveforms for operation of a midpoint rectifier with resistive load.

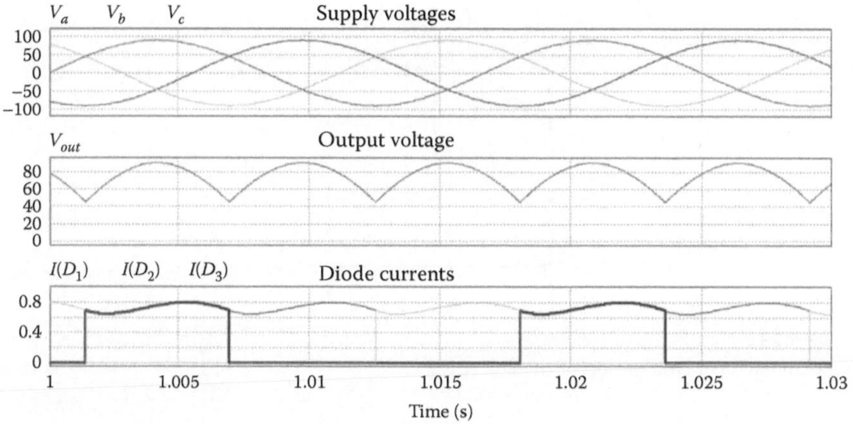

FIGURE 9.23
Waveforms for operation with a resistive-inductive load.

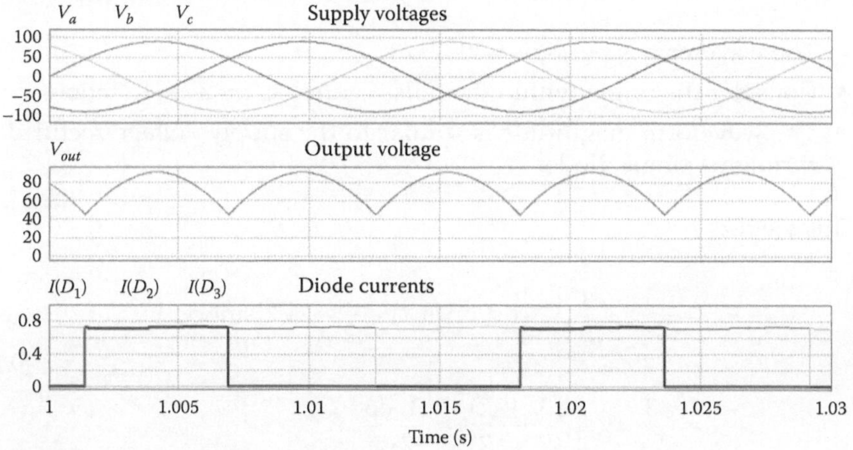

FIGURE 9.24
Waveforms for a very large inductive load.

present within the load current, and the diode currents have a quasirectangular shape. This principle is also a strong simplification idea for designing power rectifiers.

The average value of the output voltage can be considered an expression of the DC voltage and calculated with the general formula:

$$V_{dc} = \frac{1}{\omega T} \cdot \int_{0}^{\omega T} v(\omega t) d(\omega t) \tag{9.9}$$

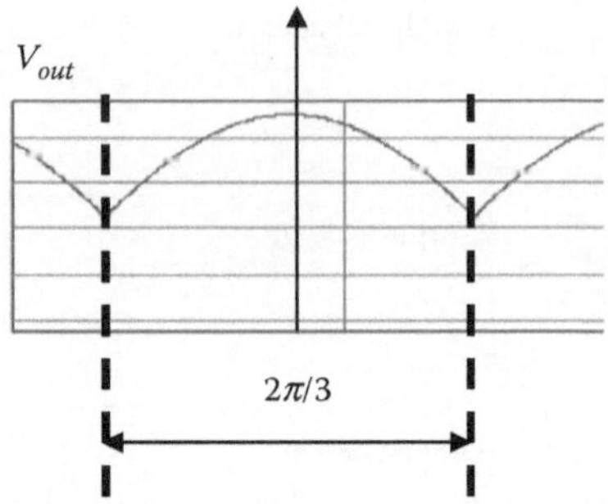

FIGURE 9.25
Waveform used for calculation of the DC voltage.

The waveform of the output voltage is considered in Figure 9.25 for a period of its ripple that equals 2π/3 at the angular frequency of the grid voltage. The following remarks help the particularisation of the general averaging relationship for this case:

- The waveform is periodic with 2π/3.
- Consider the origin within the pulse centre (as for a *cos* function).
- The waveform magnitude is similar to the supply voltage rectified through a single diode.

This yields:

$$V_{dc} = \frac{1}{\frac{2\cdot\pi}{3}} \cdot \int_{-\pi/3}^{\pi/3} V\cdot\cos(\omega t)d(\omega t) = \frac{3}{2\cdot\pi}\cdot V\cdot[\sin(\omega t)]\,\Big|_{-\pi/3}^{\pi/3}$$
$$= \frac{3}{2\cdot\pi}\cdot V\cdot\left[\frac{\sqrt{3}}{2}+\frac{\sqrt{3}}{2}\right] = \frac{3\cdot\sqrt{6}}{2\cdot\pi}\cdot E \quad (9.10)$$

where E = RMS value for the supply voltage ($V = E$ ϶ *sqrt*(2)).

The same three-phase rectifier leads to a different output voltage when a capacitor is connected at load (Figure 9.26). The operation of a rectifier with capacitive load denotes pulsed current with large magnitude (~10 times larger than the DC component), while the voltage is low-pass filtered, with a smoother waveform.

As shown for previous rectifier topologies, current pulses can have an attenuated *di/dt* with inductances placed on the current path, either on the AC or DC side. Various positions are possible for inductances with equivalent effects (Figure 9.27). Such inductances lead to the waveforms from Figure 9.28, and a different slope (*di/dt*) of the diode current waveform can be seen. This reduces EMI and distortion in the AC mains.

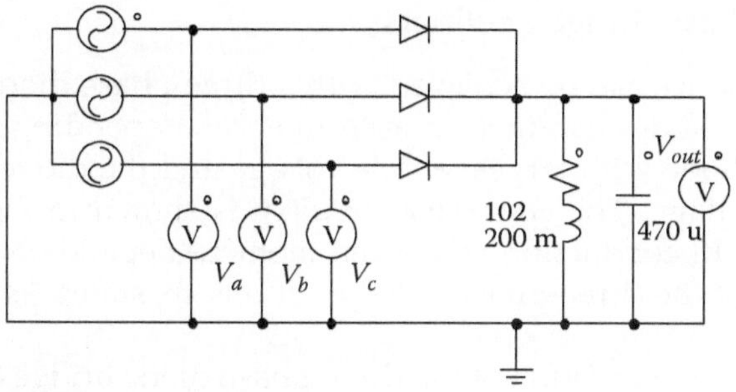

FIGURE 9.26
Three-phase midpoint rectifier with a large capacitor on the load side.

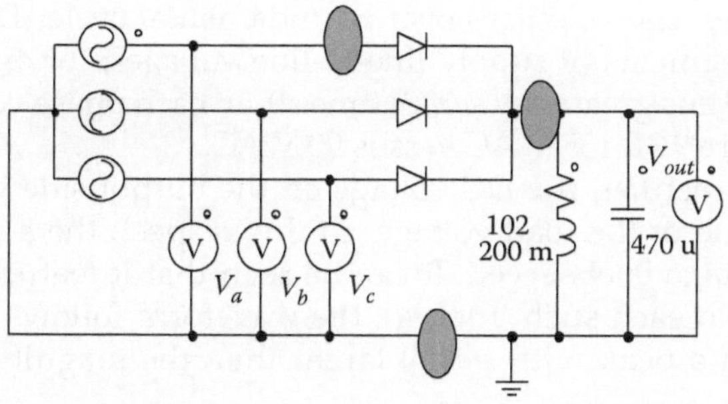

FIGURE 9.27
Possible locations for usage of commutation inductances for the midpoint rectifier.

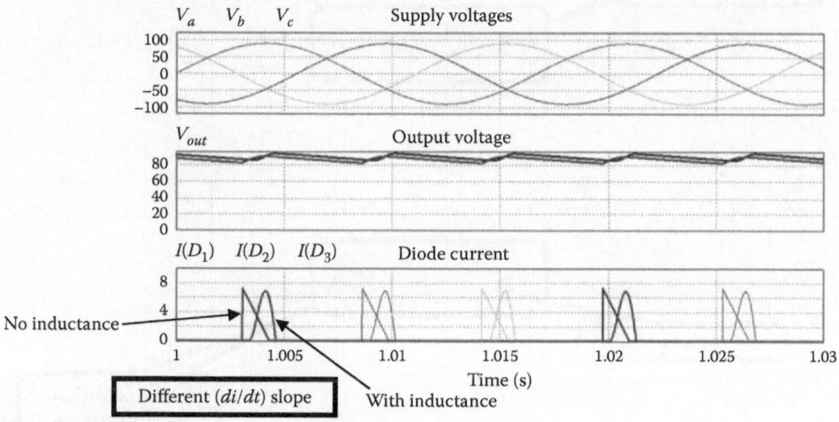

FIGURE 9.28
Comparison between the current waveforms for the case without or with commutation inductances.

9.7.2 Three-Phase Bridge Rectifiers

Better performance can be achieved with a three-phase diode bridge rectifier. Only two diodes conduct on each interval, depending on the supply voltages: the diode with largest anode voltage and the diode with the lowest cathode voltage. The converter circuitry is shown in Figure 9.29. The pair of diodes in conduction at a given moment depends on the instantaneous values of the three-phase voltages. It has six states, as demonstrated in Figure 9.30.

A special case in operation of the three-phase diode bridge rectifier is represented by a very large inductive load which is able to filter the load current and reduce its ripple. Waveforms derived in such an operation are shown in Figure 9.31.

A strongly inductive load determines a quasiconstant load current. The load voltage has six pulses over a fundamental cycle. The load voltage is formed by segments of supply line-to-line voltages; for example, $V_{phaseA}-V_{phaseB}=V_a-V_b$, thus greater by *sqrt*(3) times than each phase voltage. This can be seen in Figure 9.31: 155 V AC versus 90 V AC.

As with any rectifier, the DC voltage on the output side is calculated as an average value of the load voltage. In this respect, the actual waveform of the load voltage is observed. It can be seen that it features segments of $\pi/3$ intervals. On each such interval, the waveform follows a cosinusoidal waveform with a peak with *sqrt*(3) larger than the magnitude of the grid voltage.

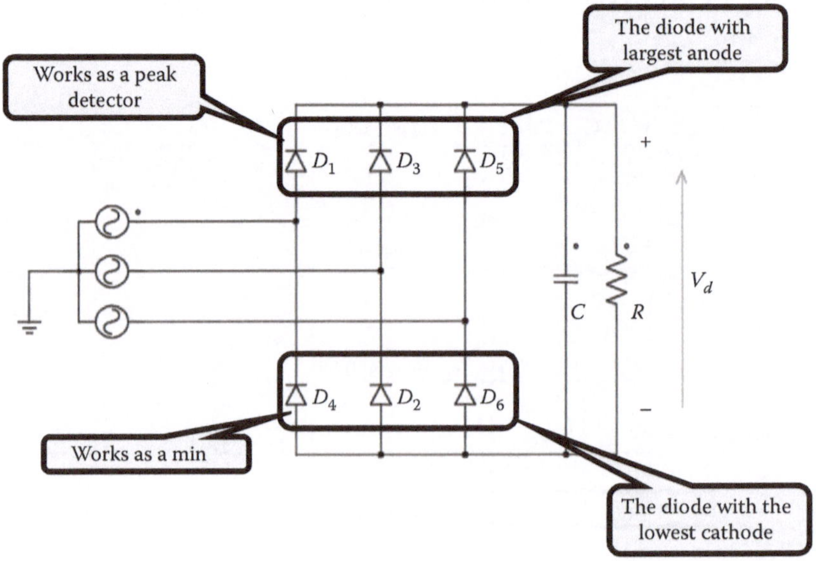

FIGURE 9.29
Three-phase diode bridge rectifier.

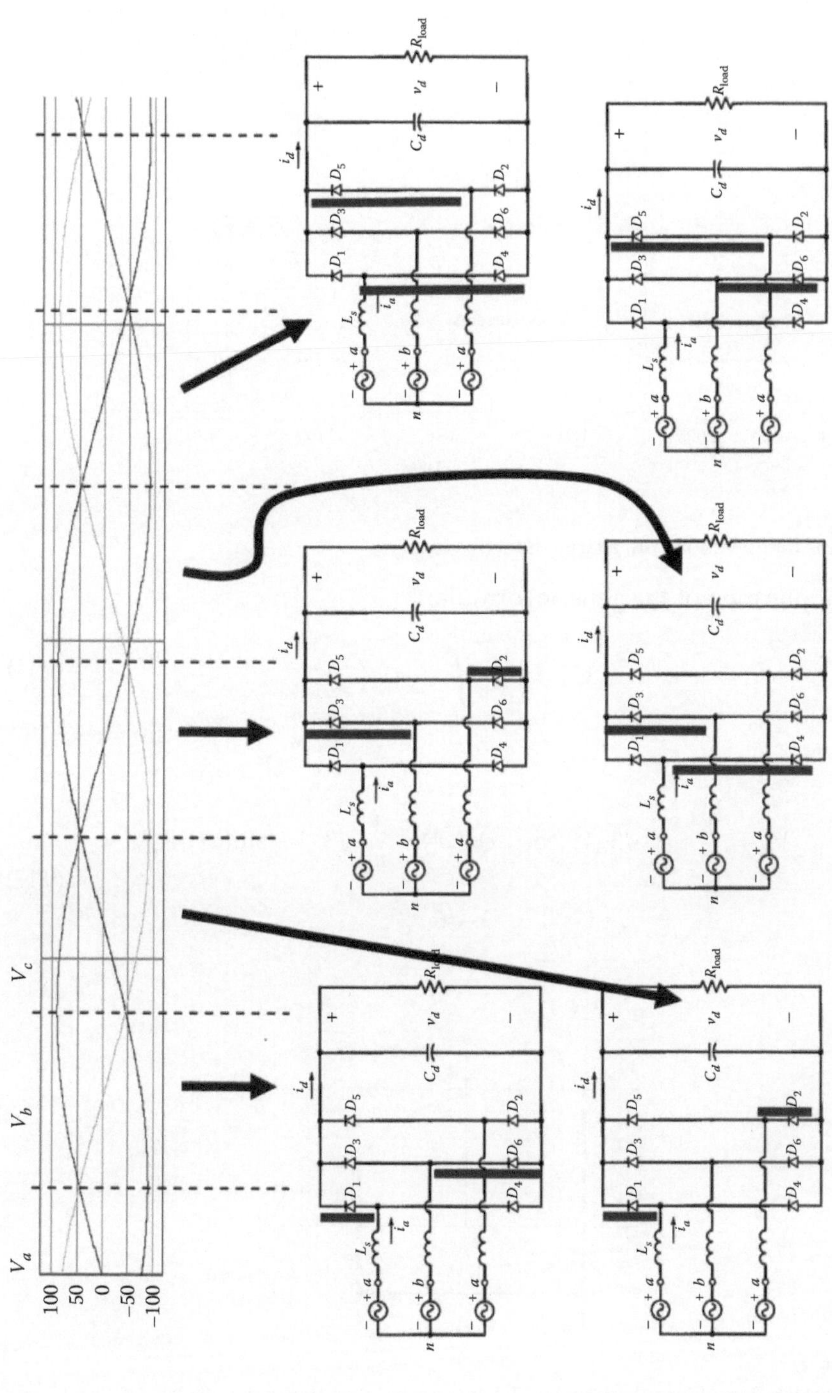

FIGURE 9.30
Explanation of the selection of the pair of diodes in conduction at a given moment.

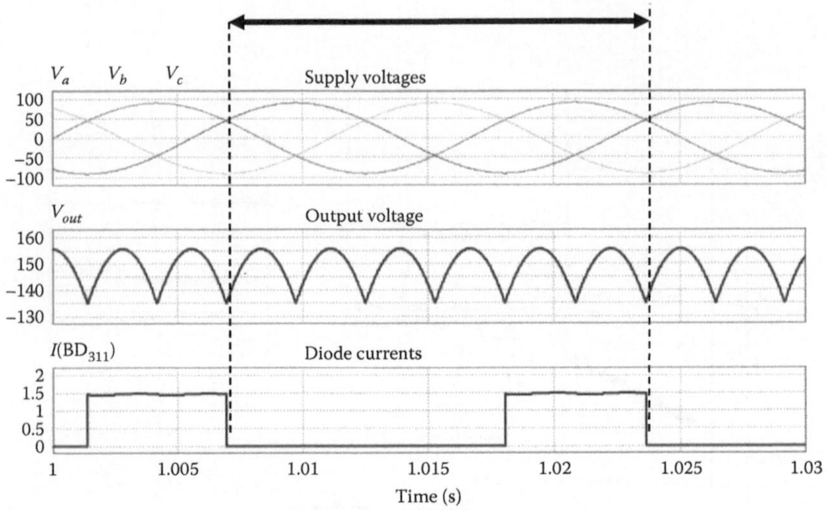

FIGURE 9.31
Waveforms for operation with a large inductive load.

The application of the generic formula:

$$V_{dc} = \frac{1}{\omega T} \cdot \int_0^{\omega T} v(\omega t) d(\omega t) \tag{9.11}$$

yields:

$$\begin{aligned} V_{dc} &= \frac{1}{\frac{\pi}{3}} \cdot \int_{-\pi/6}^{\pi/6} \sqrt{3} \cdot V \cdot \cos(\omega t) d(\omega t) = \frac{3}{\pi} \cdot \sqrt{3} \cdot V \cdot [\sin(\omega t)]\,\Big|_{-\pi/6}^{\pi/6} \\ &= \frac{3}{\pi} \cdot \sqrt{3} \cdot V \cdot \left[\frac{1}{2} + \frac{1}{2}\right] = \frac{3 \cdot \sqrt{6}}{\pi} \cdot E \end{aligned} \tag{9.12}$$

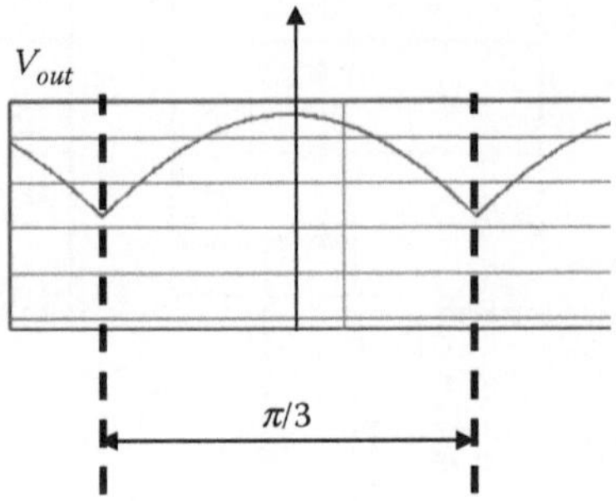

FIGURE 9.32
Waveform detail used for calculation of the average value of the load voltage.

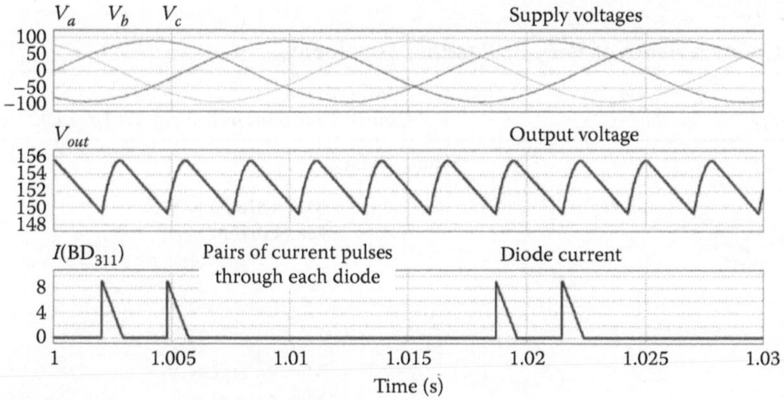

FIGURE 9.33
Waveforms for the usage of a large capacitor on the load side.

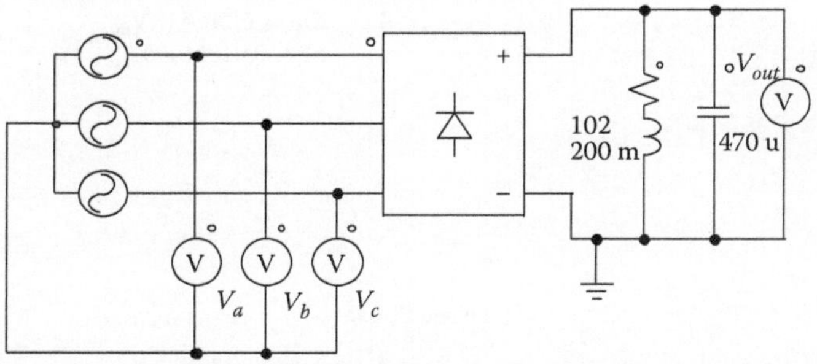

FIGURE 9.34
Three-phase bridge rectifier with a capacitive load.

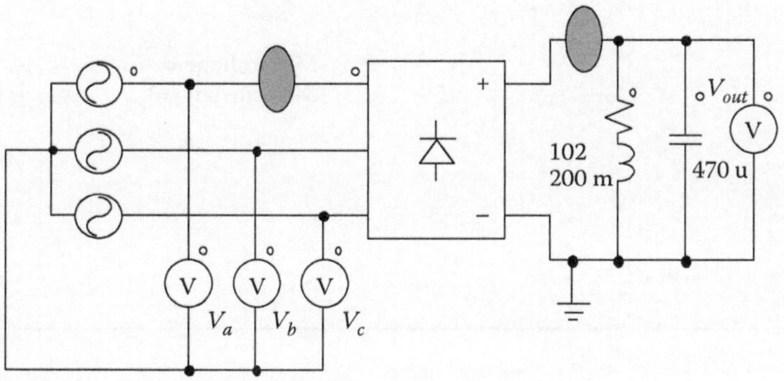

FIGURE 9.35
Possible locations for usage of commutation inductances for the full bridge rectifier.

TABLE 9.2
Main Characteristics of Various Rectifier Topologies

Topology	Diode Count	Component Rating	Output DC Voltage
Singe Phase			
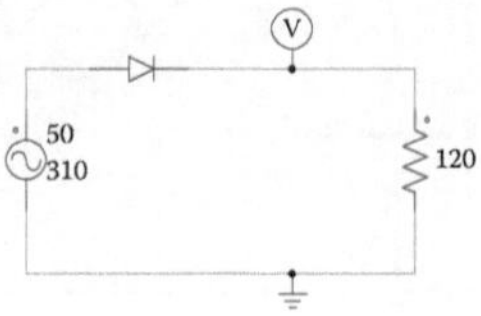	1	Max voltage = V_{in} Max current = I_{pk}	$V_{d0} = V_{ph} \cdot \frac{\sqrt{2}}{\pi}$
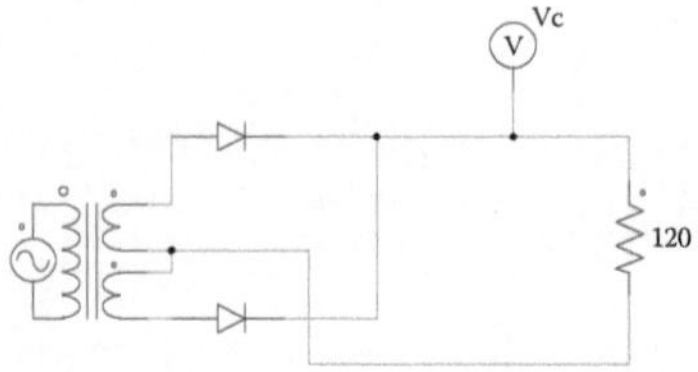	2	Max voltage = V_{in} Max current = I_{pk}	$V_{d0} = V_{ph} \cdot \frac{2 \cdot \sqrt{2}}{\pi}$
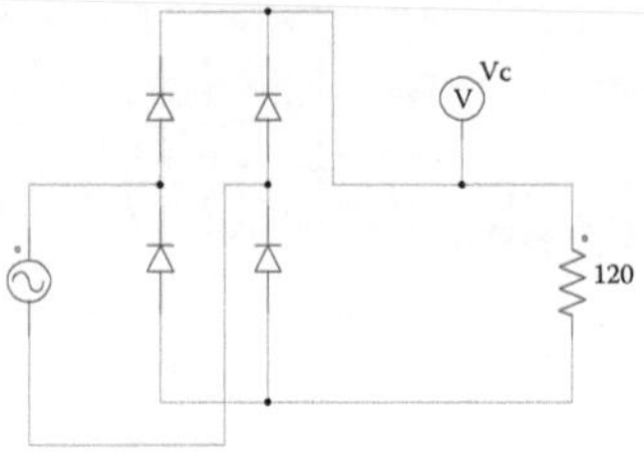	4	Max voltage = V_{in} Max current = I_{pk}	$V_{d0} = V_{ph} \cdot \frac{2 \cdot \sqrt{2}}{\pi}$
Three Phase			
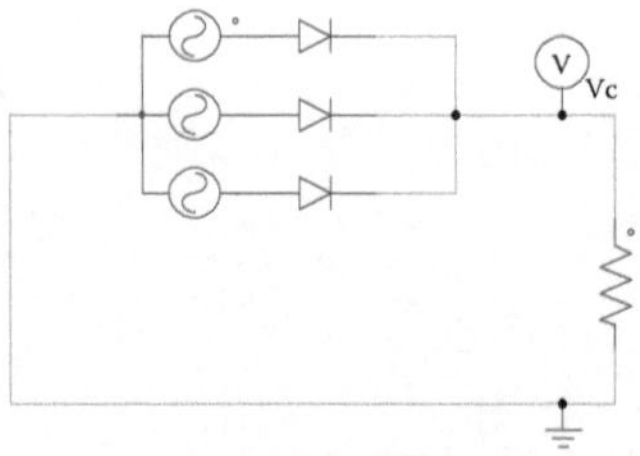	3	Max voltage = V_{in} Max current = I_{pk}	$V_{d0} = V_{ph} \cdot \frac{3 \cdot \sqrt{6}}{2 \cdot \pi}$
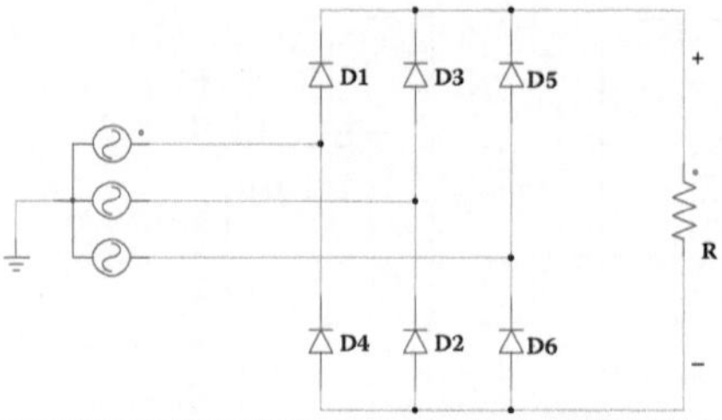	6	Max voltage = V_{LL} Max current = I_{pk}	$V_{d0} = 2 \cdot V_{ph} \cdot \frac{3 \cdot \sqrt{6}}{2 \cdot \pi}$

This average value is two times greater than the midpoint three-phase rectifier case. Basically, the three-phase bridge rectifier can be seen as two midpoint rectifiers which add their effects (Figure 9.32).

Adding a capacitor on the load side filters the load voltage and changes the shape of all waveforms as shown in Figure 9.33. The circuit used for this analysis is shown in Figure 9.34. Adding a capacitor reduces the peak-to-peak voltage and increases the currents through diodes.

A three-phase bridge rectifier with load-side capacitor produces current pulses with a large value, approximately 10 times more than the DC component, while the output voltage is low-pass filtered with a smooth waveform. The current pulses can have attenuated *di/dt* with inductances placed on the current path, either on the AC or DC side. Various positions are possible for inductances, as shown in Figure 9.34.

Summary

All diode rectifier topologies discussed within this chapter were reviewed in Table 9.2. Diode rectifiers are used less and less today since they have been replaced by *power factor correction* circuits. This is the topic of the next chapter.

References

1. Erickson, R., Maksimovic, D., 2001, *Fundamentals of Power Electronics*, 1st and 2nd editions, Springer, New York.
2. Mohan, N., Undeland, T.M., Robbins, W.P., 2002, *Power Electronics*, Wiley, New York.

10

Power Factor Correction

10.1 Harmonics and Power Factor

Since power delivery is operated on AC waveforms and most electronic loads need DC power, grid interfaces are required. These behave as AC/DC converters. Chapter 9 offers the simplest conversion solution; that is, a diode rectification. Unfortunately, such a solution always produces large current harmonics and distortion of the mains voltage [1,2].

Under the emerging pressure of harmonics standards, novel power converter interfaces are proposed to satisfy the requirements. In order to understand both the standards and the solution, we need first to look into definitions related to power transfer.

10.1.1 Sinusoidal Measures

To simplify the explanation flow, we will first consider power transferred with sinusoidal waveforms. Energy transferred on AC sinusoidal voltage can be defined with several power components:

- *Active power* (usually denoted by P) = Represents the power component which produces the desired effect ('the real work') within the load circuit. Examples = AC machine rotation, lighting, heating and so on.
- *Reactive power* (usually denoted by Q) = Represents the power component required for producing a magnetic field (inductance) or electric field (capacitor) used for good system operation.
- *Apparent power* (total power, S) = Geometrical sum of all active and reactive components.

The relationship between these components is shown in Figure 10.1.

The *power factor* is defined as a ratio between active and apparent power components:

$$\cos\varphi = \frac{P}{S} \tag{10.1}$$

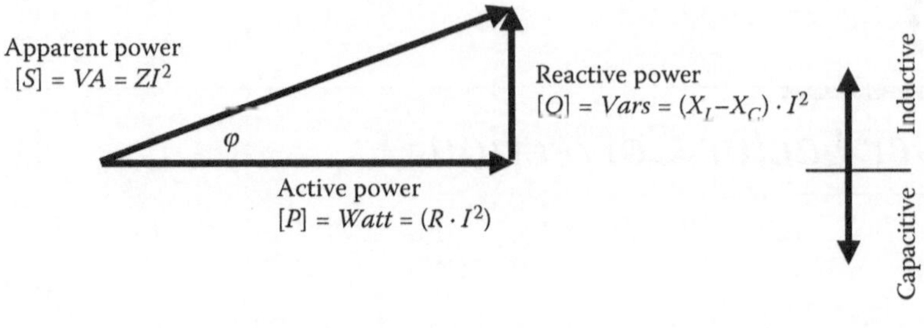

FIGURE 10.1
Relationship between power components.

and it has the meaning of showing the useful effort within the total power through the transmission and distribution system. Ideally, the power factor should equal unity (cos $\varphi = 1$, $\varphi = 0$) for the entire transmitted or processed power to produce a useful work effect.

In practice, the existence of a reactive component within a system (inductance, capacitance) or the operation of power converters through the delivery of reactive power results in a subunity power factor. This leads to additional power loss and deterioration of system efficiency, with additional effects in the decrease of reliability of power equipment.

10.1.2 Nonsinusoidal Waveforms

Switched-mode operation of grid interfaces inevitably introduces harmonics. Hence, most practical cases are characterised by sinusoidal voltage and distorted currents since the voltage corresponds to the AC distribution grid and the current comes from operation of various converters.

Two performance factors are defined to quantify both harmonics and phase shift:

- *Displacement factor* = It is usually denoted by K_φ and quantifies the angle φ between the supply voltage and the fundamental current component.

$$K_\varphi = \cos\varphi \tag{10.2}$$

- *Distortion factor* = It is usually denoted by K_d and defines the quality of the current waveforms.

$$K_d = \frac{I_1[rms]}{I[rms]} \tag{10.3}$$

where $I_1[rms]$ represents the RMS value of the current on the first harmonic (fundamental) and $I[rms]$ represents the RMS value of the entire current (fundamental + harmonics).

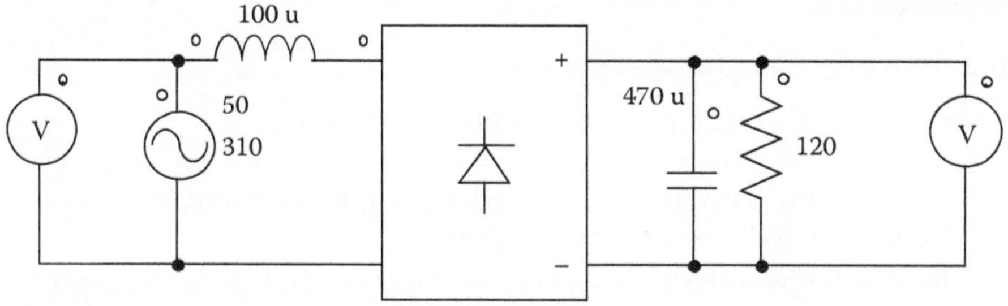

FIGURE 10.2
Grid interface with diode rectifier.

The effect of both factors can be added in order to define the power factor *PF*:

$$PF = \frac{I_1[rms]}{I[rms]} \cdot \cos\varphi = K_d \cdot K_\varphi \Leftrightarrow \begin{cases} K_d = \dfrac{I_1[rms]}{I[rms]} \\ K_\varphi = \cos\varphi \end{cases} \tag{10.4}$$

Most power supplies get power through a diode rectifier, and they do not use any form of power factor correction (Figure 10.2). This solution is mostly used for low power levels and shows some distortion of the grid current (Figure 10.3).

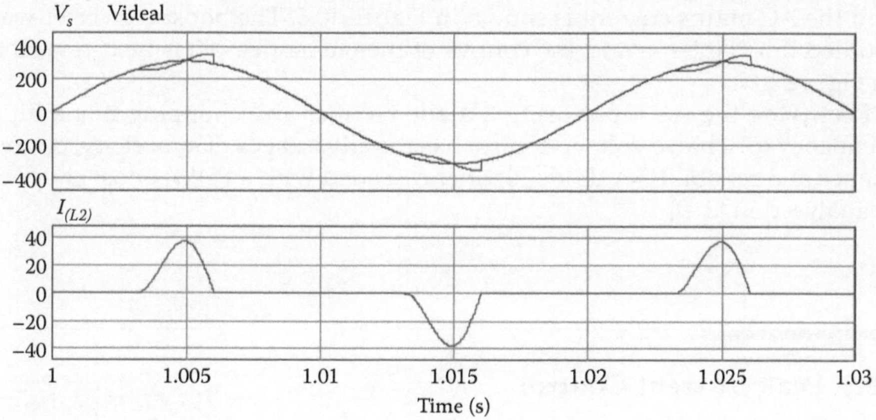

FIGURE 10.3
AC mains current distortion at diode rectifier.

10.2 Power Factor Improvement

Power factor correction (or improvement) circuits aim to

- Limit supply current and eventually bring this current to a waveform closer to the sinusoidal waveform
- Reduce the waveform distortion for the input voltage to the power supply
- Produce a balanced charging of the load capacitor for the entire cycle of the supply voltage and not for a single pulse
- Have a power supply current synchronised with the supply voltage waveform

Achieving these requirements is illustrated in Figure 10.4. This chapter discusses practical means of achieving this goal.

The quality of the current waveform is improved through direct control of the AC mains current. A topology capable of controlling the current on both polarities is required. A double-polarity rectifier (full-wave rectifier) can be used as a front-end interface, for instance, a bridge rectifier equipped with rectifier diodes. The diode rectifier has to be followed up with a switch-mode converter able to actually control the current ripple. The simplest converter using controlled switching is equipped with a single transistor in a buck- or boost-type topology.

For instance, Figure 10.5 shows a grid interface made up of a diode rectifier and a boost converter. The high-frequency switching of the power transistor controls the current ripple within the boost inductor and hence the AC mains current. The relationship between the boost inductor current and the AC mains current is shown in Figure 10.6. The boost converter was studied in Chapter 4, and the control of the current ripple is next revisited in Figure 10.7.

Rectifying the AC input using a diode rectifier and chopping it at a high frequency to achieve voltage control has the advantages of simplicity, performance and reliability. A diode rectifier cascaded with a PWM boost chopper is analysed in [3–9].

10.3 Peak Current Control

In order to explain control of the boost converter as well as control of the AC mains current, let us first consider the positive polarity of the mains current and the control circuit with sinusoidal reference tracking through

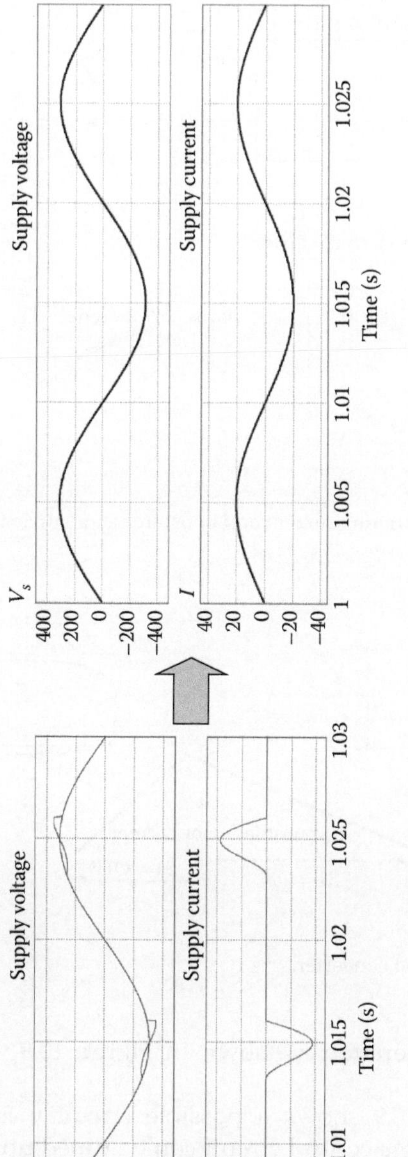

FIGURE 10.4
Transformation of grid-related waveforms from the diode-rectifier case to the power-factor-corrected case.

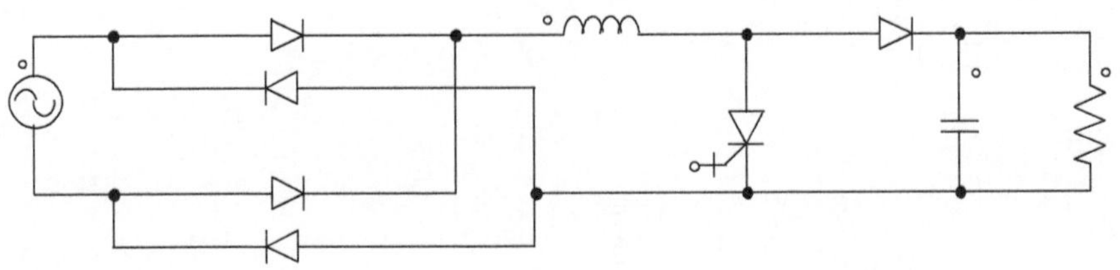

FIGURE 10.5
Power converter for AC mains current control.

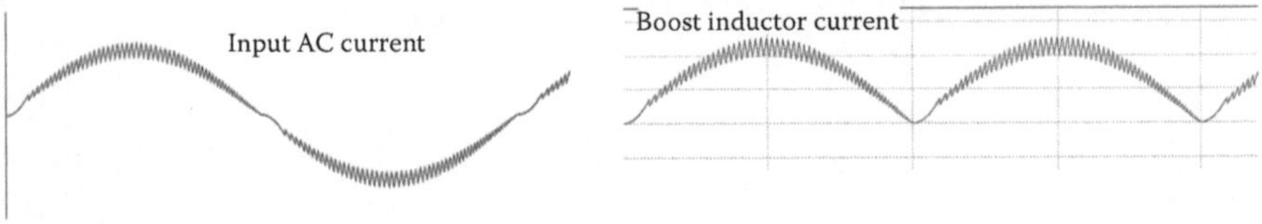

FIGURE 10.6
Relationship between the AC mains current and boost inductor current, studied with the goal of current ripple reduction.

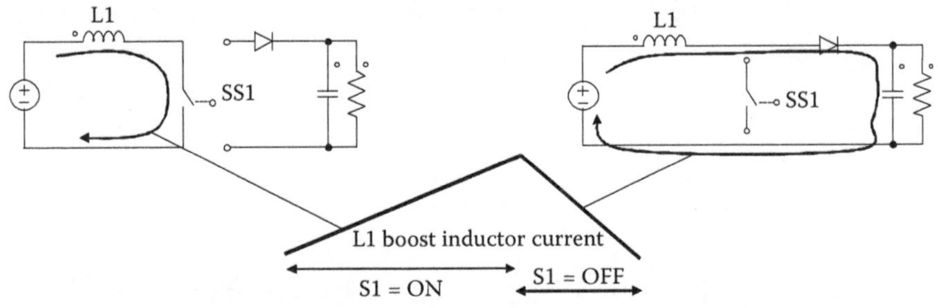

FIGURE 10.7
Current control within a boost converter.

current peaks. This operation is shown in Figure 10.8 with details of the current waveform.

Further on, Figure 10.9 shows a possible circuitry for this method of current control. The same control circuit can be tuned up for a DC/DC converter and then used for a grid interface. This produces a linear increase of the inductor current. The power transistor used as a switch is turned on at fixed time intervals. In the example considered for waveforms, the switching frequency is fixed at 5 kHz for the waveforms shown.

The input current coincides with the boost inductor current, and it is compared to a sinusoidal reference. When the inductor current reaches the reference, the transistor is turned off. This defines a variable conduction interval for the transistor.

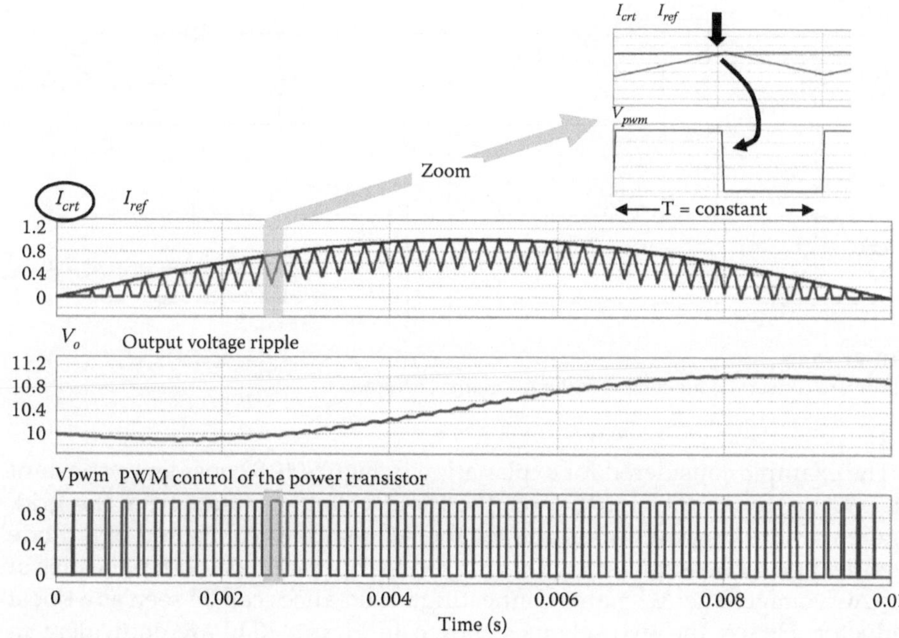

FIGURE 10.8
Details of the peak current control.

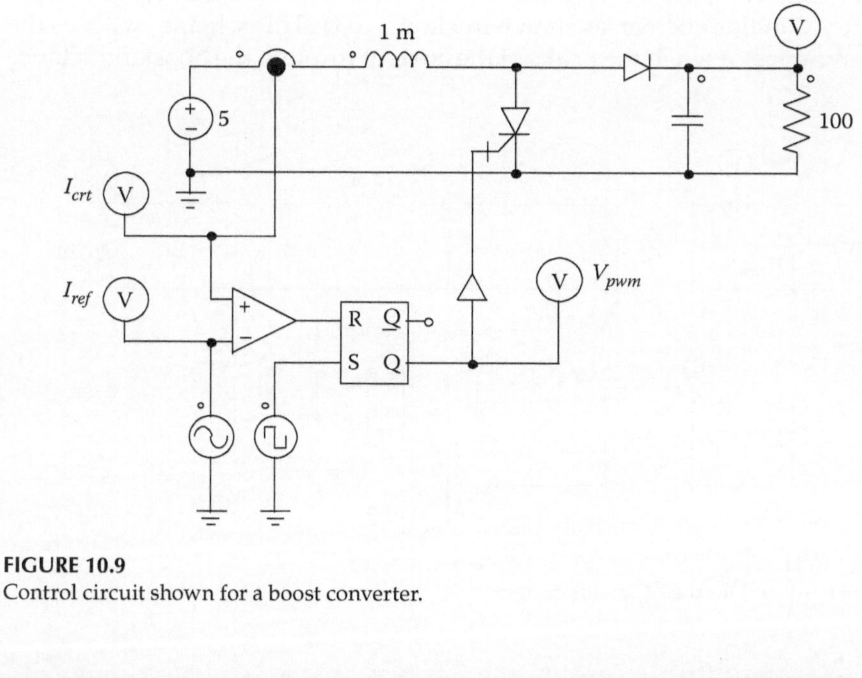

FIGURE 10.9
Control circuit shown for a boost converter.

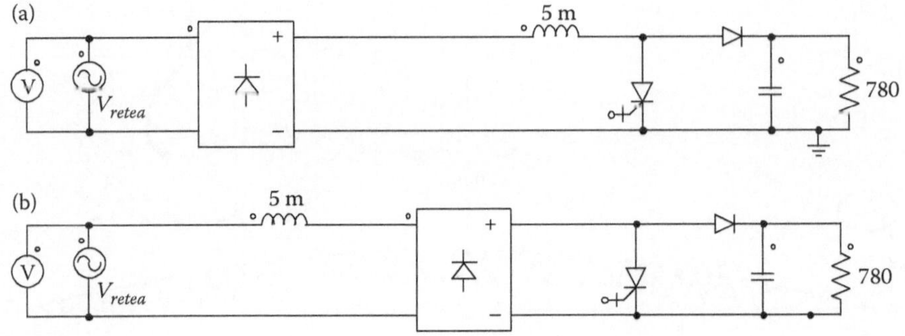

FIGURE 10.10
Hardware converters for control of the AC mains current.

The example considered for explanation in Figure 10.9 considers a constant DC voltage for supply. Further on, the input voltage is taken from the grid. Therefore, the proposed interface for the AC mains is composed of a diode rectifier and a boost converter. This is used for operation on both polarities of the AC voltage. The AC mains connecting inductances can be seen as a boost inductor. Hence, the two schemes shown in Figure 10.10 are equivalent to each other. The converter shown in Figure 10.10a is easier to follow as a boost converter, while the circuit in Figure 10.10b is closer to a physical connection to the AC mains, where the hardware converter is connected through inductances, leaving all semiconductors on the same side.

The control principles developed for a boost converter and illustrated with Figure 10.10 can now be used for the development of the control hardware for the AC mains current, as shown in Figure 10.11. This scheme switches the power transistor when the peaks of the current through the boost inductance

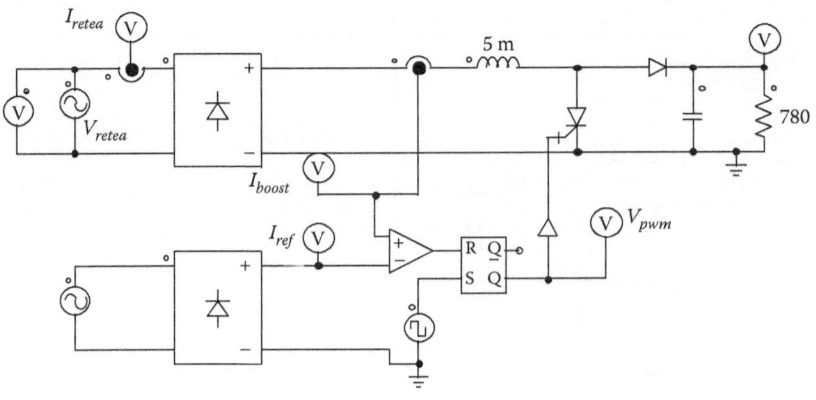

FIGURE 10.11
Control hardware for the AC mains current.

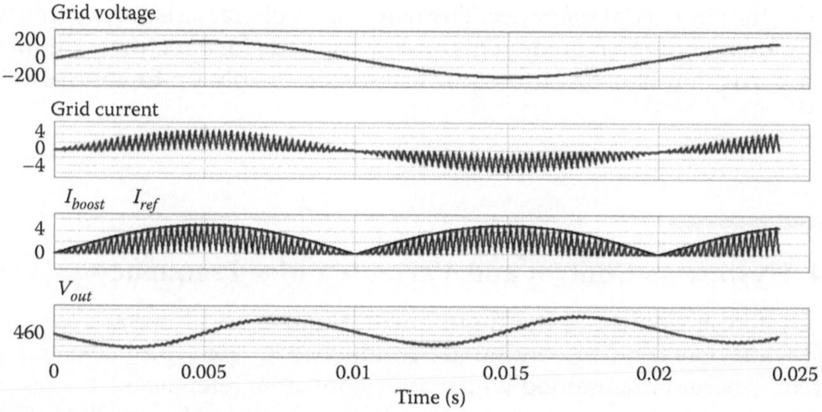

FIGURE 10.12
Waveforms for the control of the AC mains current.

reach the sinusoidal reference. Waveforms characteristic to this operation are shown in Figure 10.12.

It can be observed in Figure 10.12 that the current and voltage waveforms are in phase (synchronised). The harmonic spectrum for the mains current shows in Figure 10.13 a component on the same frequency as the AC mains voltage and another component which corresponds to the switching frequency (herein selected at 5 kHz).

The *peak current control* method (presented herein) is based on monitoring the current and switching the power transistor when the inductor current

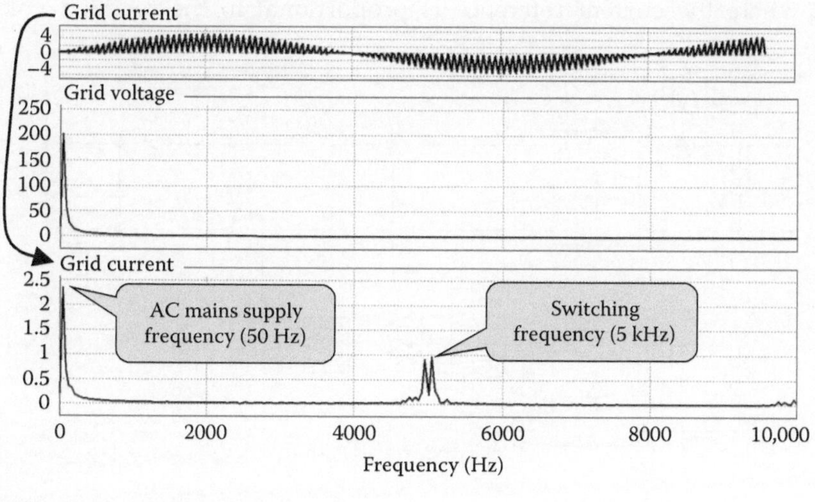

FIGURE 10.13
Spectrum of the AC mains current.

reaches the sinusoidal reference. The waveforms characterise an operation in continuous conduction mode. This operation mode for a rectifier with power factor correction is used at low power levels, possibly up to approximately 300 W.

10.4 Hysteresis Control and Variable Pulse Frequency

The hysteresis control of the input current into a grid interface can be set to provide operation in continuous conduction mode. The boost inductor current is herein maintained within two sinusoidal references.

The control scheme for such an operation is shown in Figure 10.14, and an example for the characteristic waveforms is given in Figure 10.15. Both turn-on and turn-off events result from current comparison to reference values. During an operation, the pulse frequency becomes variable. This spreads the harmonic component corresponding to the PWM operation towards higher frequencies, and a dominant harmonic component does not show up as was the case with the peak current control method. The harmonic spectrum of the AC mains current is shown in Figure 10.16.

10.5 Control with Current Compensation

The control duty cycle D is defined after a closed-loop (feedback) control, where the current reference is proportional to the rectified voltage.

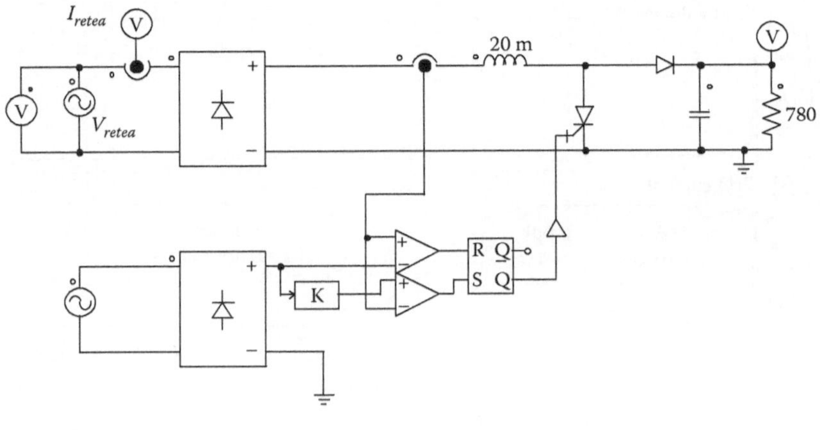

FIGURE 10.14
Control circuitry for a hysteresis controller.

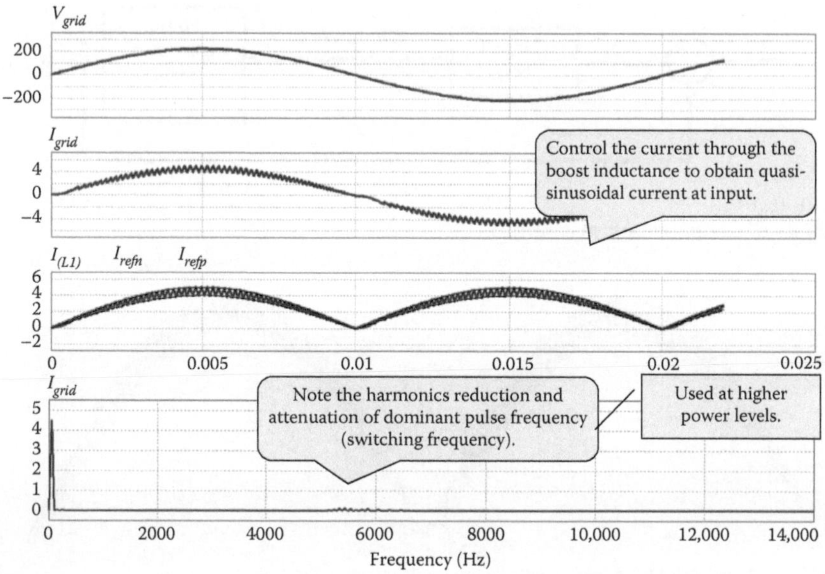

FIGURE 10.15
Waveforms for the boost current.

The principles of feedback control for power supplies were presented in Chapter 6, and they follow the generic scheme shown in Figure 10.17. This method is the most-used choice in practice. The converter operation yields a constant frequency, usually within continuous conduction mode. Results are shown in Figure 10.18 without the actual design of the compensation gains.

A control circuitry able to provide this operation mode is shown in Figure 10.19. The boost inductance current is controlled to obtain quasisinusoidal current at input. The AC mains current pulsates, with minimal ripple and high content in fundamentals, as well as the unity power factor. An example of AC current waveforms is shown with waveforms in Figure 10.20. Harmonic spectra for the AC mains current are shown in Figure 10.21, and this current meets the standards.

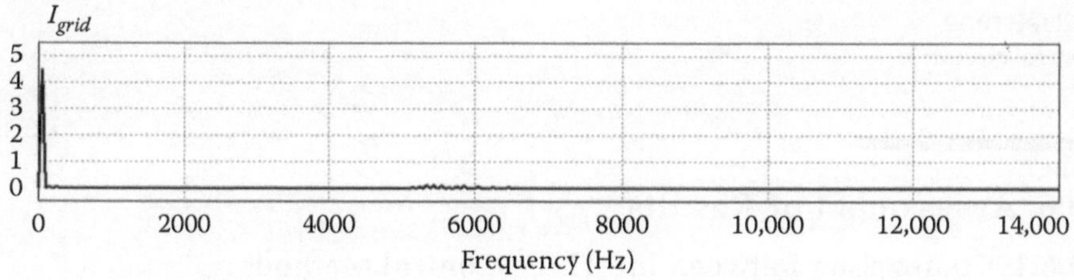

FIGURE 10.16
Harmonic spectrum for the AC mains current.

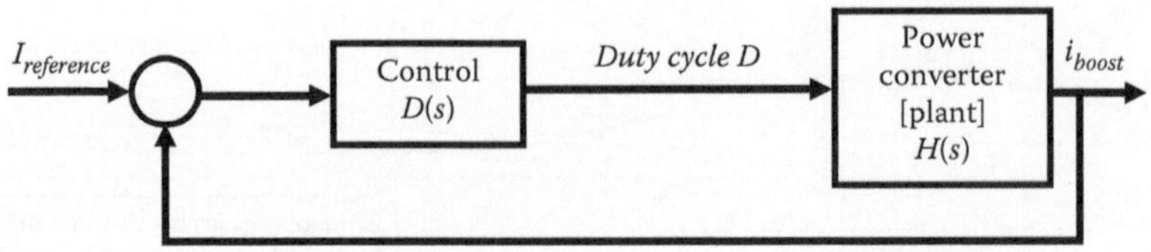

FIGURE 10.17
Principles of feedback current control.

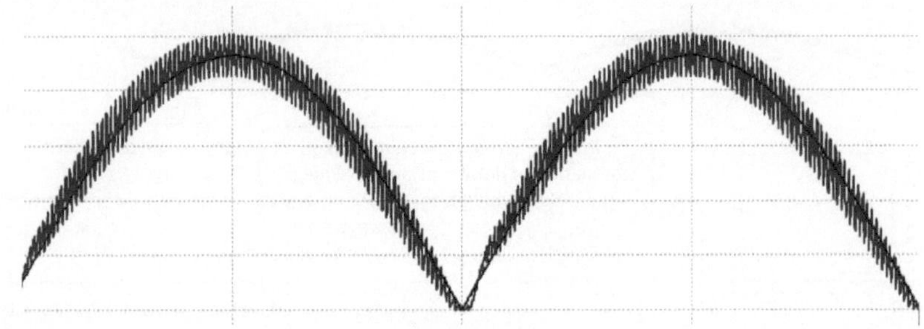

FIGURE 10.18
Generic current waveform for the inductor current.

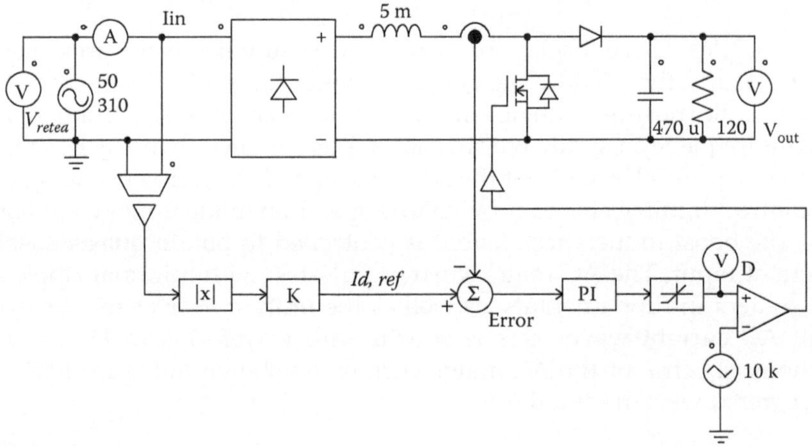

FIGURE 10.19
Control circuitry.

10.6 Assessment of Results

10.6.1 Comparison between the Three Control Methods

Several control methods have been presented for control of boost inductance after a diode rectifier in order to obtain a quasisinusoidal AC mains current.

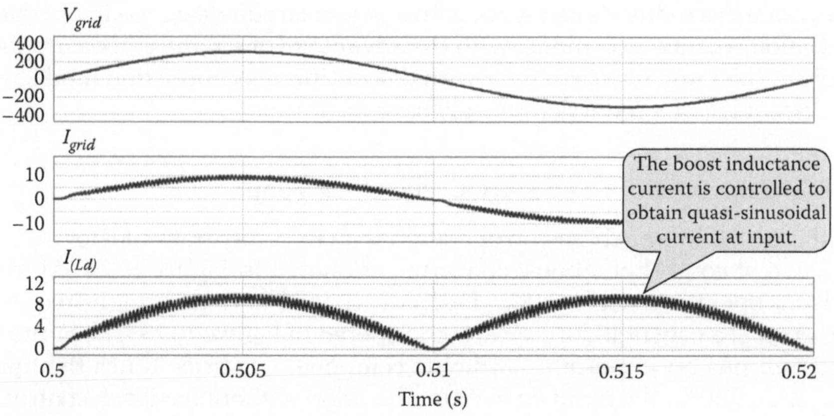

FIGURE 10.20
Waveforms for the AC mains currents.

Peak current control defines a commutation process when the current ripple's peaks reach a sinusoidal reference. *Hysteresis control with variable pulse frequency* provides a current ripple within certain boundaries. Finally, *average current compensation* requires filtering of the measured current before the actual control.

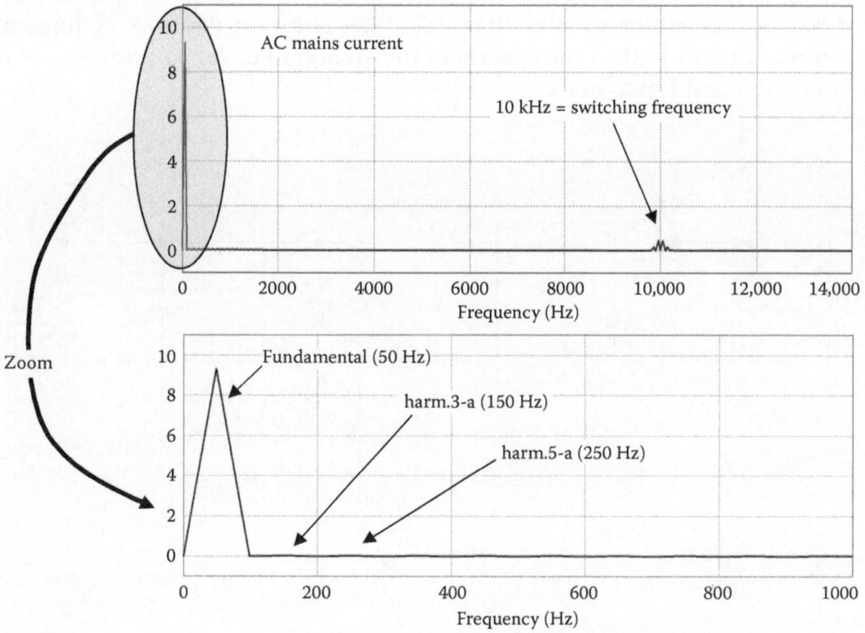

FIGURE 10.21
Harmonic spectrum for the AC mains current.

Any of these methods can work with a power circuit designed to operate in continuous conduction mode, with the advantage of a reduced peak-to-peak ripple of the inductor current, or with discontinuous conduction mode, with the advantage of a reduction in switching power loss.

10.6.2 Control with Fixed Duty Cycle, Open Loop

All these methods are based on operation with the unity power factor through DC current control of a boost converter, as shown in Figure 10.22. For comparison, operation without any feedback control and with high-frequency pulses set at a constant duty cycle is considered in Figure 10.23. The harmonic spectrum has an important harmonic component at three times the mains frequency within the input current in operation without feedback control.

This solution can be better understood when comparing the boost converter operated with a constant duty cycle (Figure 10.24a) with a conventional diode rectifier (Figure 10.24b). The operation of a diode rectifier produces many low-frequency harmonics, while the operation of the switch with a duty cycle produces a displacement of the harmonic spectra towards high frequencies.

10.6.3 Compensation Method with Second Harmonic Injection

Using this approach in high-power applications operated with low switching frequencies yields a grid-side power factor lower than unity and a current harmonics factor greater than 5%. One solution consists of injecting harmonic content within the control of the switch in order to compensate for the main current harmonics.

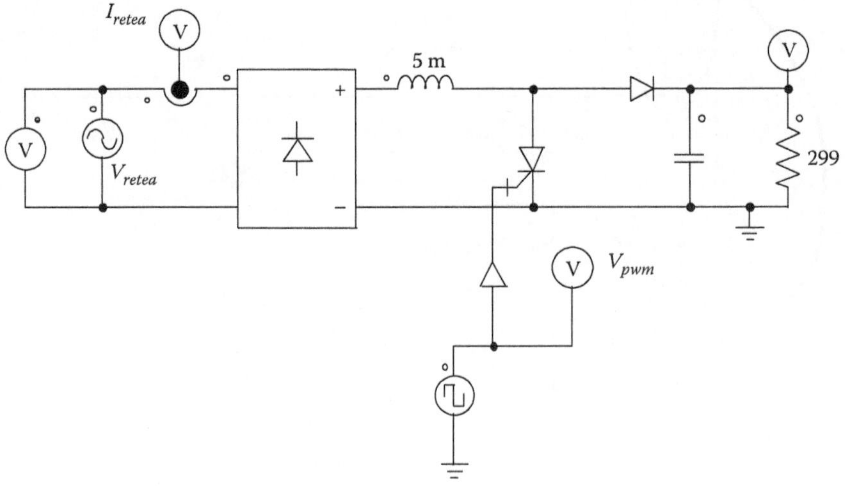

FIGURE 10.22
Power stage with unity power factor.

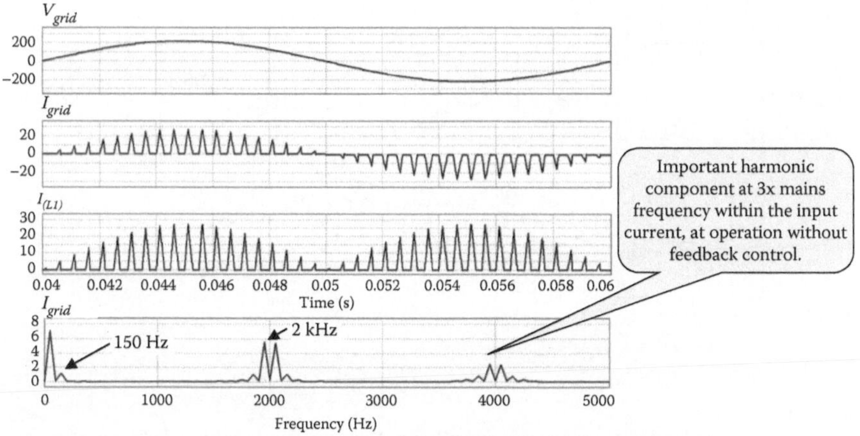

FIGURE 10.23
Waveforms from operation with constant duty cycle.

Conventional control of the boost converter grid interface was explained in the previous section without fully exploring the advantages of PWM. It has been shown in [8,9] by extensive computer analysis that the distortion of the input current for a single-phase converter is reduced when injecting a second harmonic signal within the reference (Figure 10.25).

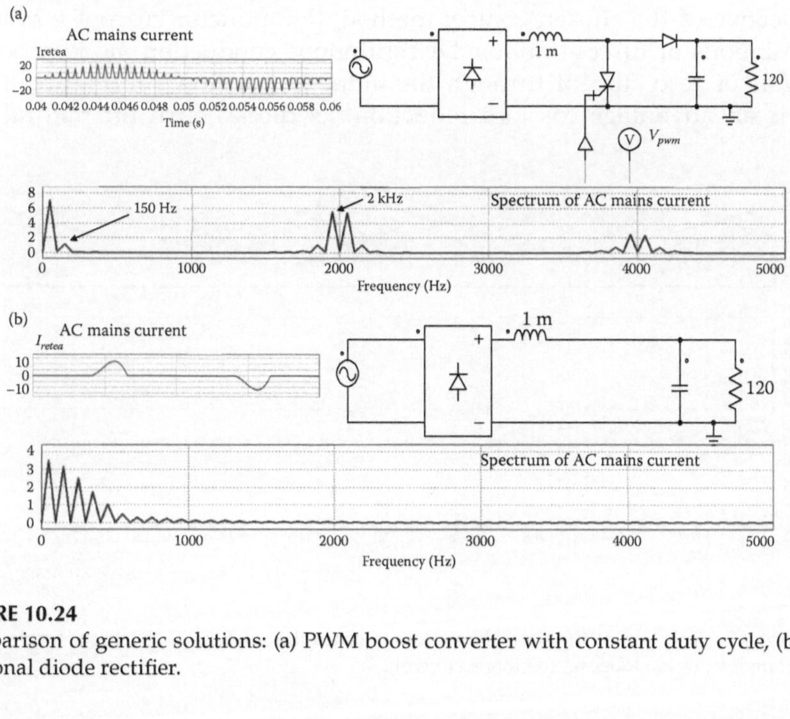

FIGURE 10.24
Comparison of generic solutions: (a) PWM boost converter with constant duty cycle, (b) conventional diode rectifier.

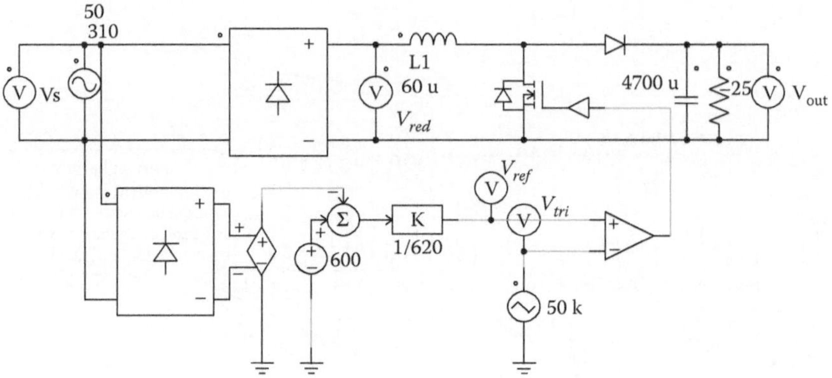

FIGURE 10.25
Circuit for injection of the second harmonic within the reference signal of a single-phase boost converter.

The reference signal for PWM generation is usually a DC value able to define the output voltage level. By injecting a second-order harmonic synchronised with the grid, the inherent variations of the output voltage and input current with the phase of the grid are reduced (Figure 10.26).

10.6.4 Discontinuous Conduction Mode versus Continuous Conduction Mode Power Factor Correction with Boundary Control

Irrespective of the chosen control method, the inductor current waveform is continuous or discontinuous. Discontinuous conduction mode produces intervals of zero current through the inductor, as shown in Figure 10.27a. This is an advantage for loss reduction, as diodes turn off naturally at

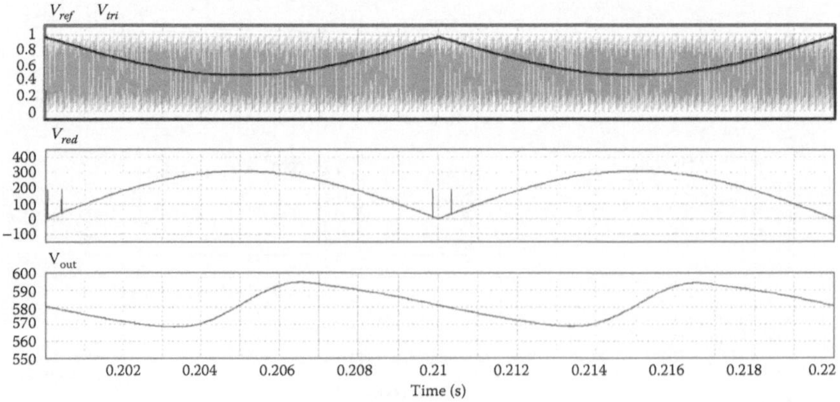

FIGURE 10.26
Waveforms for injection of the second harmonic within the reference signal of a single-phase boost converter (open loop, no feedback control).

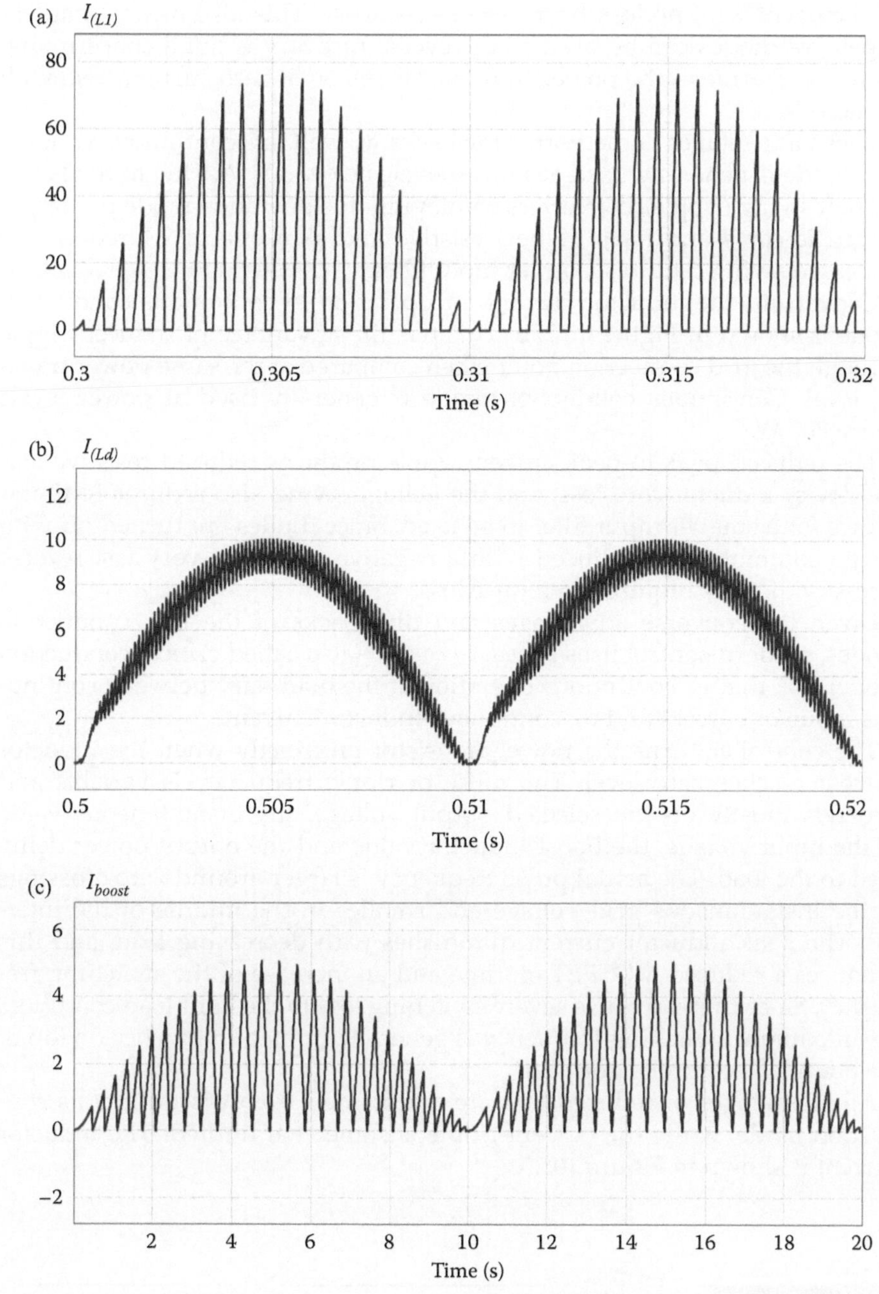

FIGURE 10.27
Inductor current for (a) discontinuous conduction; (b) continuous conduction; (c) boundary control, or critical conduction mode. Numerical values are for waveform demonstration purposes.

zero current with no loss from reverse recovery. This also means that less expensive diodes can be used since reverse recovery is not a concern anymore. Furthermore, the power transistor turns on at zero current, also with reduced loss.

These advantages come with drawbacks as well. Discontinuous conduction mode is generally used at power levels below 300 W. Due to the larger current swing (ripple), converters employing this operation mode use larger magnetic cores and have higher resistive and skin effects. Moreover, the increased swing requires a larger input filter.

Conversely, the inductor current for operation in continuous conduction mode is shown in Figure 10.27b. This has the advantage of reduced ripple through the grid connection point when compared at the same power transfer level. Continuous conduction mode is generally used at power levels above 300 W.

The reduced peak-to-peak current ripple produces reduced resistive loss and lower inductor core loss, and the voltage swing also reduces EMI and allows for a smaller input filter to be used. Since diodes are turned off with forced commutation produced with a negative voltage, a very fast reverse recovery diode is required to reduce loss.

Given the converse advantages and drawbacks of the two conduction modes, modern control uses *boundary control* (also called critical conduction mode) [10]; that is, continuous operation at the boundary between continuous inductor current and discontinuous inductor current.

The controller turns the power transistor on exactly when the inductor current reaches zero level. The pulse or ripple frequency is variable and depends mostly on the selected output voltage, the instantaneous value of the input voltage, the boost inductor value and the output power delivered to the load. The actual pulse frequency is larger around zero crossings of the instantaneous grid voltage and smaller in the middle of the interval. The peak inductor current diminishes with decreasing load, and this produces a reduced MOSFET on-time and an increase of the switching frequency. Since this can cause severe switching loss in the light-load condition, the maximum switching frequency is generally limited in practice on top of the boundary control algorithm.

All advantages of reduced loss are maintained from discontinuous conduction mode, while the current ripple is somewhat limited. The inductor current is shown in Figure 10.27c.

10.7 Three-Phase Circuit for Unity Power Factor

All the previous circuits have been presented for a single-phase conversion system. At higher power levels, three-phase systems are considered. In order

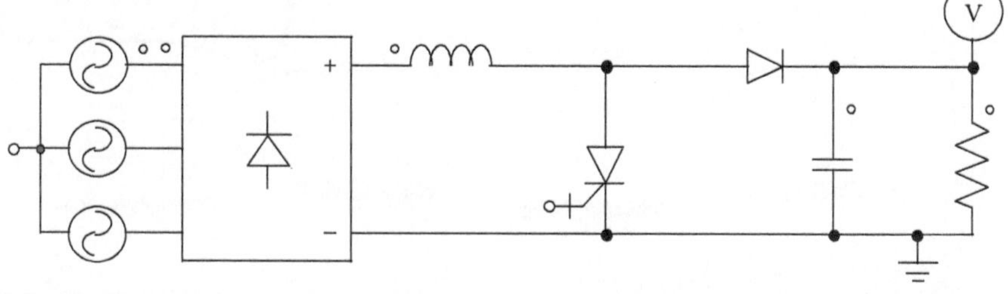

FIGURE 10.28
Three-phase boost converter.

to obtain the unity power factor, a conventional three-phase diode rectifier is followed with a boost converter (Figure 10.28). This is the simplest solution for a three-phase system. The same transistor needs to control all three phases, and a feedback control cannot be used since it is impossible to select the reference current to satisfy the requirements for all three phases. The compromise is to control the transistor with a constant duty cycle. This solution calls for third-order harmonics within the AC mains current at a single-phase rectifier. For a three-phase system, this solution calls for harmonics in all three input currents. Such results would be better than a simple diode rectification without the boost converter with a single transistor.

The distribution of the current pulses through diodes for discontinuous conduction mode are shown in Figure 10.29 for a fixed duty cycle of 0.25. A complete illustration of all waveforms for the three-phase unity power factor converter is shown in Figure 10.30. The harmonic spectrum of the AC mains current is shown in Figure 10.31. Some important components can be on the fifth and seventh harmonics.

Similar to single-phase *power factor control*, the low-frequency harmonics can be reduced with a compensation based on injection of higher-order

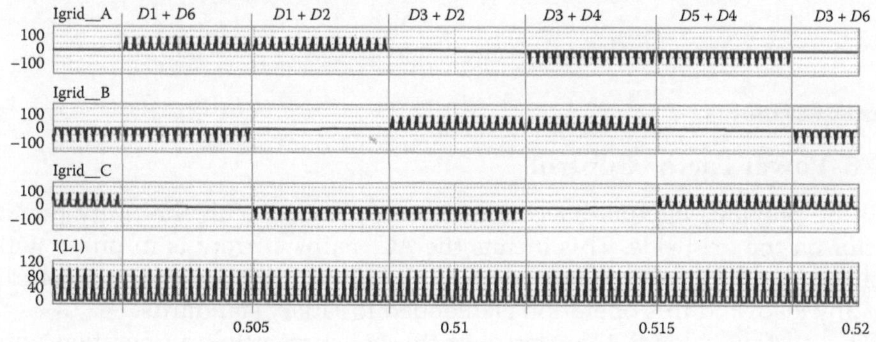

FIGURE 10.29
Pulse distribution through all six diodes.

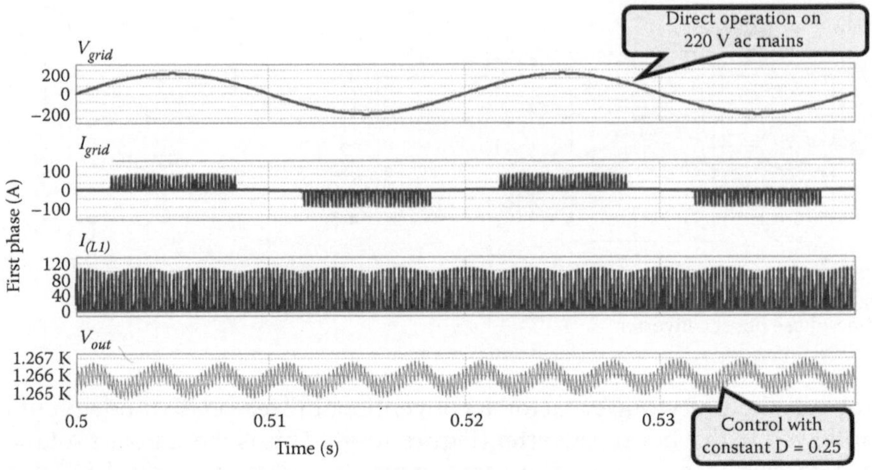

FIGURE 10.30
All significant waveforms for the unity power factor operation of a three-phase converter, operated with constant duty cycle, open loop.

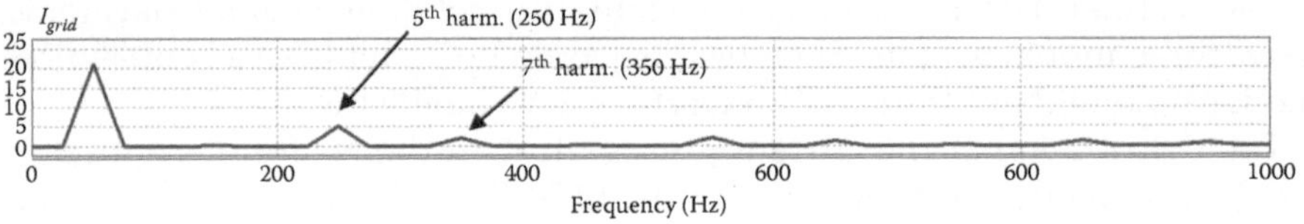

FIGURE 10.31
Harmonic spectrum for the AC mains current when operated with 220 V AC, 50 Hz.

current harmonics. A simple solution proposes the injection of a sixth harmonic in the PWM control of the single switch. Alternatively, a computer-optimized reference is made available in [11].

10.8 Power Factor Control

All the previous circuits have considered operation with the unity power factor on the grid side. This means the AC mains current is in phase with voltage (synchronised). Standards require only a unity power factor control for any load, and this operation is intended to satisfy standards.

This section extends the control of the AC current to an operation with any prescribed power factor. Control is now intended to achieve any power factor in order to correct the eventual decrease of the power factor within

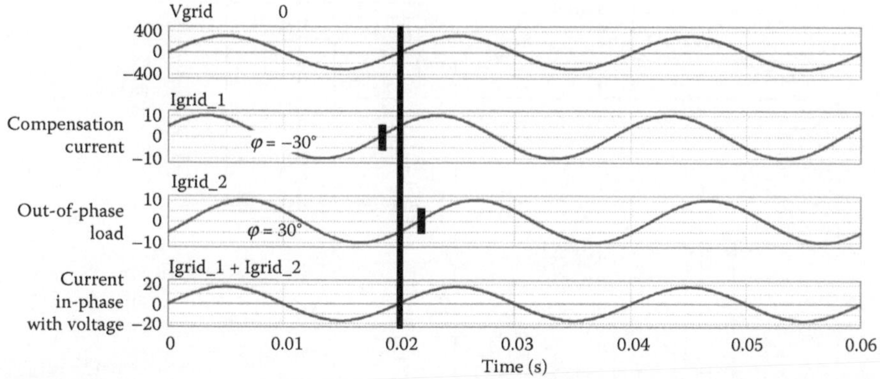

FIGURE 10.32
Operation with a variable power factor.

the connection node due to other loads at the same node. This principle is illustrated in Figure 10.32.

While the principle is easy to understand, it is not that easy to define a circuit able to achieve this operation. For instance, the previous solution involving a diode rectifier and boost converter cannot work here. If the current reference for the boost converter is phase-shifted from the voltage, the AC mains current becomes distorted. For instance, Figure 10.33 considers the previous topology with a diode rectifier and boost converter controlled with a reference at 30° out of phase. This illustrates the distortion of the waveform near zero. The diode rectifier–based topology is not recommended for unity power factor control at levels different from unity where the current is in-phase with the voltage.

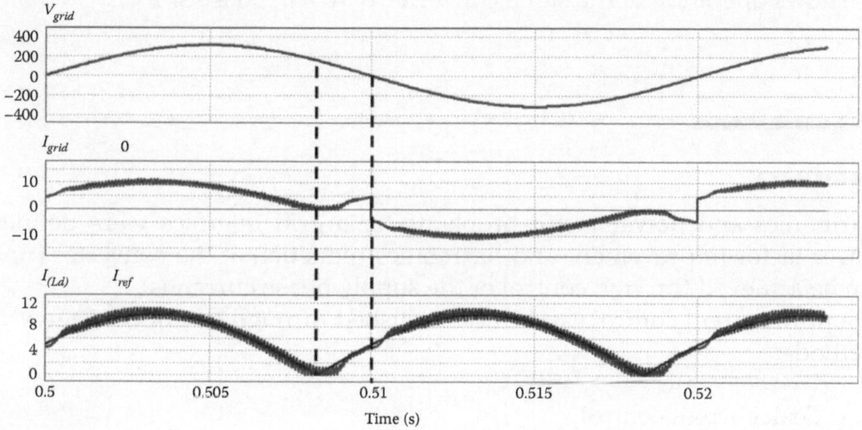

FIGURE 10.33
Distortion created by a current reference phase-shifted from grid voltage.

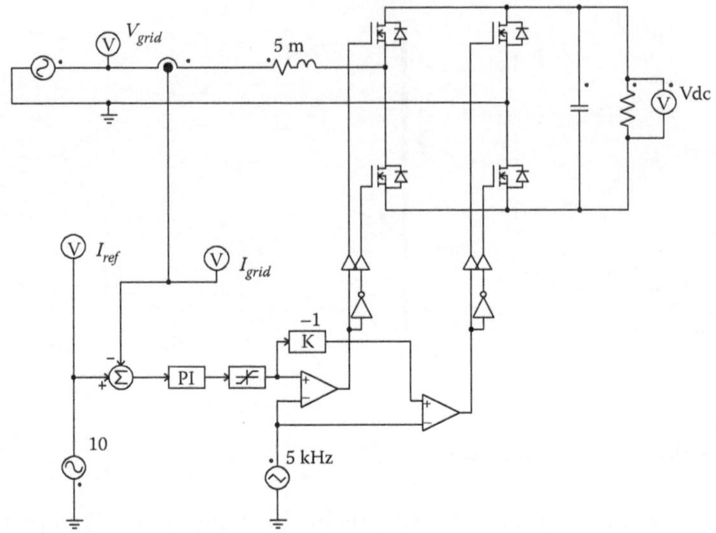

FIGURE 10.34
Full-wave controlled converter and its control circuitry switching at 5 kHz.

If the diode rectifier–boost converter does not work, what is the solution? The previous distortion of the current occurs due to diodes within the rectifier bridge which allow the current to circulate a single way during each polarity of the supply voltage. An alternative considers a topology with operation in four quadrants, which allows for polarities of both mains voltage and current. A full-wave converter built with four power transistors is shown in Figure 10.34. The operation is further illustrated in Figure 10.35, where (a) shows an operation with a 30° phase shift, while (b) shows operation of the same converter with a 0° phase shift.

Summary

Harmonics and power factor problems for a grid interface were defined. Power factor improvement and harmonics reduction at the converter input can be achieved through control of the supply phase current(s).

Supply current control can be accomplished through one of the following methods:

- Peak current control
- Hysteresis control with variable pulse frequency
- PI current control

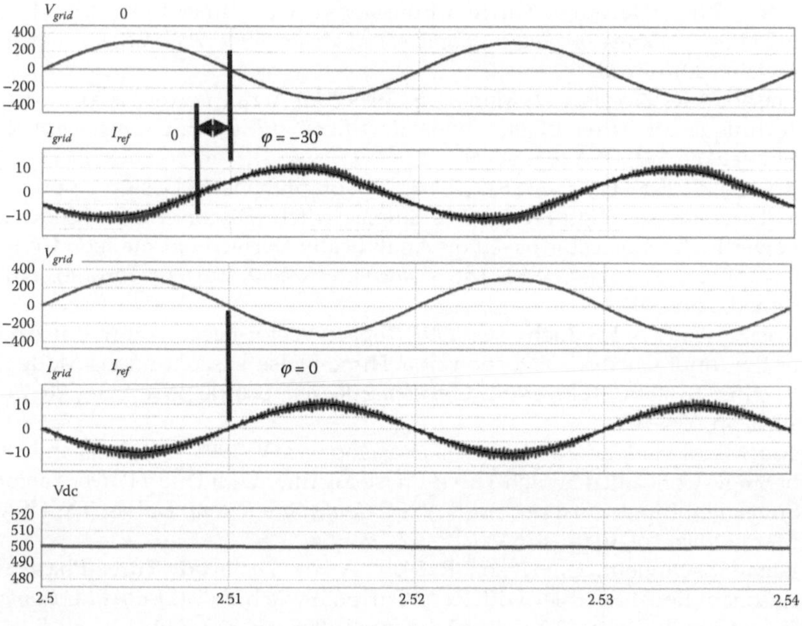

FIGURE 10.35
Waveforms for operation with various phase-shift angles.

For the sake of comparison, results for a constant duty cycle method were studied and shown without current measurement and feedback. An important third harmonic reduction was seen within the supply current.

A three-phase converter built with a diode rectifier and boost converter was analysed. There is no control possible for individual phase currents. Open-loop control with a constant duty cycle was considered, and this leads to the existence of important harmonics of the 5th, 7th, 11th and 13th orders. Some reduction in these particular harmonics can be achieved by the injection of a harmonic current: the second harmonic for single-phase converters and the sixth harmonic for three-phase converters.

The ability to achieve any value of the power factor at the grid connection point; that is, any phase for the input current, was demonstrated. Such a method can be used to compensate for other nonlinear loads connected to the same distribution node through control of the full-wave bridge converter.

References

1. Tse, C.K., 2005, "Circuit Theory and Design of Power Factor Correction Power Supplies", *IEEE Distinguished Lecture on Circuits and Systems*. Also at http://cktse.eie.polyu.edu.hk/Tse-IEEElecture2.pdf, accessed on 3 October 2017.

2. Anon, 2010, "Harmonic Current Emissions – Guidelines to the Standard EN 61000-3-2", *European Power Supplies Manufacturers Association*, Document revision 2010-11-08.
3. Prasad, A.R., Ziogas, P.D., Manias, S., 1989, "An Active Power Factor Correction Technique for Three-Phase Diode Rectifiers", *IEEE PESC Conference Record*, Milwaukee, WI, USA, pp. 58–66.
4. Kolar, J.W., Ertl, H., Zach, F.C., 1993, "A Comprehensive Design Approach for a Three-Phase High-Frequency Single-Switch Discontinuous-Mode Boost Power Factor Corrector Based on Analytically Derived Normalized Converter Component Ratings", *IEEE IAS Conference Record*, part II, Toronto, Ontario, Canada, pp. 931–938.
5. Kolar, J.W., Ertl, H., Zach, F.C., 1993, "Space Vector-Based Analytical Analysis of the Input Current Distortion of a Three-Phase Discontinuous-Mode Boost Rectifier System", *IEEE PESC Conference Record*, Seattle, WA, USA, 20–24 June 1993, pp. 696–703.
6. Tou, M., Al-Haddad, K., Olivier, G., Rajagopalan, V., 1995, "Analysis and Design of Single Controlled Switch Three-Phase Rectifier with Unity Power Factor and Sinusoidal Input Current", *IEEE APEC Conference Record*, Dallas, TX, USA, 5–9 March 1995, vol. II, pp. 856–862.
7. Ismail, E, Olivieira, C, Erikson, R, 1995, "A Low-Distortion Three-Phase Multi-Resonant Boost Rectifier with Zero Current Switching", *IEEE APEC Conference Record*, vol. II, Dallas, TX, USA, 5–9 March 1995, pp. 849–855.
8. Weng, D., Yuvarajan, S., 1995, "Constant-Switching-Frequency AC–DC Converter Using Second-Harmonic-Injected PWM", *IEEE APEC Conference Record*, vol. II, Dallas, TX, USA, 5–9 March 1995, pp. 642–646.
9. Weng, D., Yuvarajan, S., 1995, "AC–DC Converter Using Second-Harmonic-Injected PWM", *IEEE PESC Conference Record*, vol. II, Atlanta, GA, USA, 18–22 June 1995, pp. 1001–1006.
10. Anon, 2010, "Design Consideration for Boundary Conduction Mode Power Factor Correction (PFC) Using FAN7930", *Fairchild Semiconductor AN-8035*, Rev. 1.0.0, May 3.
11. Neacsu, D.O., Yao, Z., Rajagopalan, V., 1996, "Optimal PWM Control of a Single-Switch Three-Phase AC–DC Boost Converter", *IEEE PESC Conference Record*, Baveno, Italy, pp. 521–526.

11

Single-Phase Inverters

11.1 Role of Alternating Current Power Converters

Usually, energy is transmitted and distributed as AC voltages. Moreover, certain load circuits (like AC motors) require AC voltages with a certain magnitude and frequency, either in the steady state or during dynamic transients.

Dedicated power converters called *inverters* are used to achieve AC voltages with variable magnitude and frequency (Figure 11.1). Such a converter can also be called a *DC/AC converter* [1–3]. Even if this class of power converters is not directly used to supply telecom loads, it helps build uninterruptible power supplies, which are used frequently in data and telecom centres to supply loads during power failure or in order to shave power demand peaks.

Since the converter's efficiency is the most important parameter, a square-wave waveform will be generated instead of a sinusoidal waveform, followed by a low-pass filter able to depict the fundamental sinusoidal waveform.

11.2 Transistorised Inverter Leg: Resistive Load

Consider the inverter circuit shown in Figure 11.2. The supply DC voltage is delivered with a midpoint separating two equal voltages (5 V DC each for the numerical example). If two voltage sources are not available, two capacitors are used in order to divide the DC supply voltage into equal voltages.

The two transistors will be controlled alternately, producing voltages of different polarities on the load. The control pulse frequency for the transistors equals the frequency of the load voltage. Theoretical waveforms result, as shown in Figure 11.3. Note the AC square-wave voltage on the load resistance. The magnitude of the voltage waveform is half the DC supply voltage.

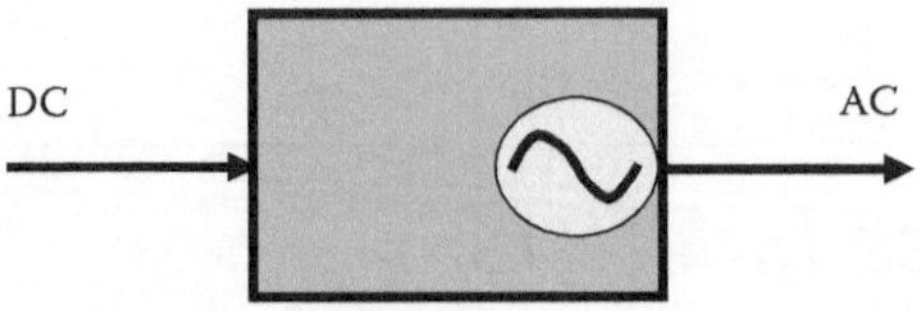

FIGURE 11.1
Principles of DC/AC conversion.

For practical considerations, dead time is used to avoid a short-circuit of the DC supply voltage with the introduction of a time interval between the turn-off of a transistor and the turn-on of the other located on the same inverter leg (Figure 11.4).

11.3 Transistorised Inverter Leg: Resistive-Inductive Load

In most cases, the load also has an inductive component, as shown in Figure 11.5. For instance, the load can be an AC motor, inductive heating, welding and so on. The inductive component in the load acts as a low-pass filter for the load current, which is also delayed from the voltage waveform (Figure 11.6). The inverter operation states are shown in detail in Figure 11.7.

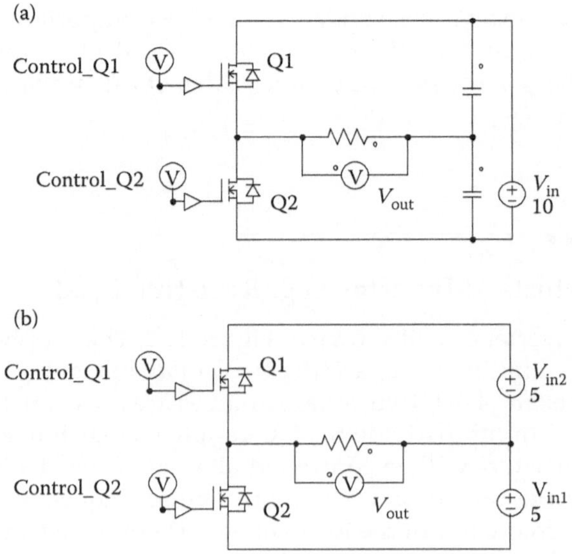

FIGURE 11.2
Inverter leg configurations: (a) midvoltage created with a capacitor divider, (b) usage of two voltage sources.

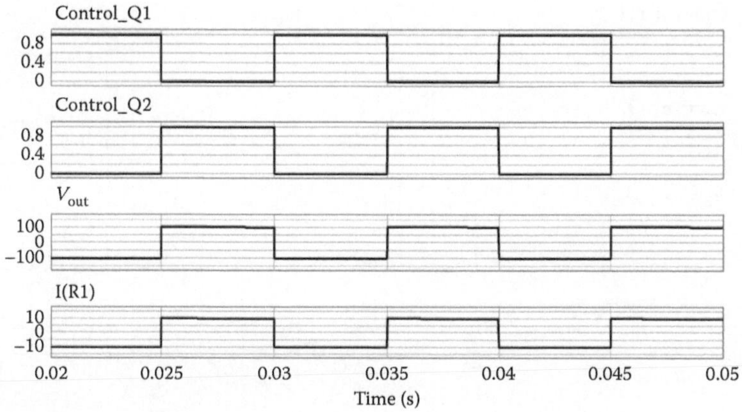

FIGURE 11.3
Waveforms characteristic to the operation of the single-phase inverter.

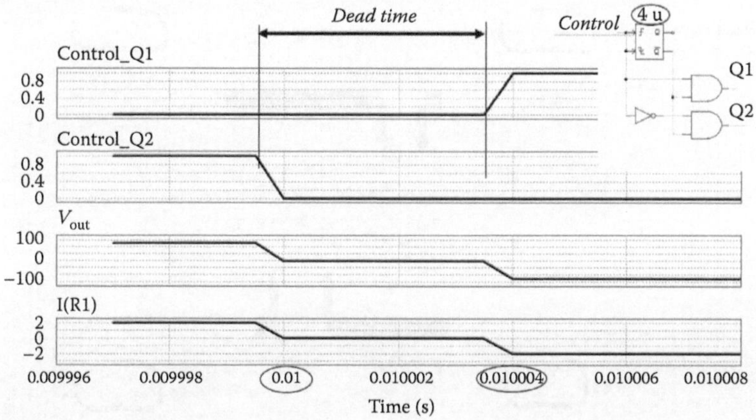

FIGURE 11.4
Definition of dead-time interval with a numerical example for 4 μsec.

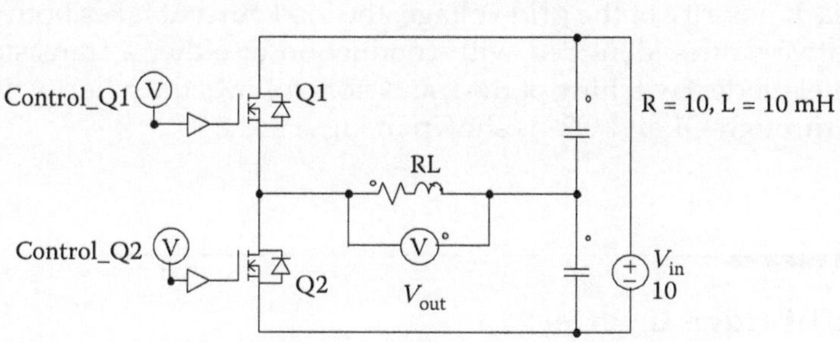

FIGURE 11.5
Inverter with R–L load.

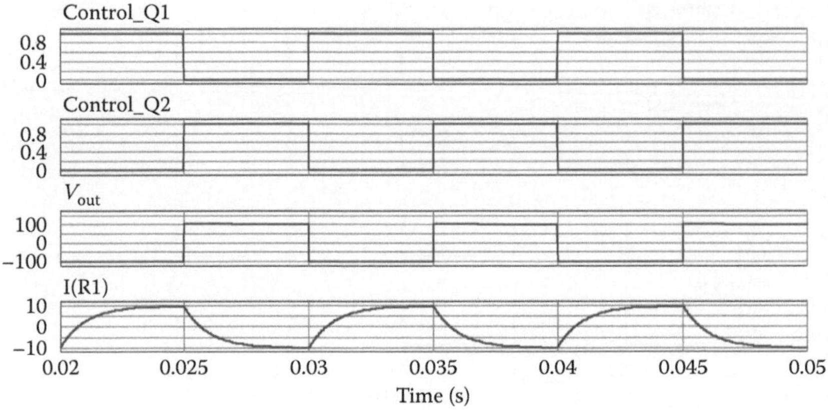

FIGURE 11.6
Waveforms for a resistive–inductive load.

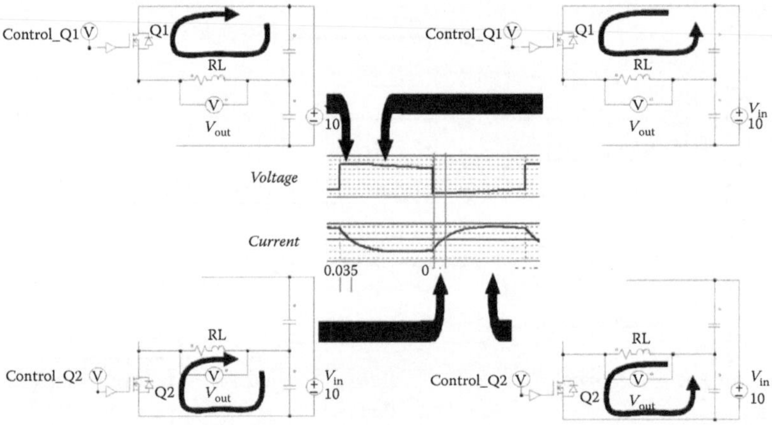

FIGURE 11.7
Inverter states.

During each polarity of the grid voltage, the load current takes both positive and negative values identified with conduction of either a transistor or its antiparallel diode. Switching of the transistors follows the polarity of the AC currents through Q1 and Q2, as shown in Figure 11.8.

11.4 Full-Bridge Inverter

The presented inverter leg can deliver a square-wave AC voltage with a magnitude equal to half the DC supply voltage. The circuit in Figure 11.9 is used

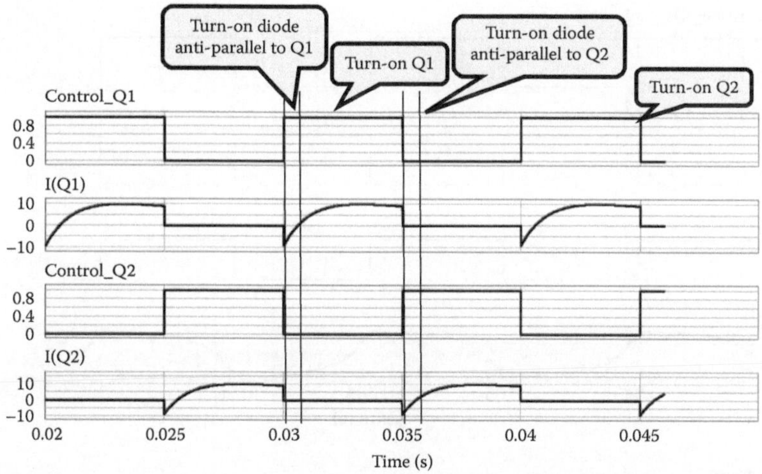

FIGURE 11.8
Waveforms for conduction of either transistors or diodes.

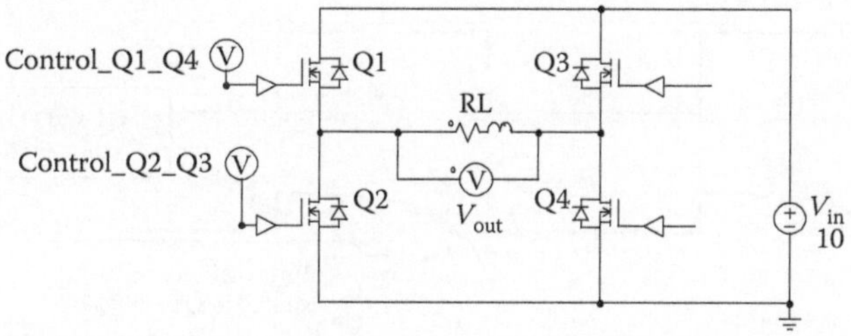

FIGURE 11.9
Full-bridge inverter.

to deliver a greater voltage on the load. This converter is also called a *full-bridge inverter.* Transistors on a diagonal of the full-bridge inverter are controlled simultaneously, producing a polarity of the load voltage, followed by control of the other two transistors to deliver the second polarity of the load voltage. The control frequency becomes equal to the frequency of the load voltage.

Waveforms for a resistive-inductive load are as shown in Figure 11.10. The inductance component of the load determines short conduction intervals for diodes connected antiparallel to transistors. This is analogous to the previous discussion of the analysis of operation for an inverter leg. There are two inverter legs operated with reversed control (180° *out of phase*), and the load is connected between the two inverter legs instead of at the capacitors' midpoint.

The commutation processes through transistors are described in Figure 11.11, where polarity of currents through Q1, Q2, Q3 and, respectively, Q4 is considered.

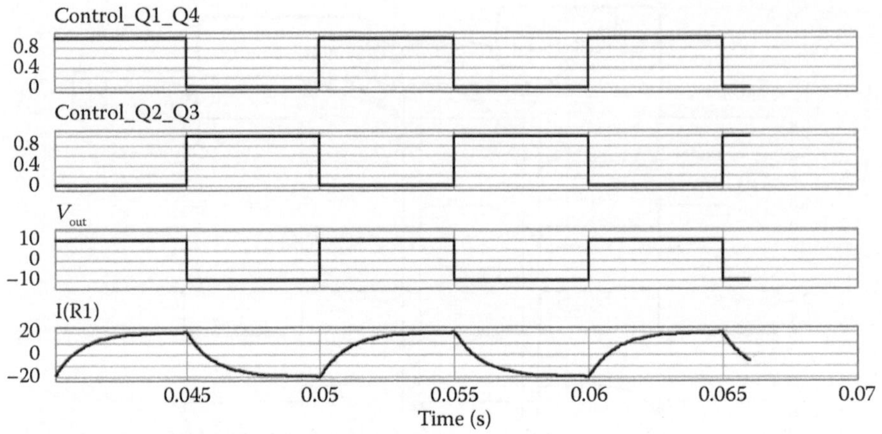

FIGURE 11.10
Operation waveforms for a full-wave inverter.

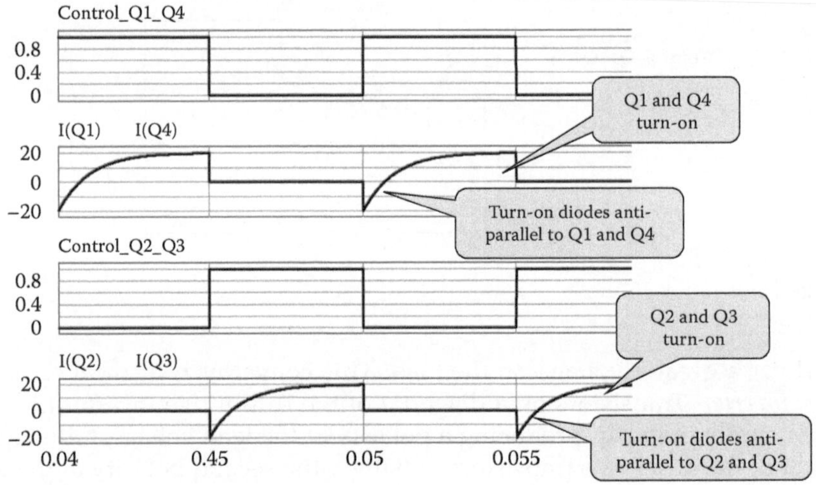

FIGURE 11.11
Details of commutation processes.

11.5 Pulse-Width Modulation Control of an Inverter Leg

11.5.1 Principle

The two solutions presented for an inverter generate a square waveform for the load voltage with a high harmonic content (see Figure 11.12). Extraction of a sinusoidal waveform with a low-pass filter (L-C) is difficult to realise. The undesired harmonics can produce power loss and undesired effects,

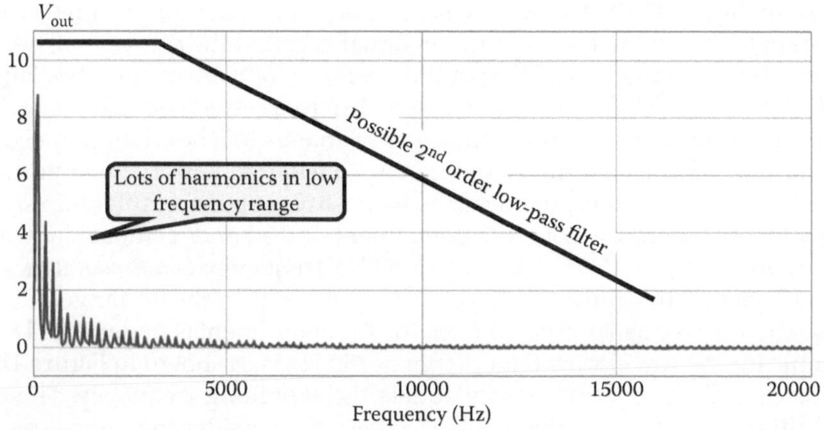

FIGURE 11.12
Harmonics for a square-wave operation of the inverter.

like vibrations or oscillations. For this reason, most modern applications use a modulation technique for control of inverters. This is called pulse-width modulation [1,4].

In order to properly introduce the PWM operation of an inverter, let us first reconsider the operation of the buck converter (Figure 11.13).

The load voltage is obtained after low-pass filtering of the train of power pulses produced by converter operation. The average value of the pulses represents the DC value on the load side. The load current comes from the load voltage and load resistance. The average value of the current through the buck inductance equals the DC component of the current through the load. The integral of the load inductance current has to be zero in steady-state mode. The control circuitry able to produce control pulses for the MOSFET transistor is used further to form the load voltage. Although there are a multitude of solutions possible for control circuitry, the most common solution is

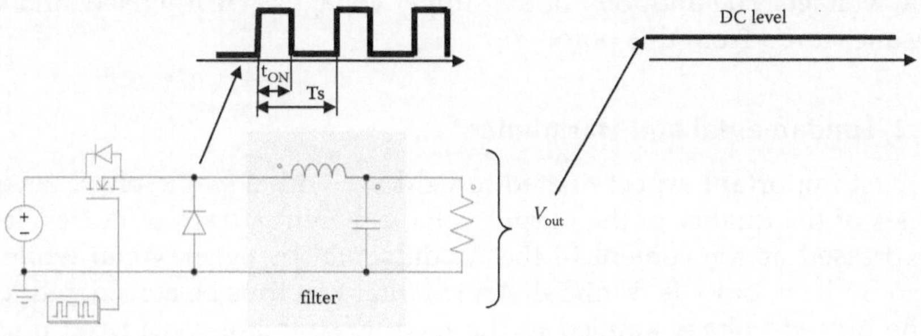

FIGURE 11.13
Buck converter circuit, waveforms and operation.

shown in Figure 11.13. A reference is compared to a triangular signal with a kHz-range frequency. This triangular signal is called the *carrier.*

For a thorough understanding of the operation of control circuitry, Figure 11.14 shows results for operation with different reference signals, while Figure 11.15 shows operation with different shapes for the triangular signal.

A variation of the reference signal can be considered within the same hardware. In order for the output voltage to be positive, a positive reference signal is considered, composed of a DC component of 0.50 and a sinusoidal waveform with a magnitude of 0.25 at 50 or 60 Hz frequency (*sinusoidal waveform with DC offset*). The entire reference signal stays within the 0:1 range.

Usually, the frequency desired from the AC component is under 100 Hz. An example for the waveforms for a carrier at 750 Hz is as shown in Figure 11.16. In this case, the carrier frequency equals the switching frequency. The output voltage is as shown in Figure 11.17, where we consider the same arrangement with 10 kHz carrier frequency for the PWM operation and an output low-pass filter.

The operation of the previous circuit provides an AC component superimposed on a DC voltage. This yields a positive current through a resistive load. This is not always enough, as many applications operate with both polarities of the load current. For this reason, the circuit is redrawn to provide operation with both polarities for the output current while considering a reversed topology able to deliver a negative current. The operation for the positive polarity of the load current is shown in Figure 11.18a, while the negative polarity is shown with Figure 11.18b.

Merging both circuits from Figure 11.18a and b yields the circuit from Figure 11.19. This circuit is similar to the one from Figure 11.2. The voltage is measured with respect to the midpoint of the supply voltage in order to define the AC voltage. The L-C components work as filters of the output voltage. The load voltage still has a DC component.

The complete schematic for a single-phase inverter is shown in Figure 11.20, where the control scheme starts off with a modulating waveform. A *modulation index* is defined as the magnitude of this sinusoidal reference. The operation can provide an output voltage at a maximum of half the DC supply voltage. The midpoint of the supply voltage is clearly fixed, and voltage is measured from this point on.

11.5.2 Fundamental and Harmonics

The most important aspect related to working with a single-phase inverter consists of the quality of the output voltage, herein a train of pulses. It can be expressed as the content in the fundamental being maximal while the content in harmonics is minimal. An LC filter can thus be considered to filter the pulsed voltage applied to the load. Results are considered both in time and frequency and are illustrated in Figure 11.21. The harmonic spectra clearly show a strong fundamental component at 50 Hz and a dominant

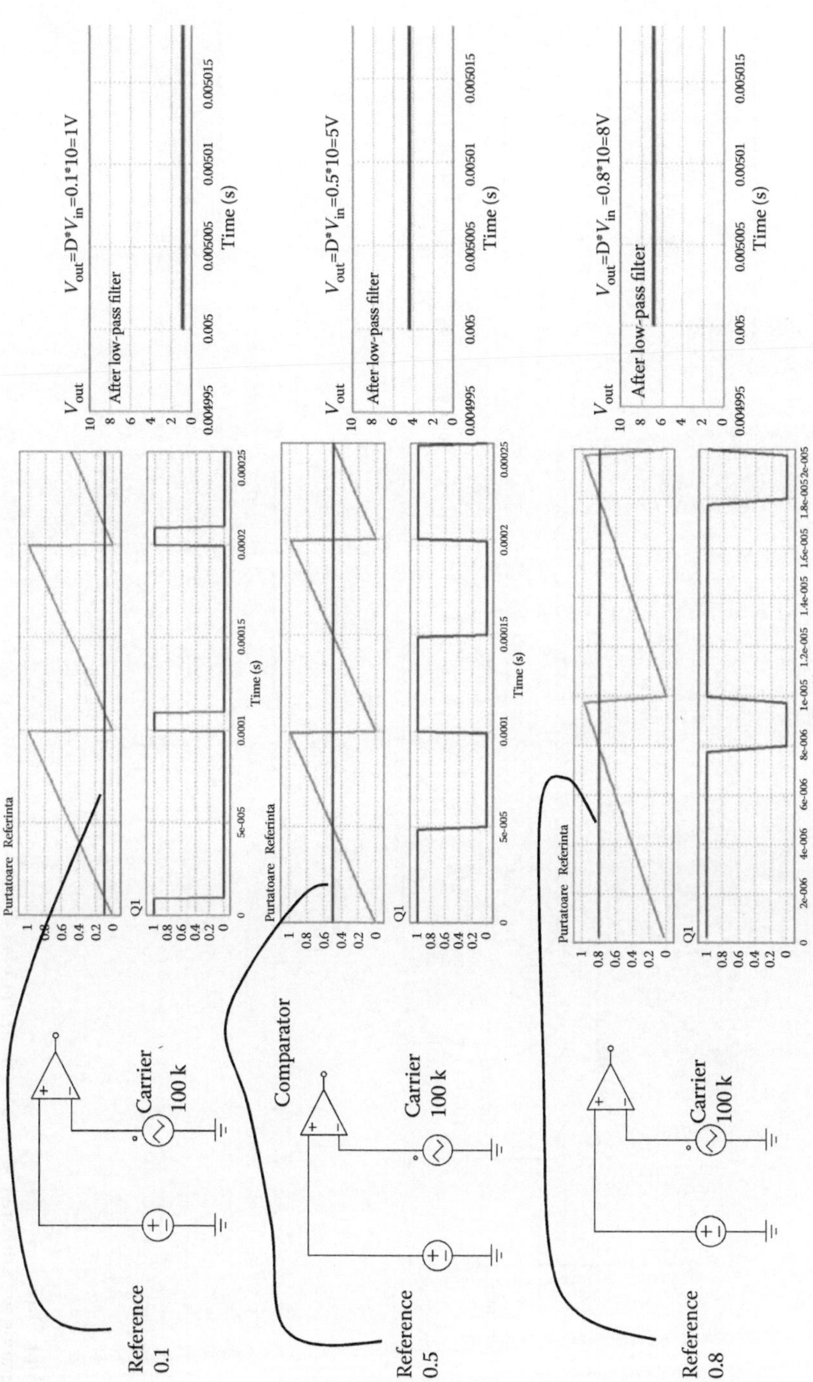

FIGURE 11.14
Operation with different reference signals.

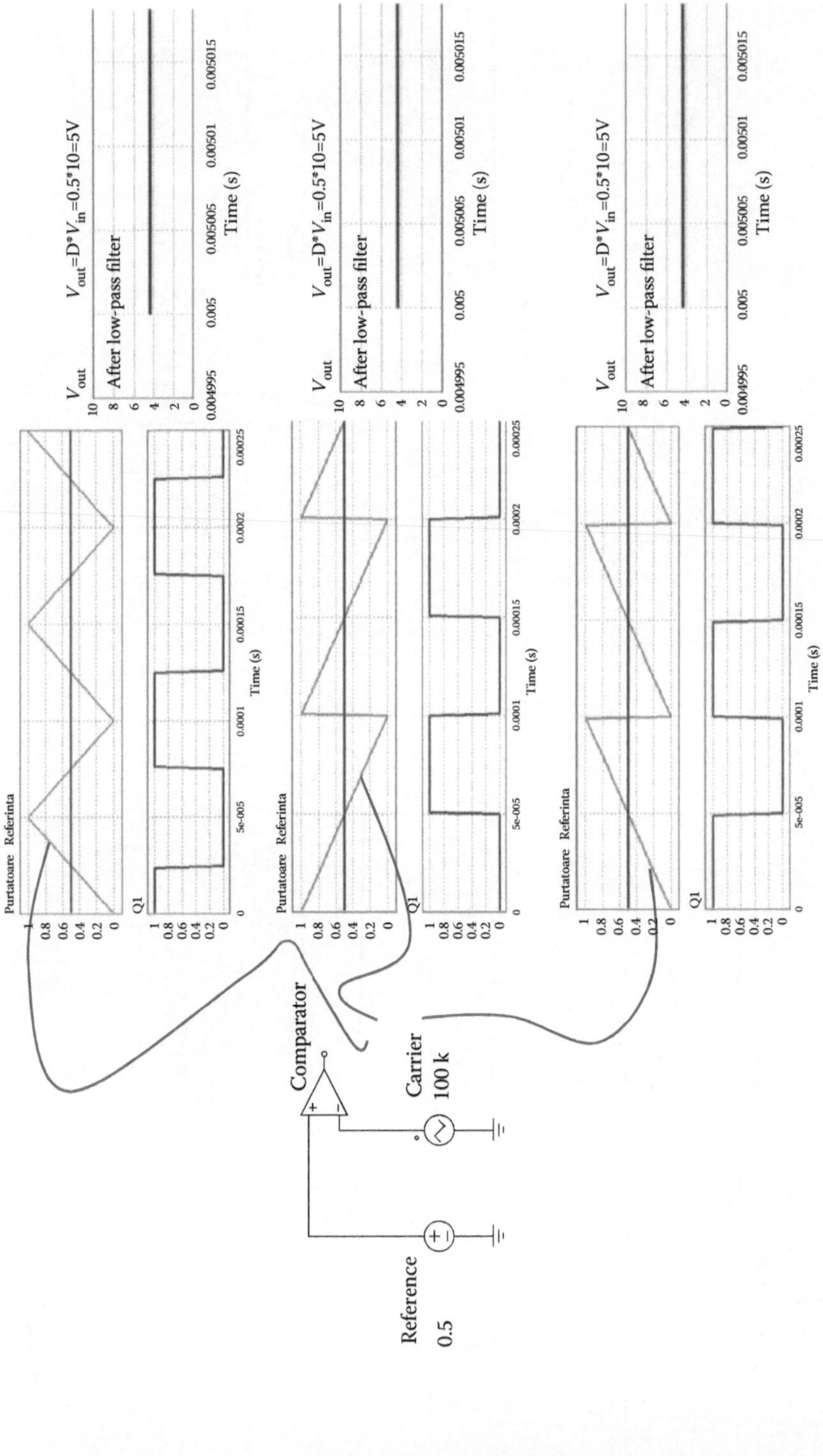

FIGURE 11.15
Operation with different shapes for the triangular signal.

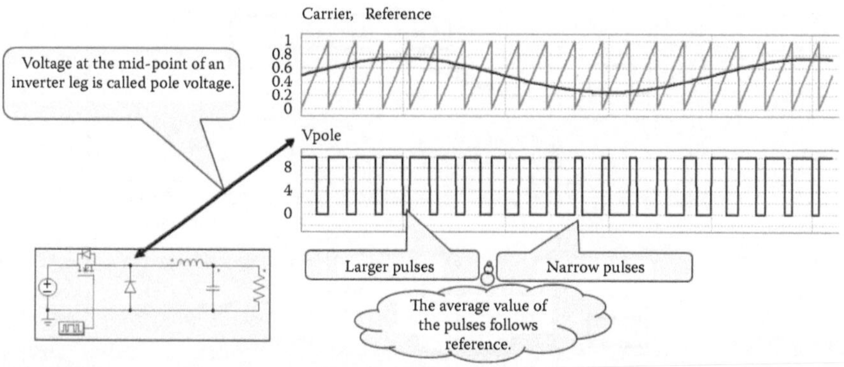

FIGURE 11.16
Introducing an alternative reference signal.

harmonic component at 10 kHz that corresponds to the carrier frequency from the PWM algorithm. This fast Fourier transform (FFT) calculation of the harmonic spectra also shows components at integer multiples of the carrier frequency.

Figure 11.22 illustrates the benefits of using a PWM method. The harmonic content for the output voltage generated by a square-wave inverter is shown in Figure 11.22a. There are many harmonics in the low-frequency range, close to the fundamental frequency. These harmonics are nearly impossible to filter with a conventional low-pass LC filter. An LC filter provides a 40 dB/decade slope. The requirements for the harmonics are to reduce the most important harmonics by 60–80 dB; that is, by 1000 times from the level of the

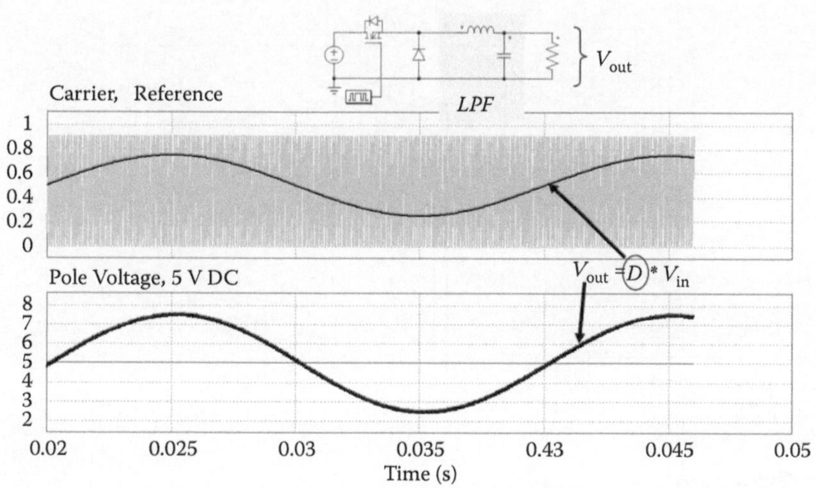

FIGURE 11.17
Achieving an AC output voltage with a low-pass filter.

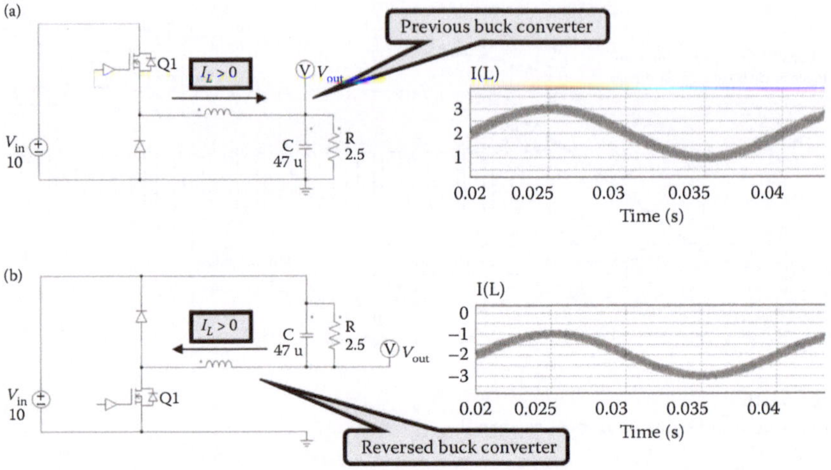

FIGURE 11.18
Circuits for positive and negative polarities of output voltages.

fundamental component. When operating with grid voltages at 110–220 V AC, the voltage ripple needs to be reduced in the range of 0.1 V while the fundamental stays unaltered. There is not enough room to fit such a filter between the fundamental component and the low harmonics generated in square-wave operation.

This situation changes with a PWM inverter, where harmonics move towards high frequency.

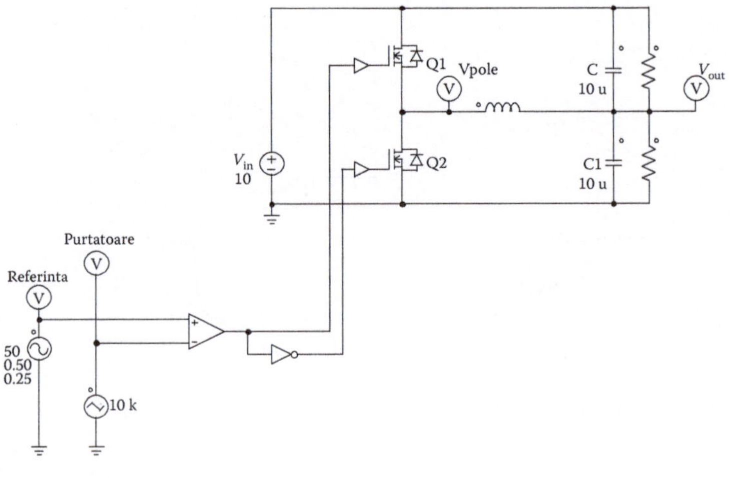

FIGURE 11.19
Circuitry for both polarities in output voltage.

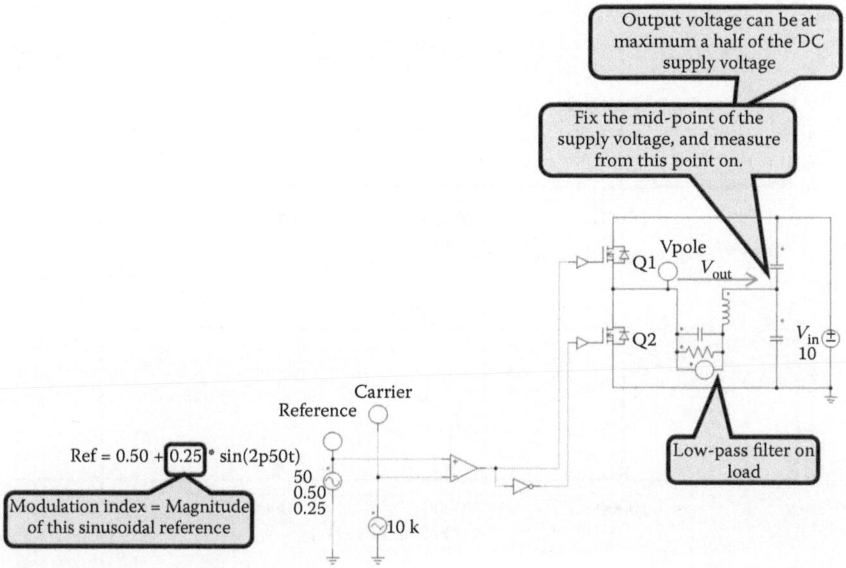

FIGURE 11.20
Control scheme for a single-phase inverter.

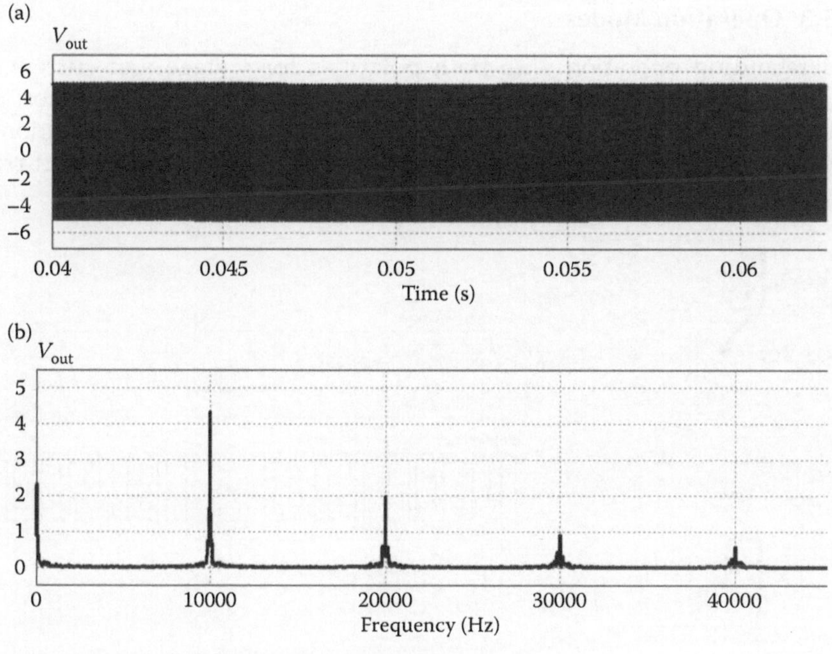

FIGURE 11.21
Waveform and spectra for the operation of the single-phase inverter.

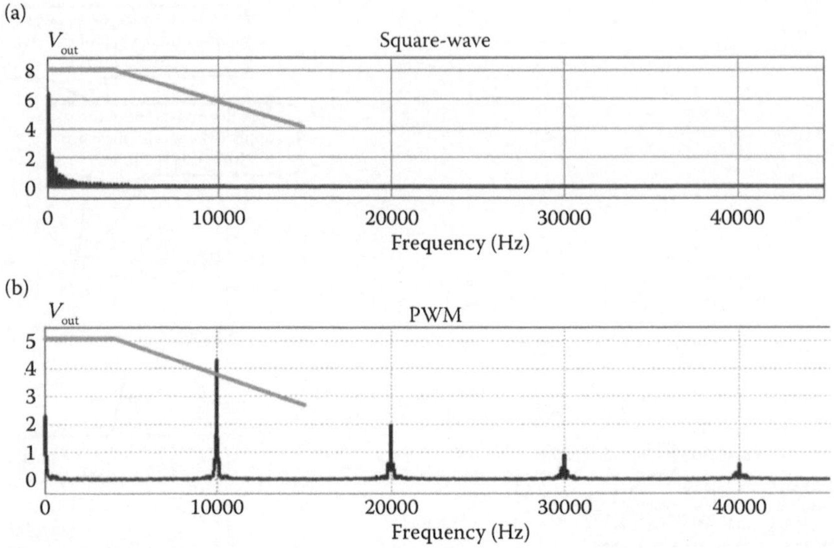

FIGURE 11.22
Output voltage harmonic comparison between a square-wave operation (a) and a PWM operation (b).

11.5.3 Operation Modes

Understanding operation with both polarities for voltage and current on the load side requires a careful analysis of device conduction states; see Figure 11.23. All four combinations compose a four-quadrant operation of the power converter. During each polarity of the load current, the current

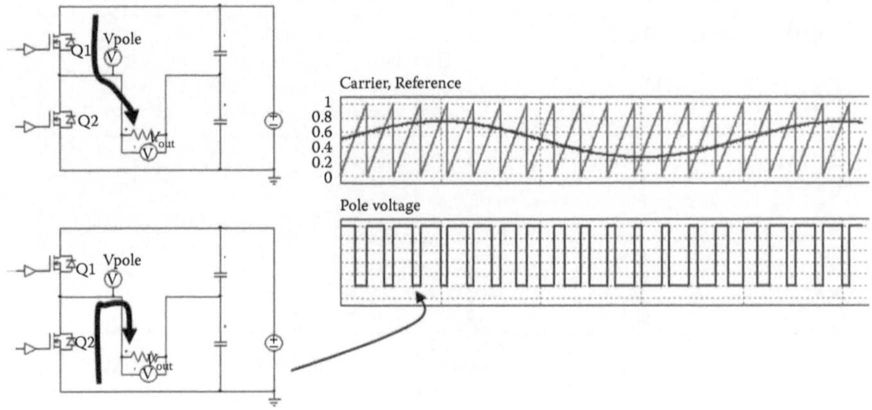

FIGURE 11.23
Two conduction states across a carrier period for the same polarity of the load current.

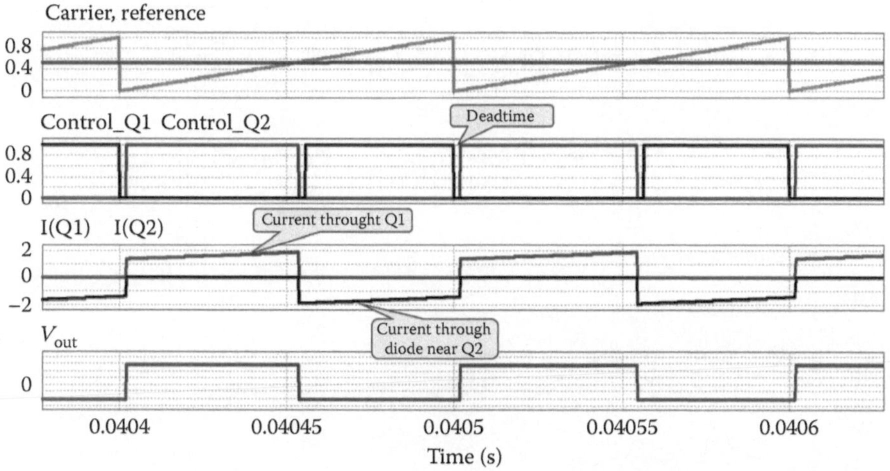

FIGURE 11.24
Details of operation.

circulation switches between a transistor and a diode connected antiparallel with the other transistor. Since the period of the carrier is very short, the same polarity of the current can be considered during a switching period. This is shown in Figure 11.24.

11.6 Full-Bridge Inverter with Pulse-Width Modulation

11.6.1 Control Circuit

The previous converter can provide sinusoidal voltages with a magnitude of a maximum of half the whole DC supply voltage. In many applications, a full-bridge inverter is preferred for delivering a sinusoidal voltage with a maximum magnitude equal to the DC supply voltage; that is, twice that of a single inverter leg. The PWM operation of a full-bridge inverter is achieved with a control scheme, as shown in Figure 11.25. The waveforms for the gate control signals and output voltage are similar to Figure 11.26, where a low carrier frequency of 750 Hz has been chosen for illustration purposes. Variation of the reference magnitude in Figure 11.25 determines variation of the fundamental of the output voltage. Herein, the modulation index equals the magnitude of this reference $m \in (0,1)$.

Switching processes through transistors can be observed, along with the polarity of the currents through the transistors Q1, Q2, Q3 and Q4. Switching details are illustrated in Figure 11.27 for a carrier frequency of 10 kHz.

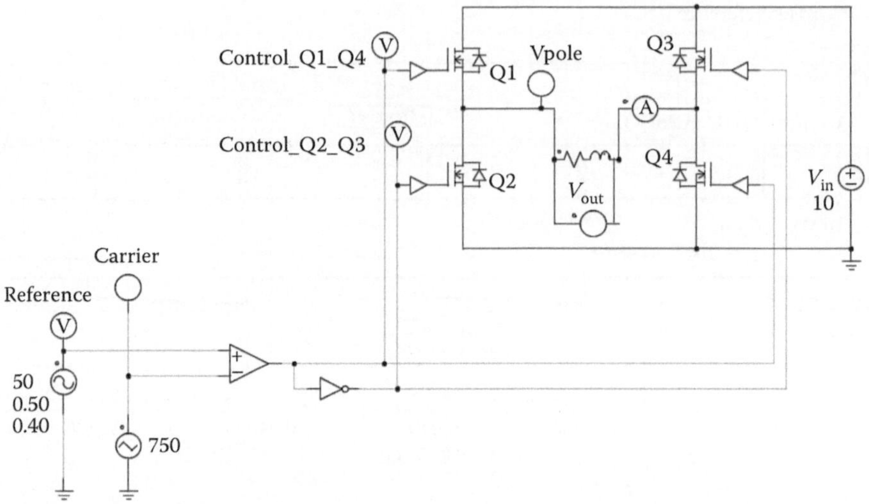

FIGURE 11.25
Power stage and control circuitry for a full-bridge inverter.

A low-pass filter of the output voltage has been considered in Figure 11.28, where the PWM waveform has a carrier of 10 kHz. The high carrier frequency guarantees the same polarity of the current during a period of this signal. These results show the correspondence between the reference waveform and the output voltage.

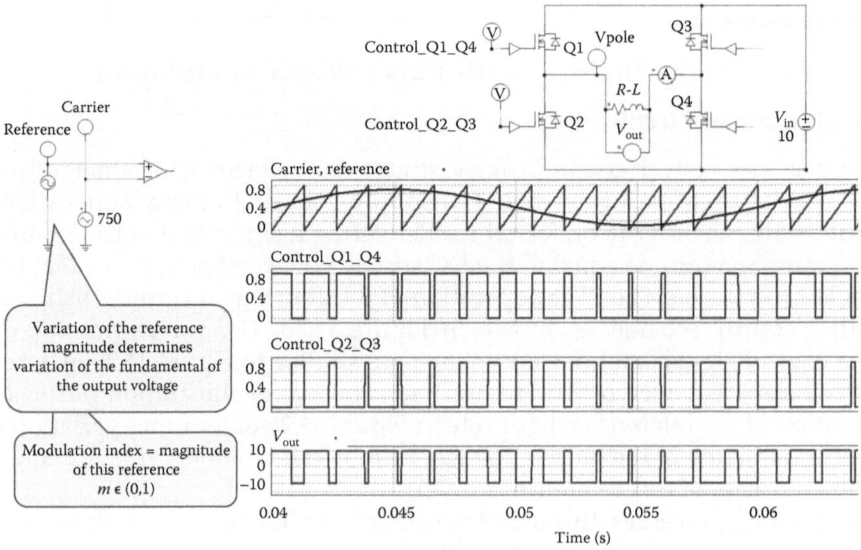

FIGURE 11.26
Waveforms for the circuit shown in Figure 11.25.

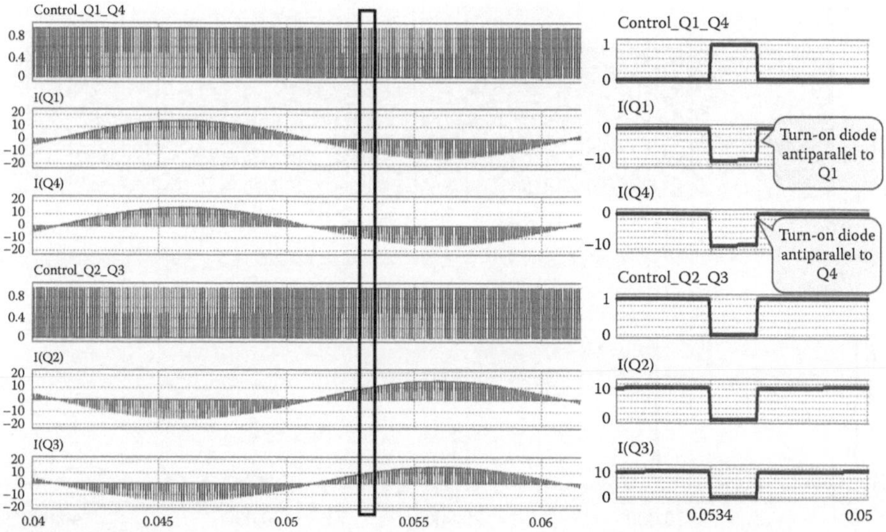

FIGURE 11.27
Details of switching waveforms.

11.6.2 Fundamental and Harmonics for the Full-Bridge Inverter with Pulse-Width Modulation

When the load is an AC motor, the quality of the output voltage does not matter that much, just the content in the fundamental. In other words, a high content in the fundamental with harmonics at higher frequencies would work successfully. In such a situation, the inverter can work without the LC filter, considering the pulse voltage applied directly to the load.

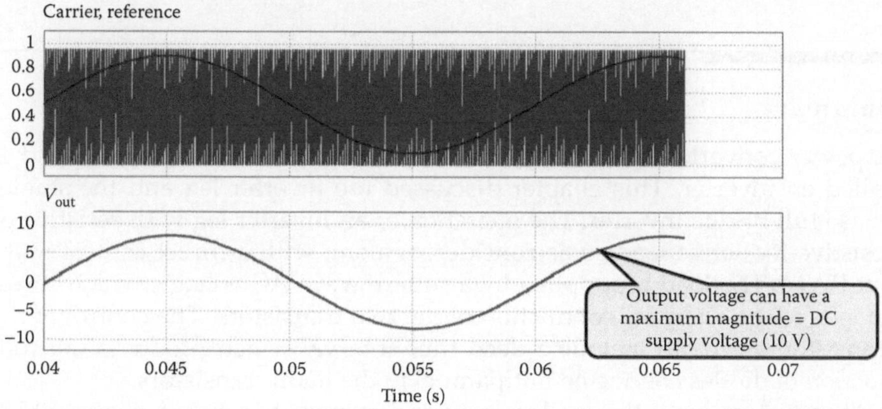

FIGURE 11.28
Correspondence between the reference waveform and output voltage.

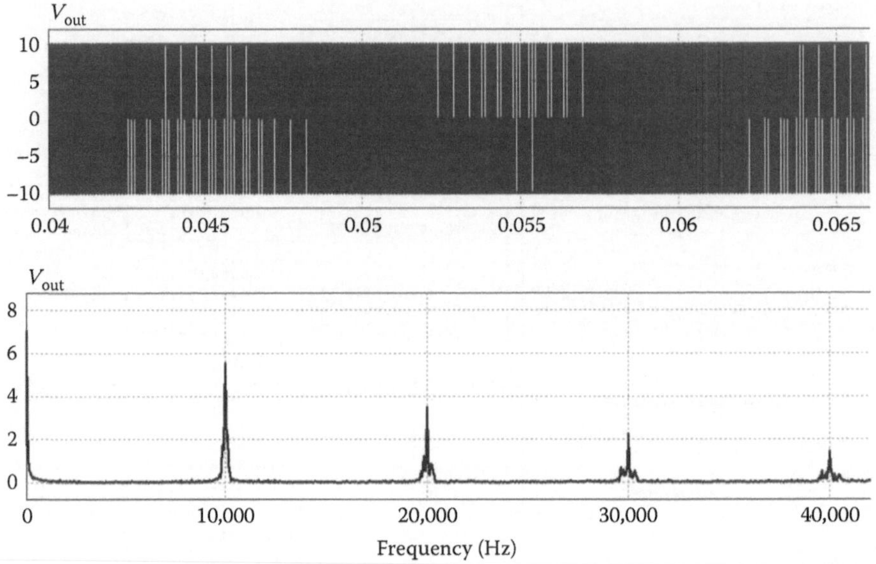

FIGURE 11.29
Harmonic spectrum for the output voltage of a generic single-phase full-bridge PWM inverter switched at 10 kHz.

Figure 11.29 shows the actual output voltage and its spectra. Important content in the fundamental can be seen. The other important harmonics are grouped at high frequencies. A large harmonic can be seen at the carrier frequency, and other important harmonics are at multiples of the carrier frequency; that is, herein at 10 kHz.

Summary

A power converter used to convert a DC energy source to an AC load is called an inverter. This chapter discussed the inverter leg and the single-phase full-bridge inverter. The operation of an inverter leg with resistive or resistive-inductive loads was considered, along with a direct control without PWM. The load is supplied by a square-wave AC voltage, which corresponds to the alternate conduction of the two transistors. The control of the transistors needs to include a dead-time interval which produces the conduction of diodes connected antiparallel to the main transistors.

At inductive load, the load voltage and current are out of phase, which defines the conduction of the two diodes connected antiparallel with transistors instead of the actual transistor being controlled. Reduction of low-frequency harmonics within the load voltage and an easier filtering of the load voltage are achieved when using a pulse-width modulation algorithm.

Particulars of using a PWM algorithm were presented. During a polarity of the load current, a transistor and the antiparallel diode connected near the second transistor conduct sequentially. The harmonic spectrum for the output voltage shows a fundamental component at the same frequency as the reference waveform, no significant low-frequency harmonics and high-frequency harmonics at multiples of the carrier frequency. When compared with square-wave operation, the harmonic spectra seem to move towards a higher frequency range.

Single-phase inverters are used as part of uninterruptible power supplies.

References

1. Neacsu, D.O., 2013, *Switching Power Converters – Medium and High Power*, CRC Press, Taylor and Francis, Boca Raton, FL.
2. Erickson, R., Maksimovic, D., 2001, *Fundamentals of Power Electronics*, 1st and 2nd editions, Springer, Berlin.
3. Mohan, N., Undeland, T.M., Robbins, W.T., 2002, *Power Electronics*, Wiley, New York.
4. Neacsu, D., 2001, "Space Vector Modulation", *IEEE IECON Tutorial*, Denver, CO, USA, 29 November–2 December 2001.

12

Three-Phase Inverters

12.1 Definitions Related to Three-Phase Power Inverters

Energy delivery on a single-phase line is limited in power. For instance, residential circuits use breakers up to 20 A/circuit in most of Europe. Certain countries limit the electronic loads which can be supplied on residential low-voltage power lines (4.6 kW in Germany or Austria, 5 kW in Great Britain or Italy and 10 kW in Australia) [1].

At higher power levels, energy transmission can be done on three-phase systems. Three-phase low-voltage power circuits can transmit energy as AC voltage with 380 V rms line-line, 50 Hz, or 208 V rms line-line, 60 Hz. Standards for power limitation are different in different countries. For instance, electronic power converters can interface with three-phase low-voltage power delivery systems if their power is under 25 kW in Mexico, 30 kW in Australia or 100 kW in Portugal. Higher power level converters can be connected to higher voltage three-phase delivery power systems [2].

It is thus important to review the terminology related to power systems. In this respect, Figure 12.1 illustrates the energy flow from generation to final distribution. After the energy is generated in large power generators, a transformer steps up the voltage level into a high-voltage transmission line. The long-distance high-power transmission of energy can be done on 500 kV, 345 kV, 240 kV or 138 kV lines. This is called *high voltage* in power system lingo. As the energy gets closer to the customer sites, step-down transformers create a *medium voltage* distribution system, usually at voltages like 28 kV or 69 kV. Some customers, such as steel industry factories, take and use energy at this level. Most of the energy is further delivered at lower voltage levels like 13 kV or 4 kV, where industrial customers or residential distribution companies operate. Local (street)-level transformers ensure energy delivery to various residences or small businesses with 220 V rms AC in Europe or two-phase 120 V rms AC in the United States.

The connection of loads on three-phase systems can be on star or delta circuits (Figure 12.2). A star connection features access to a median point of the connection, called neutral, and the entire connection is therefore made on four wires. Conversely, the delta connection is carried out on three wires.

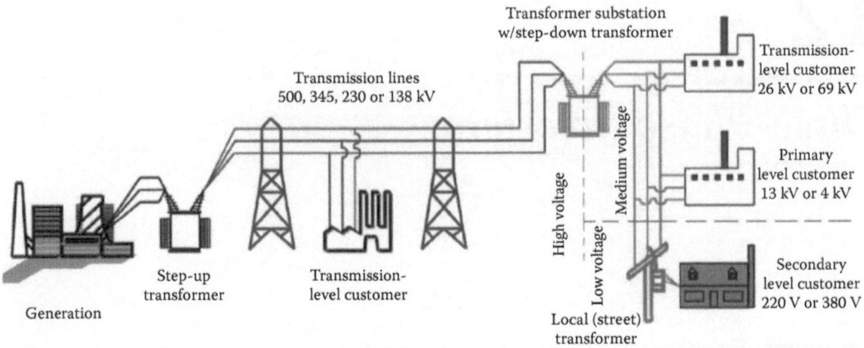

FIGURE 12.1
Energy flow from generation to final distribution.

Numerical values in Figure 12.2 are provided for the European system (220 V rms/phase at 50 Hz).

Equations for transformation from delta to star measures are:

$$\begin{cases} V_{LL\_AB} = V_{ph\_A} - V_{ph\_B} \\ V_{LL\_BC} = V_{ph\_B} - V_{ph\_C} \\ V_{LL\_CA} = V_{ph\_C} - V_{ph\_A} \end{cases} \tag{12.1}$$

Transmission of energy on three-phase power lines determined the creation of specific loads with operation under three-phase AC voltages. If variation in or control of voltage parameters like frequency and magnitude is sought for these systems, three-phase electronic circuits called inverters are used. These transform energy from DC form into three-phase loads.

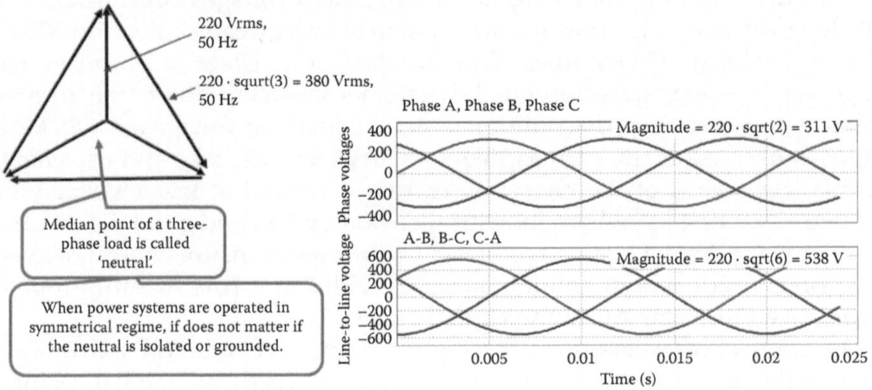

FIGURE 12.2
Star or delta connection of loads.

12.2 Building a Three-Phase Inverter

The operation of an inverter branch with an R-L load is shown in Figure 12.3, while characteristic waveforms are shown in Figure 12.4 as an example. It can be seen that the train of pulses follows the reference.

Three identical inverter branches are next considered for building a three-phase inverter. References for the three branches need to be phase-shifted by 120° so that they will produce a symmetrical three-phase system. The three load impedances are connected to the median point of the applied voltage (also called neutral) so that a star-connected load is formed (Figure 12.5).

The waveforms characteristic to operation of this power converter at a modulation index of:

$$m = 0.777 = \frac{220 \cdot \sqrt{2}}{400} \tag{12.2}$$

are shown in Figure 12.6. Generally, the modulation index shows how much the reference varies around the unmodulated signal. The exact definition depends on application and inverter topology. The waveforms in Figure 12.6 illustrate the relationship between the reference signals and the output phase voltages. The output voltage follows the reference even if it is actually produced as the average of a train of pulses. The waveforms are shown

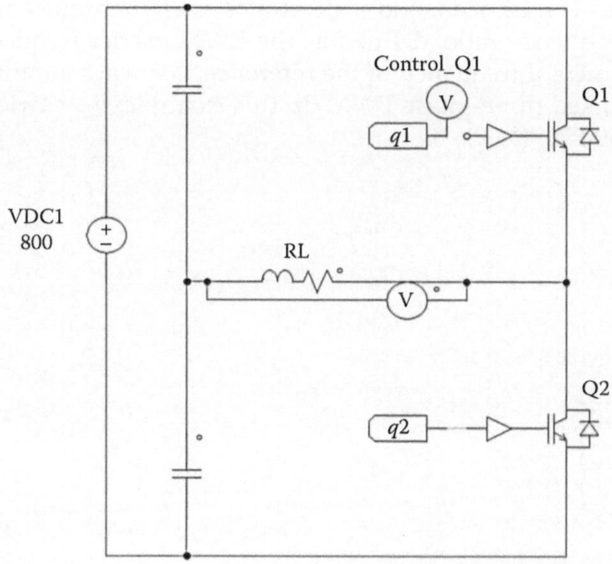

FIGURE 12.3
Inverter branch with R–L load.

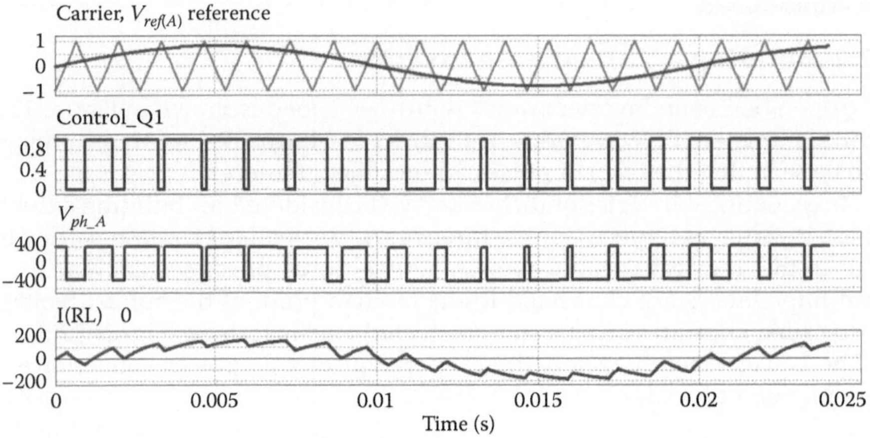

FIGURE 12.4
Waveforms for operation of an inverter branch with modulation index 0.90, and for a 750 Hz carrier waveform and a 50 Hz fundamental.

for a 750-Hz carrier waveform and a 50-Hz fundamental. This is a non-synchronised PWM technique, which means that the control pulses are not symmetrical across the fundamental.

Conversely, Figure 12.7 shows waveforms for a synchronised PWM technique. Waveforms are shown for a 750-Hz carrier waveform and a 50-Hz fundamental. The relationship between the reference signals and the output phase voltages can be seen, and it illustrates the synchronised PWM technique. The frequency ratio, defined as the PWM carrier frequency divided by the fundamental frequency of the reference, is always a multiple of 6 for any synchronised three-phase PWM. In this example, the carrier frequency is 900 Hz $= 50$ Hz $\cdot (3 \cdot 2) \cdot 3$.

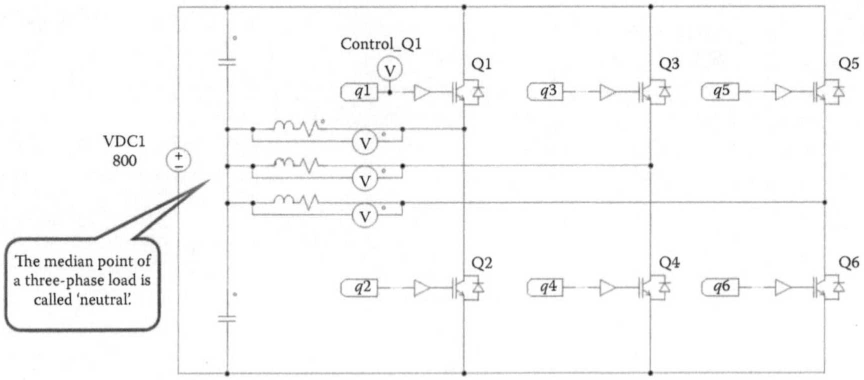

FIGURE 12.5
Building a three-phase inverter from three identical inverter branches.

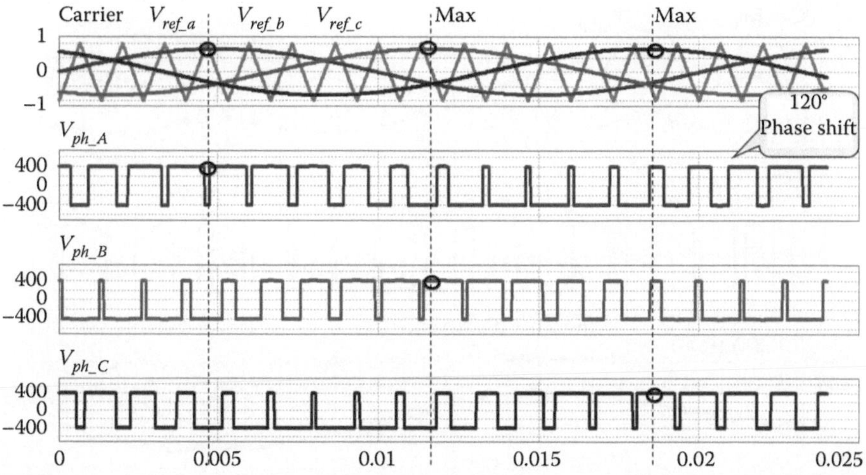

FIGURE 12.6
Sample waveforms for operation of the three-phase converter built with three identical inverter branches at 0.77 modulation index and a carrier frequency of 750 Hz (non-synchronised case).

Figure 12.8 allows us to observe the relationship between a phase voltage and a phase current for a load with a strong inductive character. The current waveform is smoother and is phase-shifted from the voltage waveform.

The inverter built with three identical branches and the load connected to the median point of the DC-side capacitors have an educational role, less used in practice, since most three-phase loads do not offer neutral access.

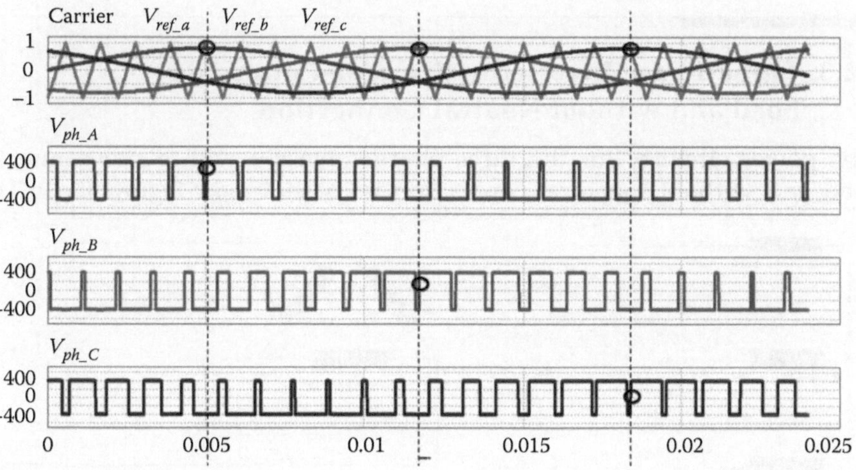

FIGURE 12.7
Sample waveforms for operation of the three-phase converter built with three identical inverter branches at 0.77 modulation index and a carrier frequency of 900 Hz (synchronised case).

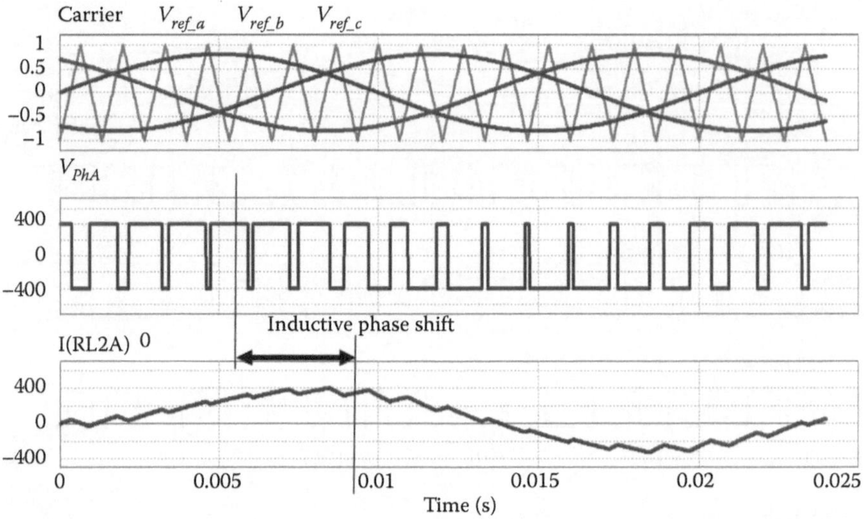

FIGURE 12.8
Load voltage and current.

In most cases, three-phase loads are connected in a star or delta (Figure 12.9) configuration. It is very rare that the load neutral is also connected to the midpoint of the DC capacitor bank, as previously discussed. For this reason, the following analysis focuses on the operation of a three-phase inverter with one of the more common load configurations (star or delta).

12.3 Three-Phase Inverter with a Star-Connected Load and without Neutral Connection

The power circuitry for the three-phase converter with a star-connected load and without a neutral connection is as shown in Figure 12.10. This

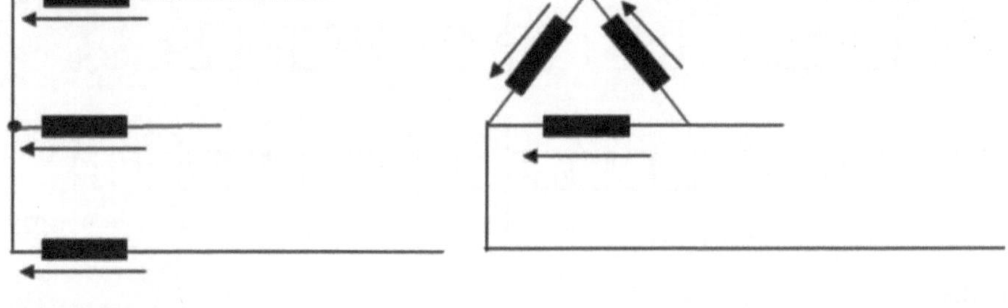

FIGURE 12.9
Star or load connection of the three-phase load.

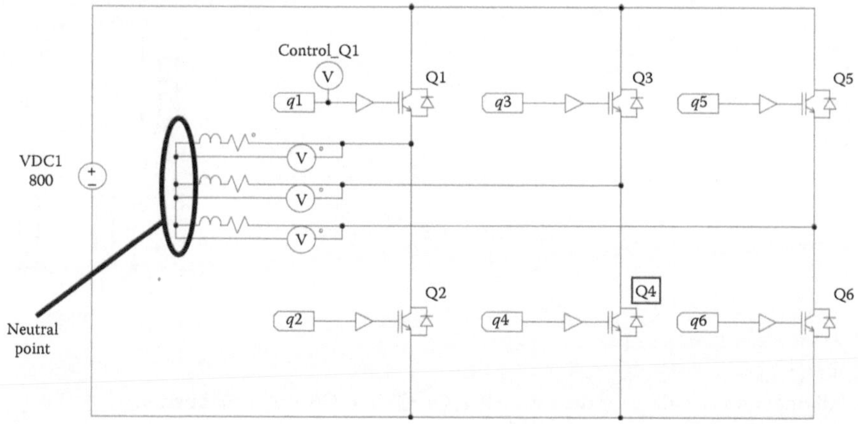

FIGURE 12.10
Inverter schematics.

three-phase inverter is controlled with the same PWM technique, and the load voltages are different. The waveform of the output phase is shown in Figure 12.11, and it is substantially different from the previous case. These output voltages feature three levels at 0, 0.33 V DC and, respectively, 0.66 V DC during the positive polarity of the voltage and three similar levels for the negative polarity. Details of this waveform can be understood from the analysis of all the possible operation states.

If we neglect the effect of the dead time on the phase voltages, each inverter branch has two possible states: one for high-side transistor turn-on and one for low-side transistor turn-on. Since the neutral point is not connected to the

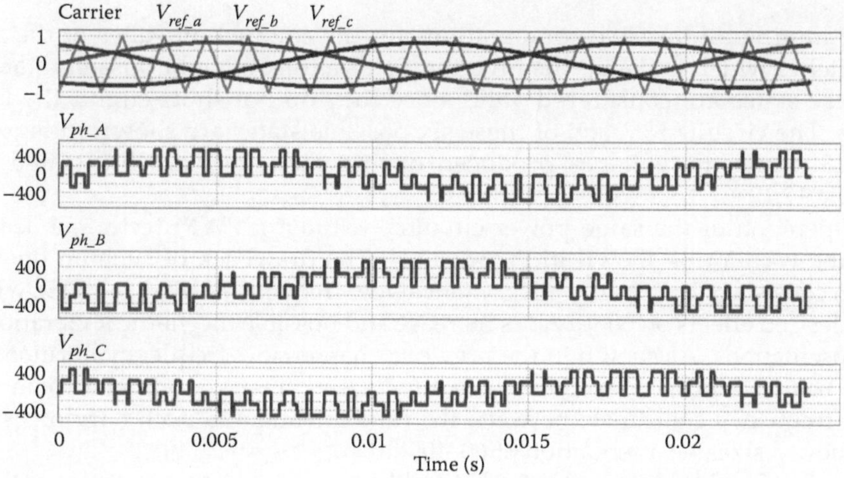

FIGURE 12.11
Inverter waveforms.

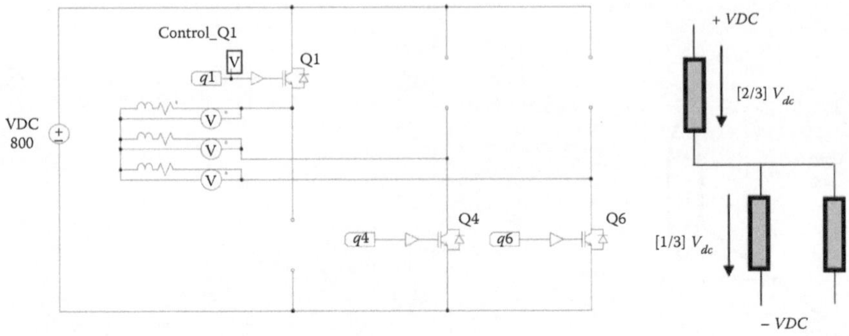

FIGURE 12.12
Load divider and circuitry for the case when Q1, Q4 and Q6 are in conduction.

capacitor bank midpoint, an impedance divider is formed among the three load impedances. For instance, if we consider Q1, Q4 and Q6 turned on, the equivalent circuitry for the impedance divider is as shown in Figure 12.12.

Reading the phase voltages across each phase as measured from the inverter branch towards the neutral point:

$$\begin{cases} V_1 = \dfrac{2}{3} \cdot V_{dc} \\ V_2 = \left(-\dfrac{1}{3}\right) \cdot V_{dc} \\ V_3 = \left(-\dfrac{1}{3}\right) \cdot V_{dc} \end{cases} \tag{12.3}$$

Counting all possible combinations for the conduction states of the six switches which make up the inverter yields a total of eight possible states. These states are considered valid since they do not short-circuit the DC bus. The circuits for each of these six possible states are shown in Figure 12.13, while the voltages produced during each state are illustrated in Table 12.1.

Operation of the same power circuitry without a PWM technique leads to the waveforms shown in Figure 12.14. The drawback of such an operation consists of high harmonic content at low frequencies (Figure 12.15), with undesired effects on energy loss increase and possible mechanical vibrations or oscillations. As shown in the previous chapter, low-frequency harmonics are nearly impossible to filter since any passive filter would need room on the frequency coordinate between the fundamental and such harmonics to achieve a sizeable attenuation of 60–100 dB.

A PWM technique similar to that used for single-phase converters can be used for three-phase converters [3,4]. All three references for three-phase PWM are generated using three sinusoidal waveforms as references, followed

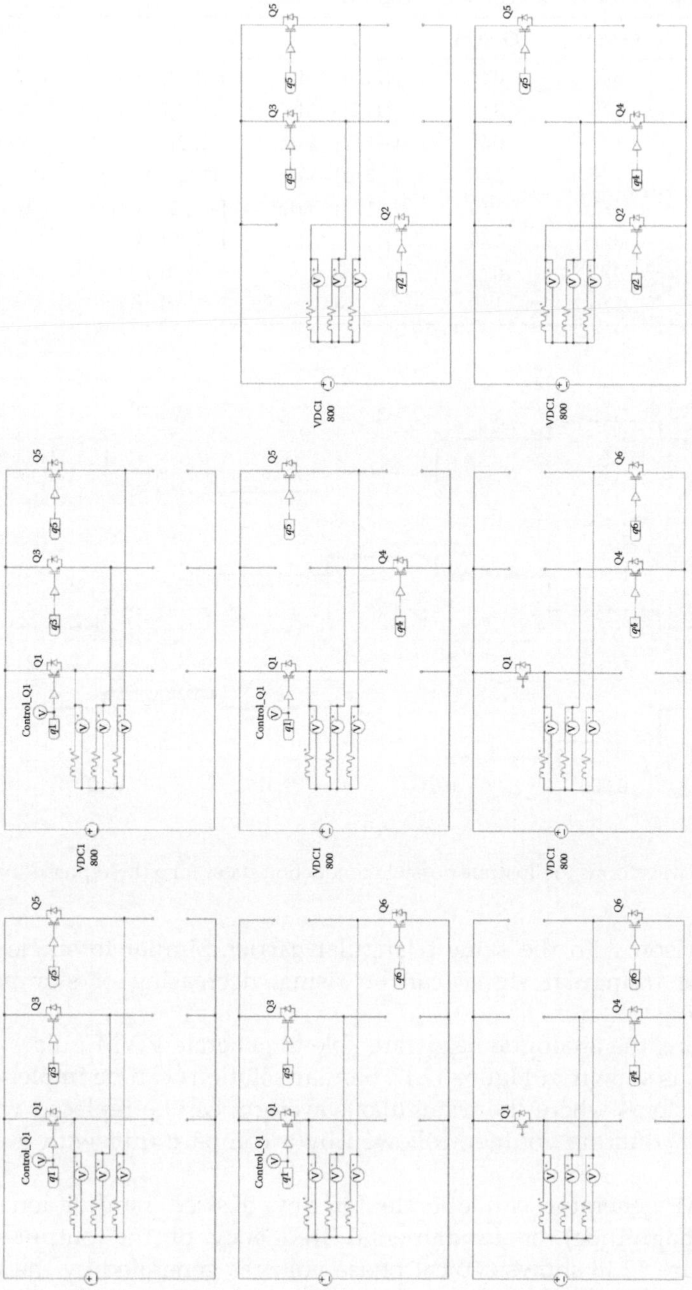

FIGURE 12.13
All possible states for the operation of a six-switch inverter.

TABLE 12.1
Voltages and States for All Possible Conduction States

	Q1/Q2	Q3/Q4	Q5/Q6	V_{ph_A}	V_{ph_B}	V_{ph_C}
2	1/0	0/1	0/1	[2/3] · Vdc	[−1/3] · Vdc	[−1/3] Vdc
3	1/0	1/0	0/1	[1/3] · Vdc	[1/3] · Vdc	[−2/3] · Vdc
4	0/1	1/0	0/1	[−1/3] · Vdc	[2/3] · Vdc	[−1/3] · Vdc
5	0/1	1/0	1/0	[−2/3] · Vdc	[1/3] · Vdc	[1/3] · Vdc
6	0/1	0/1	1/0	[−1/3] · Vdc	[−1/3] · Vdc	[2/3] · Vdc
1	1/0	0/1	1/0	[1/3] · Vdc	[−2/3] · Vdc	[1/3] · Vdc
Z1	0/1	0/1	0/1	0	0	0
Z2	1/0	1/0	1/0	0	0	0

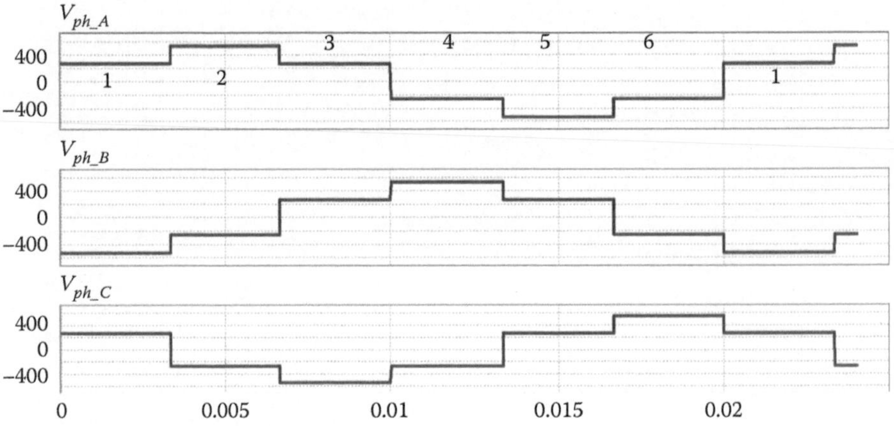

FIGURE 12.14
Output voltage waveforms to illustrate possible conduction states for a three-phase six-switch inverter.

by a comparison with the same triangular carrier. Similar to single-phase inverters, the triangular signal can be rising, decreasing or symmetrical (Figure 12.16).

Furthermore, the analogue hardware able to generate PWM pulses for all three phases is shown in Figure 12.17. Similar solutions can be implemented on digital systems where the triangular waveform can be replaced with an up- or down-counting counter, followed by a compare unit with a digital reference.

This PWM generator can operate for any desired modulation index (reference magnitude) or fundamental frequency of the output phase voltage. Figure 12.18 shows typical phase voltages generated by the circuit from Figure 12.17, now for a modulation index of 0.77. Furthermore, it can be seen that each pulse comes from two neighbouring levels in the waveform for the unmodulated inverter, also shown at the top of Figure 12.18.

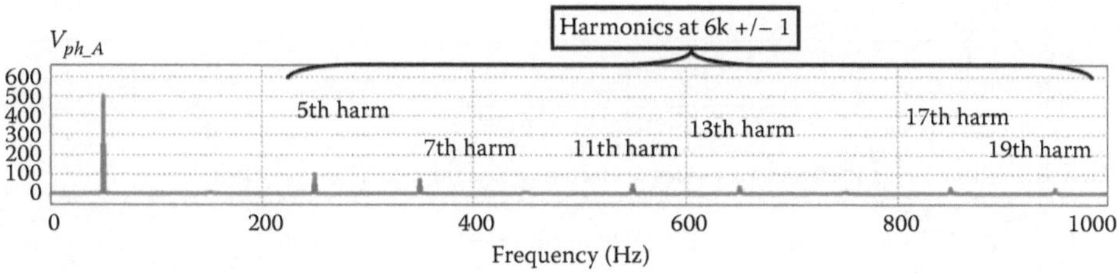

FIGURE 12.15
Harmonic spectra for the output voltage when PWM is not used.

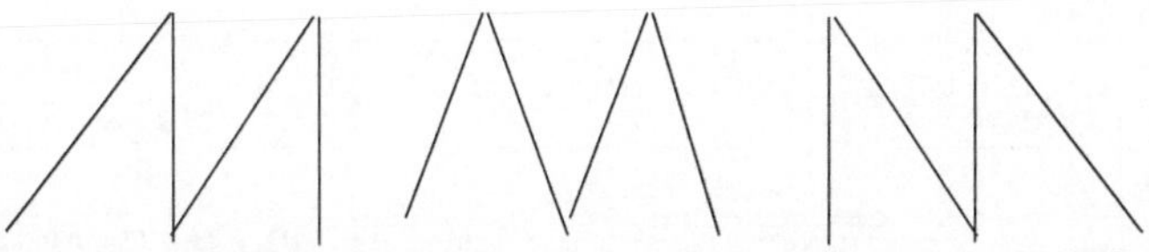

FIGURE 12.16
Possible waveforms for the triangular carrier signal.

The average calculated over a constant sampling period should follow the reference, herein sinusoidal.

The main reason for and advantage of using PWM consists of harmonic reduction in low frequency ranges. Actually, it can be said that harmonics other than the fundamental move towards a higher-frequency domain. For instance, the output phase voltage's harmonics for the case with a modulation index of $m = 0.777$, a pulse frequency of $f_p = 5$ kHz and a fundamental component of 50 Hz are shown in Figure 12.19. Pair components near the carrier frequency and its multiples $(n \cdot f_p)$ are seen.

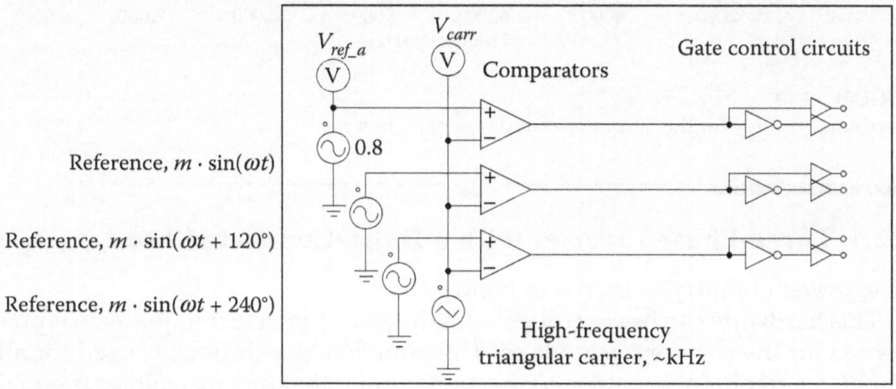

FIGURE 12.17
Hardware for three-phase PWM generation.

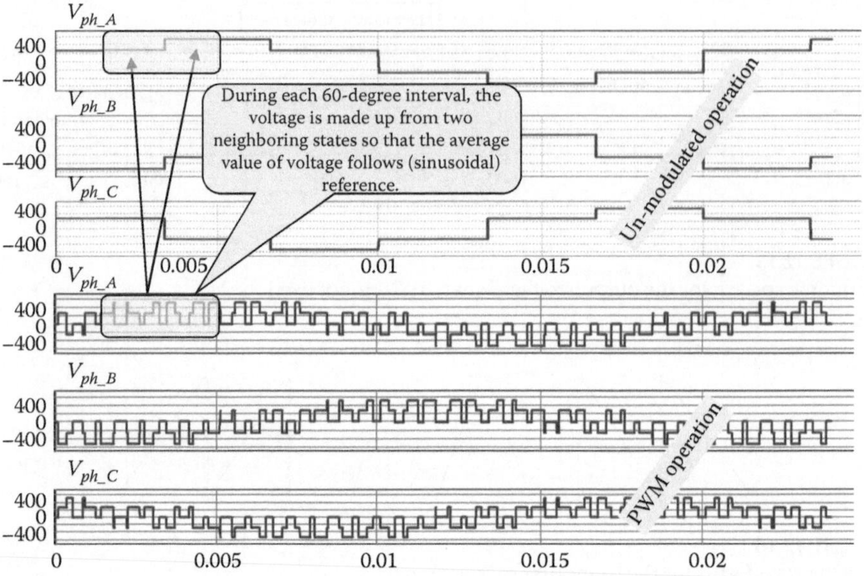

FIGURE 12.18
Waveforms for output phase voltages for either a unmodulated or modulated inverter.

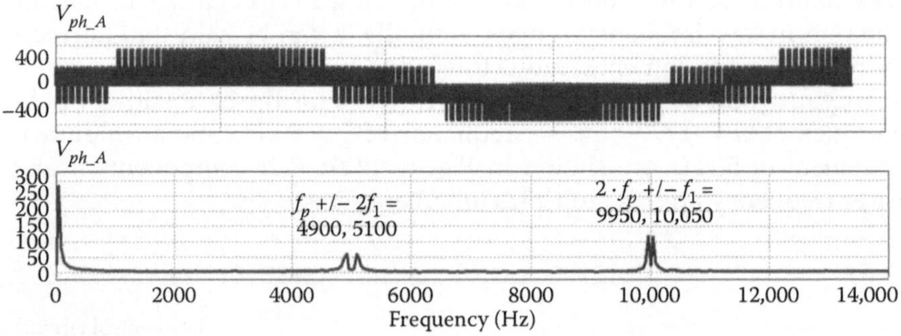

FIGURE 12.19
Harmonic content for the output phase voltage of an inverter.

12.4 Three-Phase Inverter with a Delta-Connected Load

The power circuitry is shown in Figure 12.20.

This hardware can be controlled with the same PWM technique as that presented for the star-connected load. The same PWM technique is used for all three inverter branches. This connection implies that the line voltage is equal to the difference between two pole voltages (inverter branches) and also that the line voltage equals the difference between the two phase voltages.

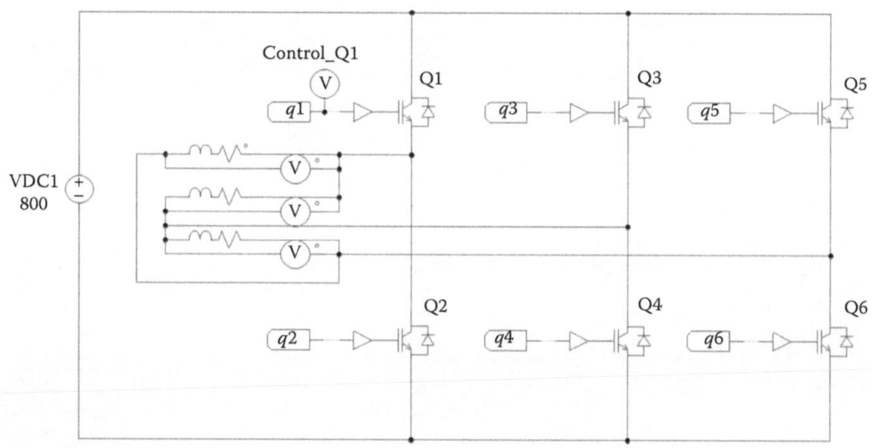

FIGURE 12.20
Inverter with load connected in delta.

The waveforms measured this way feature only two levels (0 or V DC). The modulation technique provides *sqrt*(3) more voltage when using a delta-connected load than a star-connected load since each line voltage represents the difference between two phase voltages. Figure 12.21 shows waveforms for a carrier of 750 Hz and a modulation index of $m = 0.777$ when the DC bus voltage equals 800 V DC. Pulses of 800 V are seen within the line voltages. It can be shown that a single switching occurs for each power transistor during each period of the carrier waveform when an up- or down-counting carrier is considered. That means that the switching frequency equals the carrier frequency.

The output voltage has a harmonic spectrum composed of the fundamental and harmonics around the carrier frequency and its multiples (Figure 12.22).

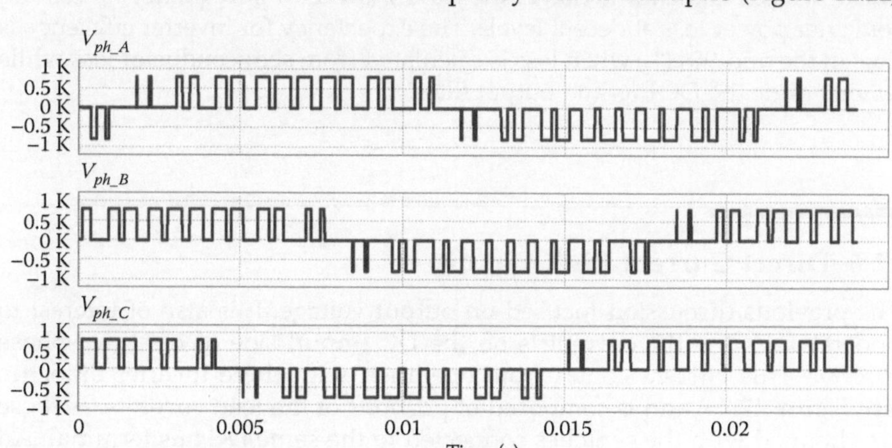

FIGURE 12.21
Waveforms for PWM operation with a load connected in delta.

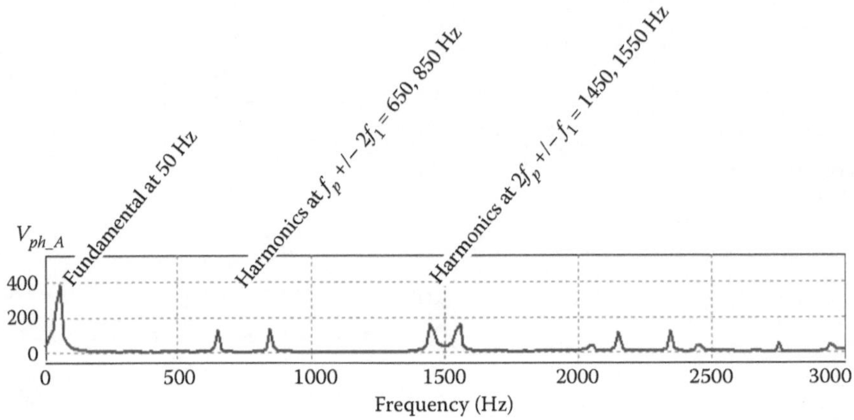

FIGURE 12.22
Harmonic components for the output phase voltage with an inverter operated with PWM.

Details of the spectral composition depend on the selection made for the carrier waveform – triangular signal with rising, decreasing or symmetrical waveforms; such details are not important for the design of a passive filter – and it is important to observe we have harmonics around the carrier frequency and its multiples far enough from the fundamental. There is enough room between the fundamental and first significant harmonic for a second-order filter with 40 dB/decade fallout. From 50 to 650 Hz, we can possibly get 40 dB attenuation, which is about 8 V peak-to-peak ripple at an 800 V DC bus. When the switching and carrier frequencies are selected as higher, even more attenuation is possible. This is the most important advantage of using the PWM method. The downside is the limitation of the carrier (switching frequency) to contain converter power loss at decent levels. The expectancy for inverter efficiency is thus in the mid-90s (%) when loss is calculated from semiconductor loss while leaving aside the DC bus and output filter.

12.5 Direct Current Link Circuit

The previous discussion focused on output voltage. It is also of interest to understand what the current is on the DC (input) side of the three-phase inverter. This current comes from operation with three inverter branches (see Figure 12.23) and is composed of portions of the load currents, as these are chopped with the switches connected to the same DC bus terminal. An example is shown in Figure 12.24.

The harmonics of the DC-side current (input current) are required for sizing the DC filter. The previous shape of the DC current can be investigated

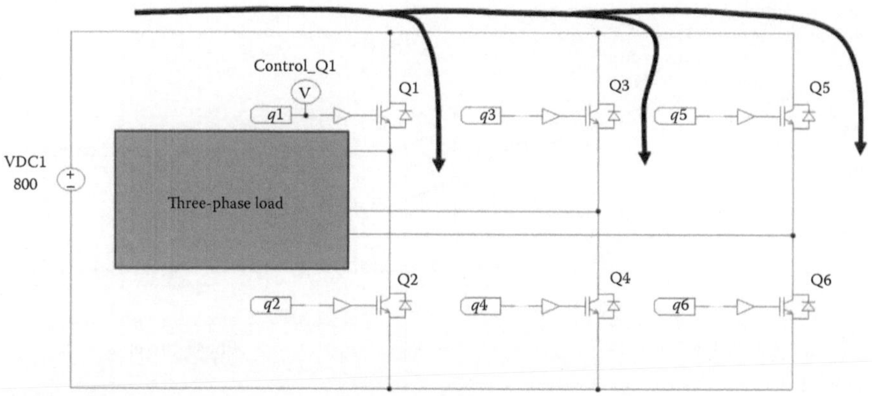

FIGURE 12.23
Formation of DC current from chopped load currents.

with an FFT for understanding the current harmonics. The outcome is shown in Figure 12.25.

12.6 Output Voltage Variation: Overmodulation

The previous analysis concerned the operation for a certain modulation index of $m = 0.777$, able to produce a voltage of 220 Vrms when the three-phase inverter is supplied from 800 Vdc. However, the modulation index can be modified within a large domain, and this is mainly used in motor drives to

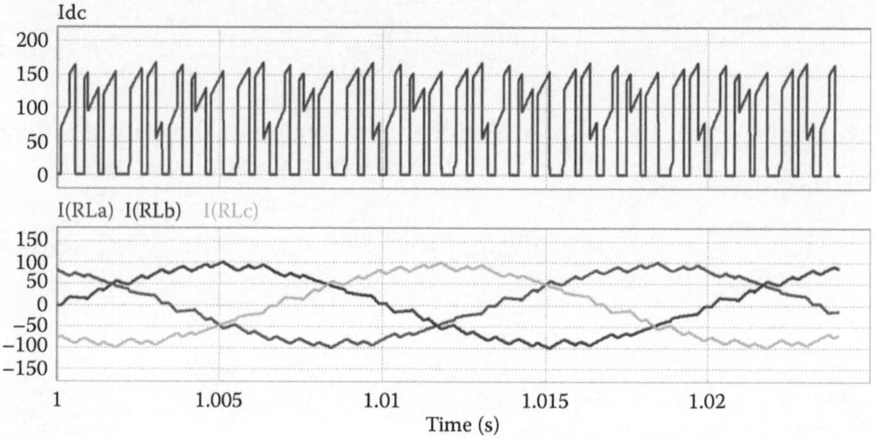

FIGURE 12.24
Waveforms pertaining to the formation of the DC current for a three-phase inverter.

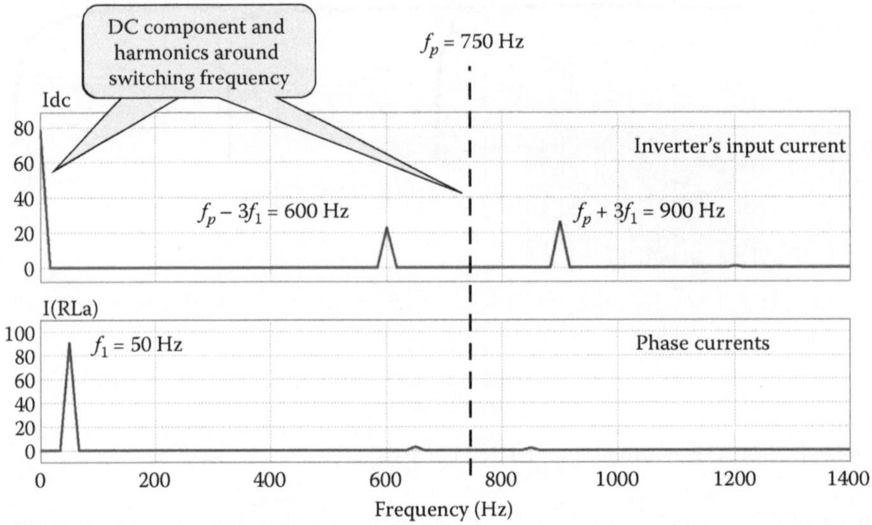

FIGURE 12.25
Harmonics within the DC-side current and the appropriate AC currents.

adjust the operation point of the load. Alternatively, AC power supplies use the modulation index variation to adjust for the DC (input) voltage ripple or decay, as well as for any voltage drop induced by operation with large currents.

The modulation index can be adjusted through the magnitude of the reference waveform. The magnitude of the reference waveform needs to be lower than the magnitude of the triangular waveform. This constraint can also be expressed from the DC bus voltage (Figure 12.26).

The maximum is:

$$V_{phase[rms]}^{(1)} = \frac{V_{dc}}{2 \cdot \sqrt{2}} \tag{12.4}$$

This voltage can be compared with the *rms* value of the output voltage waveform for the case without modulation, which was also discussed.

$$V_{phase[rms]}^{(1)} = \frac{V_{dc}\sqrt{2}}{\pi} \tag{12.5}$$

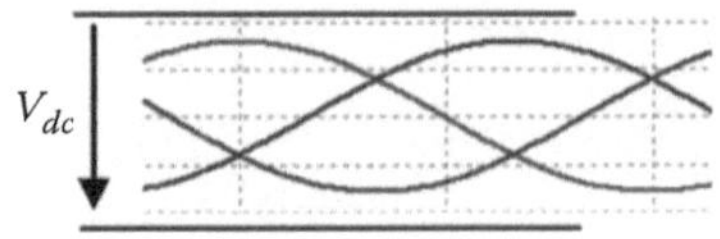

FIGURE 12.26
Depiction of the maximum for the modulation index.

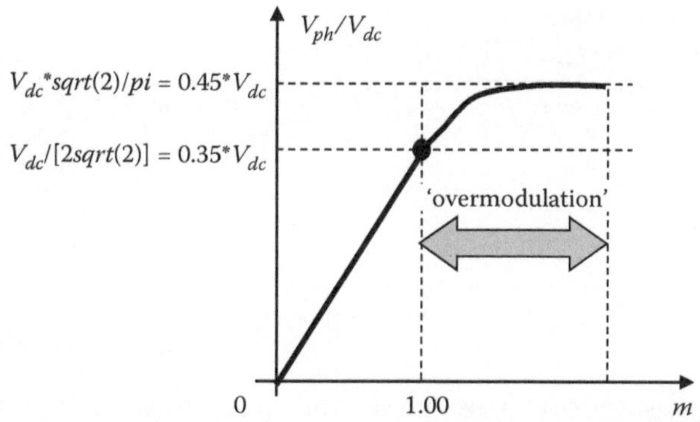

FIGURE 12.27
Dependency for the entire range of operation of a three-phase inverter.

Operation at any modulation index can be mapped into a dependency between the reference signal and the output phase voltage. This dependency is shown in Figure 12.27. It continues the previously demonstrated linear range for the PWM generator into the so-called overmodulation region. During operation in the overmodulation range, the dependency is nonlinear and the inverter simply tries to push more voltage than the conventional generator can provide. This interval can also be seen as a link between the linear PWM mode and the square-wave operation of the same hardware.

Also note that the line-to-line voltage is *sqrt(3)* larger than the phase voltage. In order to force the same hardware to operate in this overmodulation region, the designer should allow the reference waveform to go beyond the peak value of the triangular waveform. This is shown with Figure 12.28. The inverter is now controlled in the overmodulation regime. The principle illustrated in Figure 12.28 explains the nonlinearity of the transfer characteristic as the interval with PWM varies over the fundamental.

12.7 Other Three-Phase Topologies

Three-phase inverters are usually built with IGBT devices, which need to be selected for a voltage at least equal to the DC supply voltage. Operation at medium voltage is therefore difficult to equip with conventional IGBT devices, which are rated at lower voltages. Various more complex topologies have been considered. Obviously, series connection of devices can somewhat solve the problem. However, a better solution is provided with multilevel converters, where each IGBT sees a lower voltage across its collector-emitter terminals. This is also currently a very trendy research field.

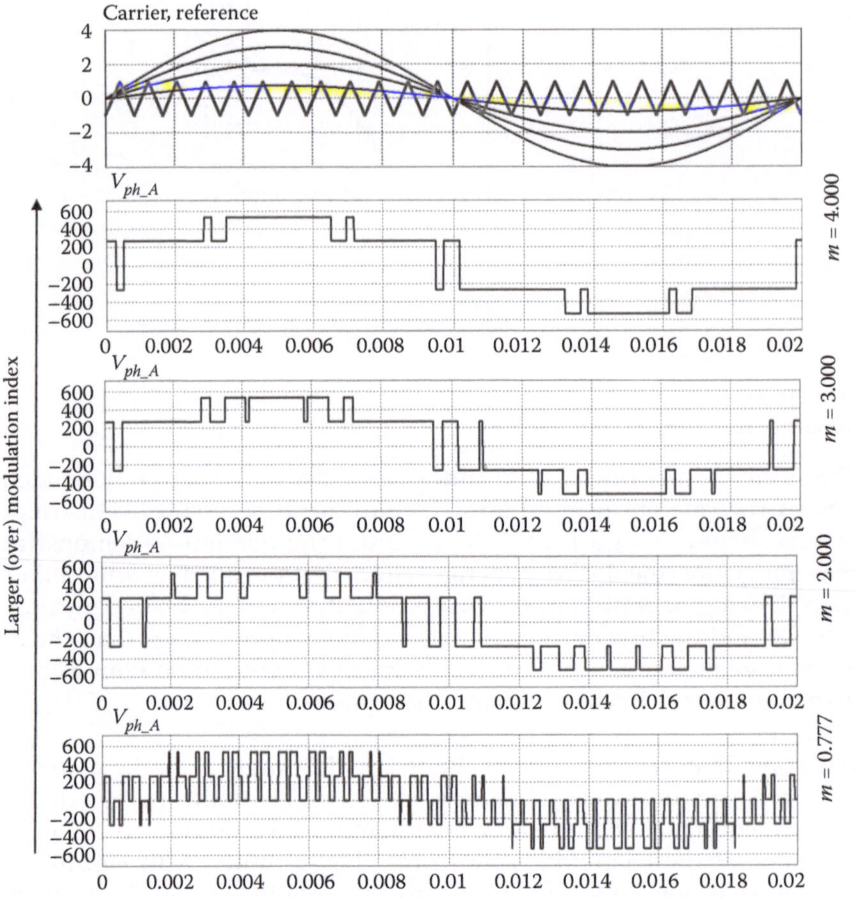

FIGURE 12.28
Waveforms for the inverter operation with various modulation indices, including overmodulation regime.

12.7.1 Voltage Three-Phase Inverter, with Multiple Levels (Multilevel)

Sometimes the output voltage is built through modulation between multiple voltage levels. Multilevel inverters allow the selection of power semiconductor devices rated at lower voltages with the advantage of building three-phase inverters at large voltages, for instance, at DC bus voltages of 4500 Vdc. While there are a multitude of multilevel inverter topologies, one example is considered herein for information. The principle of these inverters relies on the existence of multiple levels able to participate in building the output voltage, like a staircase.

The example considered in Figure 12.29 considers a staircase output voltage, as shown on the left side of the figure. The power structure is shown on the right side for an inverter branch (single phase). Each voltage level can be produced with appropriate control of the switches, allowing connection of more or fewer local DC-side capacitors or voltage sources into the phase voltage. For each individual structure, if the high-side IGBT is turned on, the entire DC-side voltage contributes to the pole voltage formation. A turn-on of the low-side IGBT produces just zero voltage contribution.

This principle is improved further with PWM operation of each individual structure. Any intermediary level between two consecutive levels can be created with time-appropriate averaging between these levels. This principle is illustrated further with Figure 12.29 for a transition between the kth and $(k+1)$th levels.

The principle illustrated with Figure 12.29 is also called an *N-level cascaded multilevel converter*, where N is the number of levels, $N = 2 \cdot m + 1$, and m is the number of leg or bridge inverters supplied with m independent sources.

A different multilevel converter can be built as a neutral-point clamped multilevel converter (Figure 12.30). A set of diodes is used to clamp different voltages from the inverter to the midpoint of the capacitor bank. Often, this also represents the neutral of the three-phase voltage system. An example with four IGBT devices on each leg is shown. Transistors are controlled in pairs, as the control scheme suggests.

The two topologies for multilevel converters discussed in this section are the most-used principles for hardware generation of a multilevel output voltage.

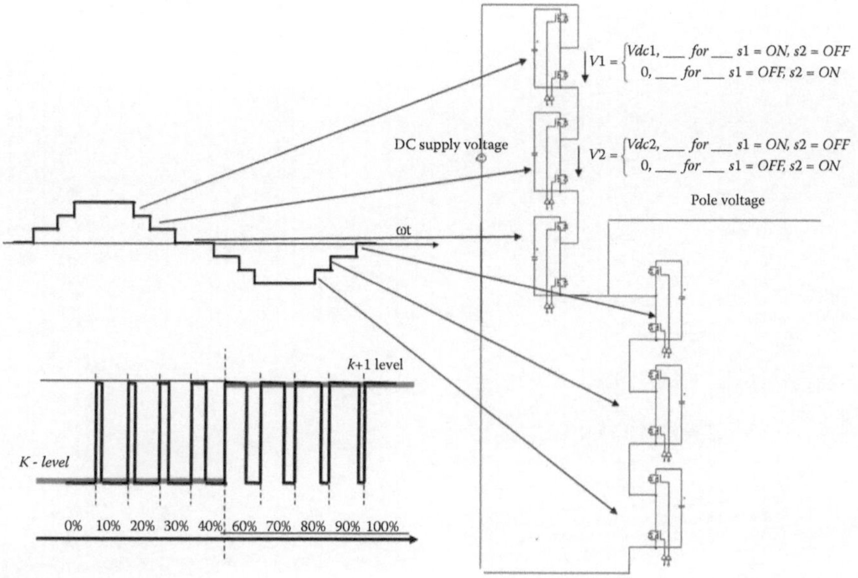

FIGURE 12.29
Building a multilevel inverter.

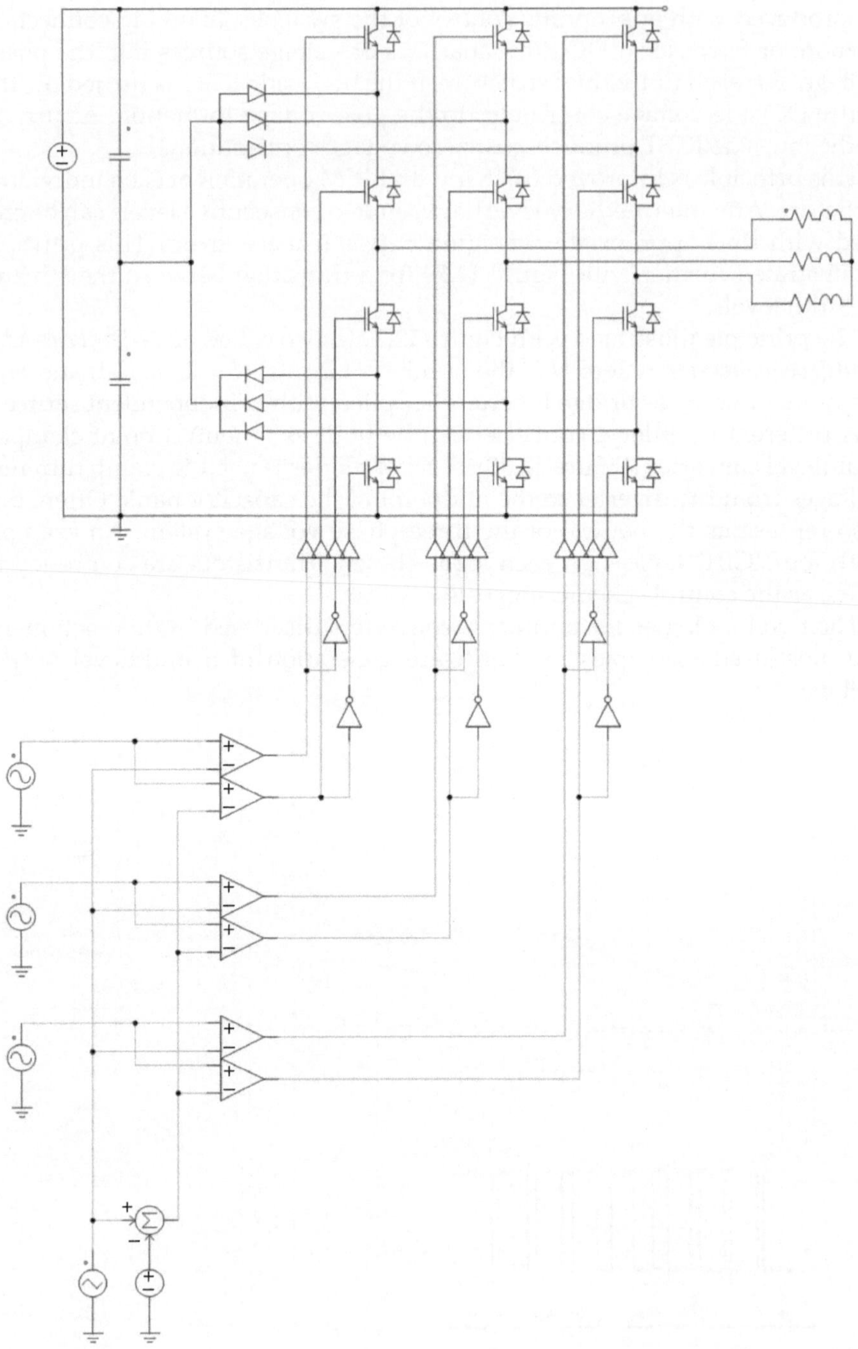

FIGURE 12.30
Neutral point clamped multilevel inverter.

Summary

Three-phase systems are used at higher power levels. They can provide energy to loads connected with a star or delta connection. Three-phase voltage systems can also be created with electronic converters.

A three-phase inverter can be created with three inverter branches controlled at the same modulation index and a phase shift of 120°. This configuration is not used very often since the large majority of practical loads do not offer access to the neutral point. The same converter topology with star or delta loads was observed within this chapter.

Low harmonics from the output voltage are observed – a fundamental component and the harmonic components around the multiples of the carrier frequency. The supply current has an important DC component and harmonics around the multiples of the carrier frequency.

The fundamental component of the output voltage can be modified with a variation of the modulation index. A linear dependency is achieved up to a certain modulation index when this is defined as up to half the supply voltage. The fundamental component of the output voltage can be increased through a nonlinear dependency up to unmodulated (six-step) operation.

Multilevel converters can be used at large voltages (400–10,000 V DC), where they can be built with low voltage–rated transistors.

Three-phase inverters were discussed herein as an alternative to AC voltage generation, and this will be used further in Chapters 13–14 for uninterruptible power supplies.

References

1. Neacsu, D.O., 2013, *Switching Power Converters – Medium and High Power*, CRC Press, Taylor and Francis, Boca Raton, FL.
2. Abur, A., Gomez Esposito, A., 2004, *Power System State Estimation: Theory and Implementation*, CRC Press, Taylor and Francis, Boca Raton, FL.
3. Neacsu, D.O., 2001, "Space Vector Modulation," *Seminar IEEE-IECON*, Denver, CO, USA, 29 November–2 December 2001, pp. 1–130.
4. Homes, D.G., Lipo, T.A., 2003, *Pulse Width Modulation for Power Converters: Principles and Practice*, Wiley-IEEE Press, New York.

13

Converters with AC Reference

13.1 Review of Inverter Operation

Single-phase inverter operation was studied in Chapter 11 for both square-wave and PWM inverters with either resistive or resistive-inductive load. In this chapter, the same hardware is used for bidirectional energy transfer between the AC and DC sides of the inverter. In this respect, an AC source e_A is considered in the load circuit, with the same frequency as the inverter reference, so that energy transfer is possible either from the power converter to the AC source or from the AC source to the power converter.

Since we want to perform this energy transfer between the inverter and the AC source (Figure 13.1), an impedance is needed to take over the instantaneous difference between the two voltage sources. Moreover, in order to reduce the energy loss with the voltage drop across impedance, a pure inductance is considered as having zero resistance.

In Chapter 11, a 90° phase shift between the current and voltage at the output of an inverter with inductive load was demonstrated. The case shown in Figure 13.1 is more complex since the current depends on the amplitude and phase of both the AC source and the inverter phase voltage.

In order to understand this relationship, a review of PWM operation is necessary herein. The inverter branch is controlled with PWM. This was deemed necessary in Chapter 10 for current harmonics reduction.

Figure 13.2 reviews the difference between low-frequency current harmonics when using a diode rectifier as a grid interface and current harmonics at the switching frequency and its multiples when using PWM converters. Even these harmonic components are very reduced in magnitude at the switching frequency and multiples.

Due to the connection inductance, a current–voltage phase shift under 90° has previously been considered for the single-phase inverter. An AC voltage source e_A is added to the load circuitry in Figure 13.1 to form the new circuit shown in Figure 13.3. This voltage source has the same frequency as the inverter control; that is, 50 Hz in our example.

This setup allows a bidirectional energy transfer between the AC source and the inverter. Since a transfer between two voltage sources (the inverter

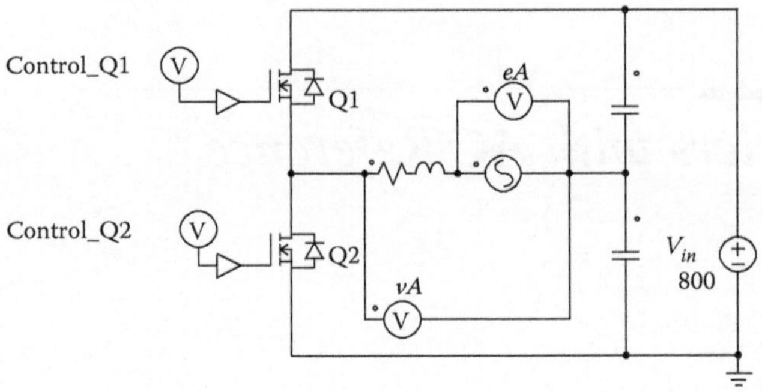

FIGURE 13.1
Inverter branch with AC source on the load circuit.

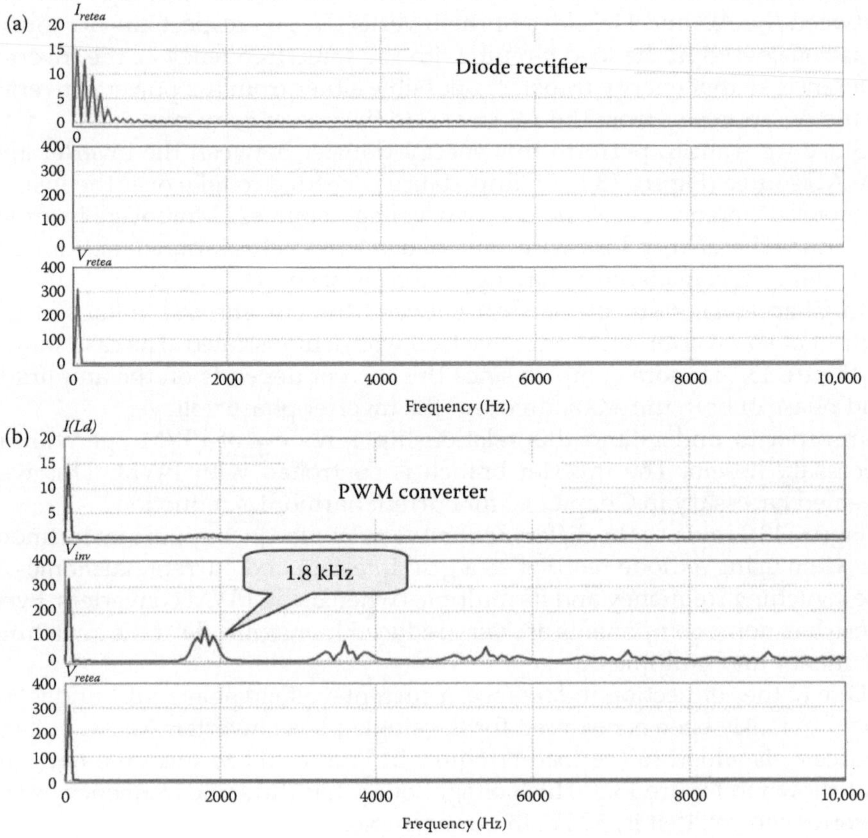

FIGURE 13.2
Current harmonics for a grid interface with (a) diode rectifier, (b) PWM power converter.

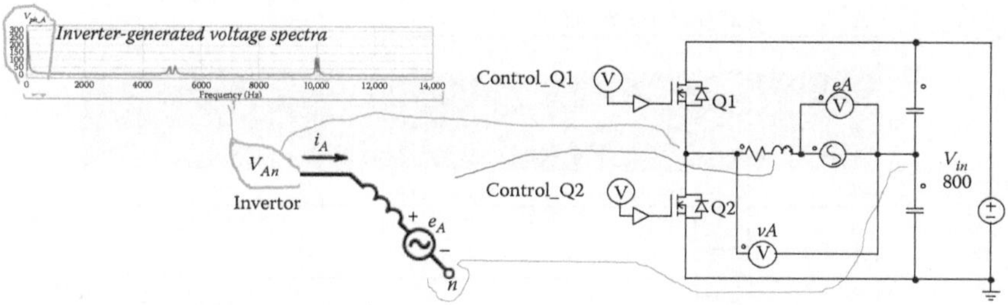

FIGURE 13.3
Inverter branch with voltage source within the load circuitry.

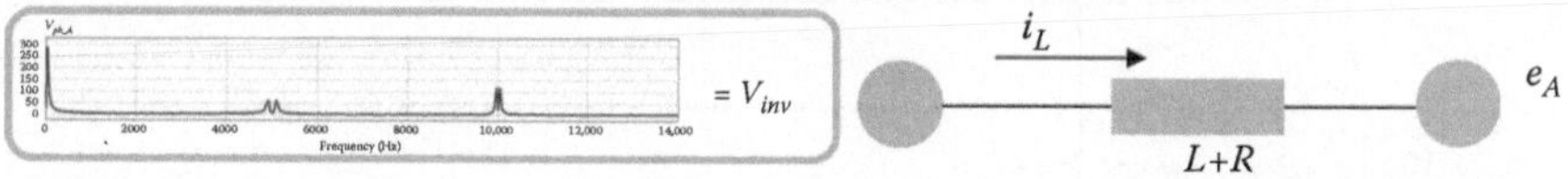

FIGURE 13.4
Simplified load circuit diagram.

and the newly introduced voltage source) is sought, an impedance is needed to overtake the instantaneous voltage difference between the two sources. Further on, considering an inductance with low resistance will reduce loss. Despite PWM operation of the inverter, only the low-frequency component (50 Hz herein) is of interest, and the equivalent circuit on the fundamental frequency is shown in Figure 13.4.

This circuit can be analysed for each harmonic. First, at the fundamental frequency (herein 50 Hz), the voltage equation is:

$$V_{A-n} = L \cdot \frac{di}{dt} + R \cdot i + e_A \tag{13.1}$$

while at any high frequency, the source $e_A = 0$ and the impedance becomes very large (ωL), with a very small current i_{harm}:

$$V_{A-n} = L \cdot \frac{di_{harm}}{dt} + R \cdot i_{harm} \tag{13.2}$$

Although a calculation is possible for this equivalent circuit to define positioning of all the sinusoidal measures from the circuit, examples for operation with switching at 10 kHz are next provided for better understanding of this discussion.

Case A (Figure 13.5)

$$\text{Reference} = 0.50 + 0.39 \cdot \sin(2 \cdot \pi \cdot 50 \cdot t) \tag{13.3}$$

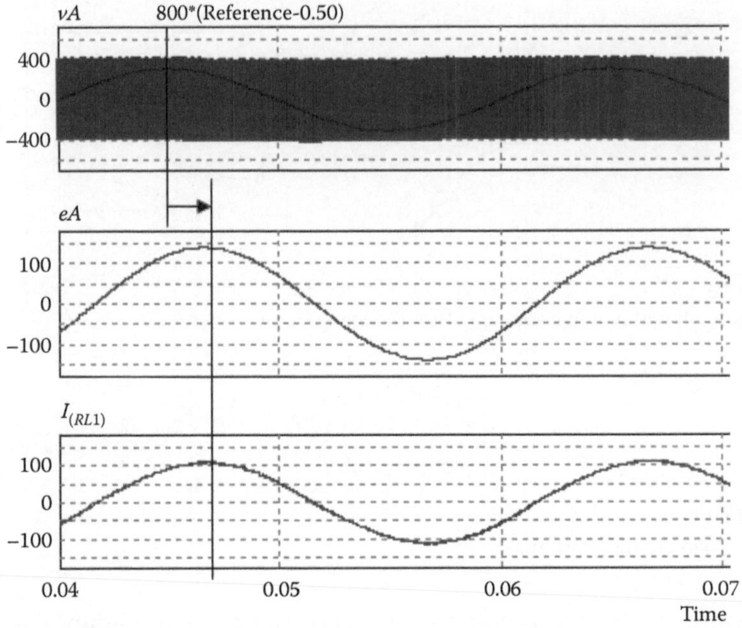

FIGURE 13.5
Case A for discussion of the load circuitry.

$$\text{Source} = 140 \cdot \sin(2 \cdot \pi \cdot 50 \cdot t \; - \pi/6) \tag{13.4}$$

Case B (Figure 13.6)

$$\text{Reference} = 0.50 + 0.39 \cdot \sin(2 \cdot \pi \cdot 50 \cdot t) \tag{13.5}$$

$$\text{Source} = 180 \cdot \sin(2 \cdot \pi \cdot 50 \cdot t - \pi/3)\,\text{v} \tag{13.6}$$

13.2 Rectifier Operation Mode of the Inverter

The analysis can be completed with a phasor diagram representation of voltages and currents from the load circuitry. Sinusoidal measures with time variation can be associated with phasors. Phasors are different from vectors, which represent physical measures defined in a bi- or tridimensional physical space. Magnetic induction is an example of vector. The magnitude, frequency and phase of the sinusoidal function are considered time invariable. Figure 13.7 shows the definition of a phasor for a sinusoidal measure.

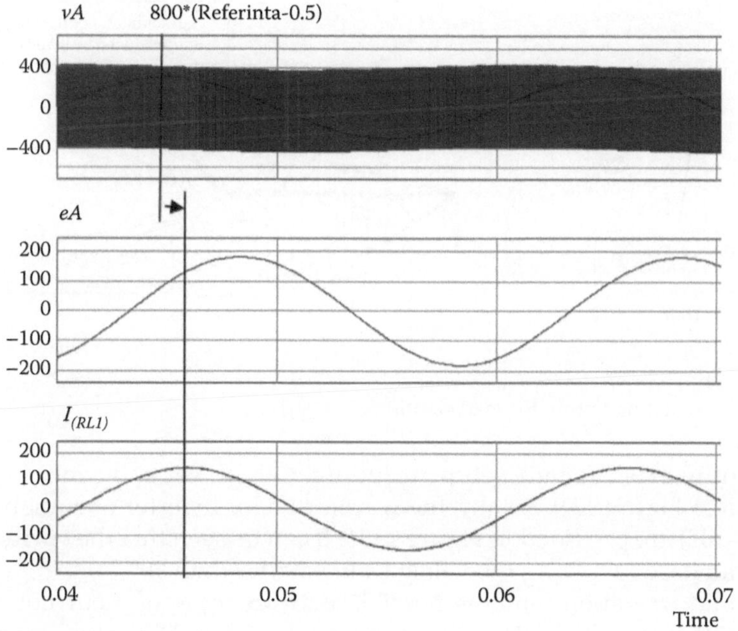

FIGURE 13.6
Case B for the load discussion.

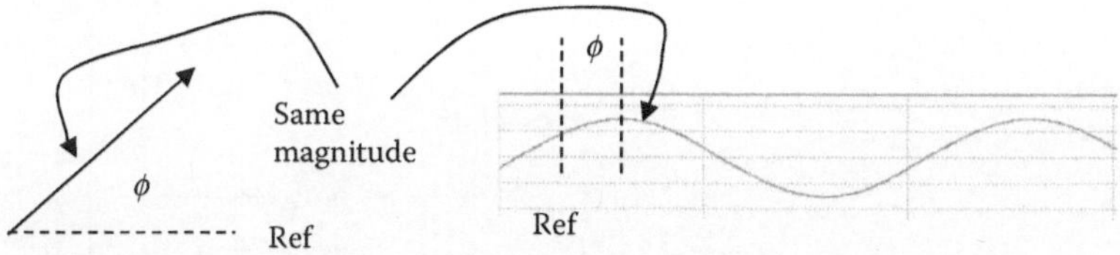

FIGURE 13.7
Definition of a phasor.

The complete phasor diagram for voltages and currents within the load circuitry is shown in Figure 13.8. For an inverter operation, the energy passes from inverter to source, so the current i_A has a phase coordinate following the inverter voltage, but smaller than 90°. The current i_A can be decomposed into two components:

- One aligned with the newly introduced voltage e_A (this will produce an active power component P in the source)
- One orthogonal to the newly introduced voltage (this will produce the reactive power component Q)

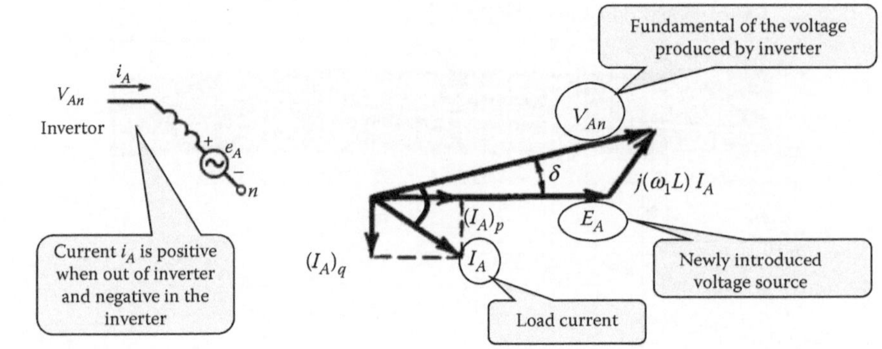

FIGURE 13.8
Phasor representation within the load circuitry.

The entire experimental setup, including both power and control circuits, is shown in Figure 13.9. Finally, the waveforms for inverter operation (transfer DC→AC) are provided in Figure 13.10. It can be seen that the current i_A (or I_{RL}) has a phase following the voltage, but smaller than 90°.

A second operation mode is possible with the case of a current–voltage phase shift larger than 90° within the load circuitry. The active component (from P = active power) of the current has a reversed direction in this case (I_{AP} *versus* E_A). The energy transfer is made from the source e_A towards the inverter. The phasor diagram representation changes, as shown in Figure 13.11.

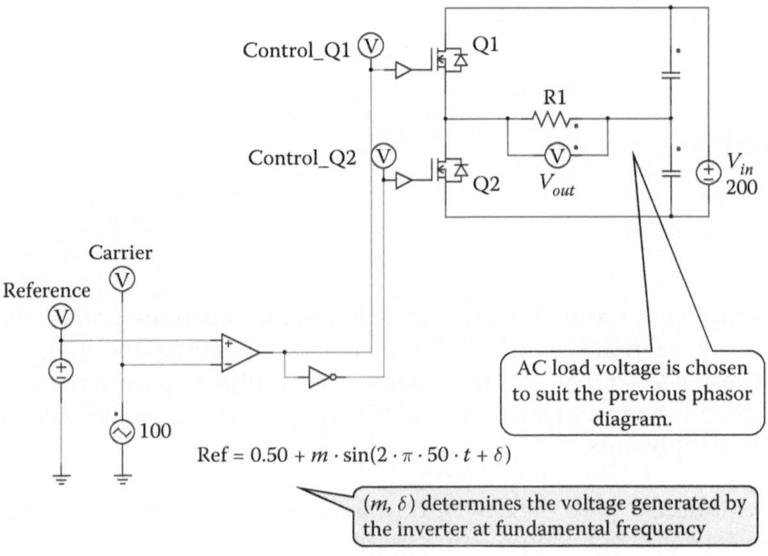

FIGURE 13.9
Complete circuitry with both the power and control circuitry.

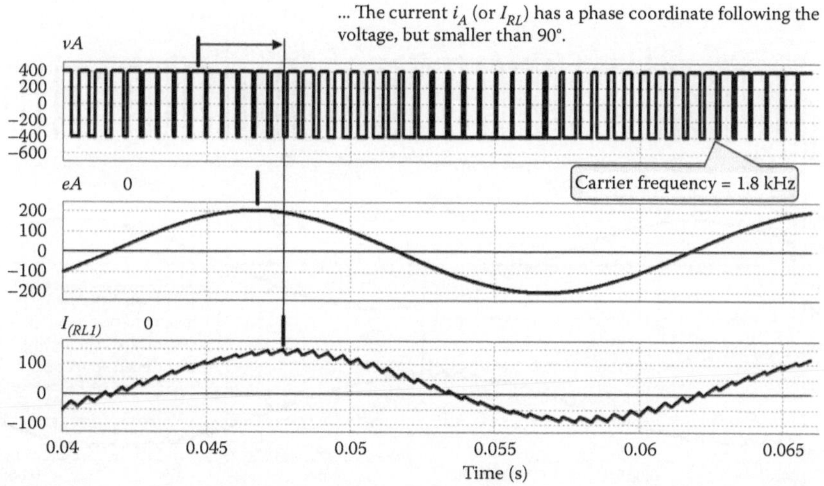

FIGURE 13.10
Waveforms for inverter operation mode.

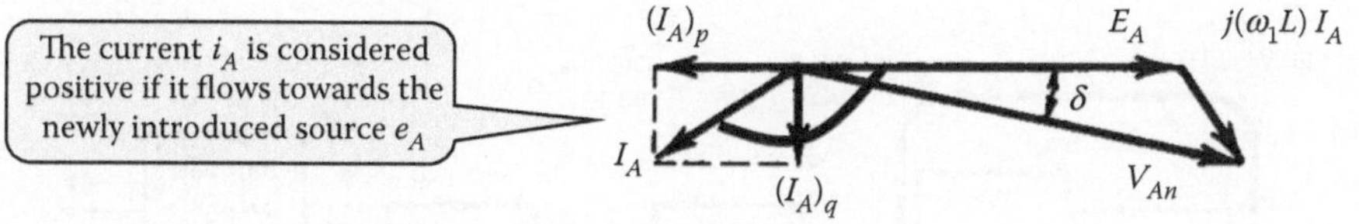

FIGURE 13.11
Phasor diagram for rectifier operation mode.

This operation mode is called rectifier mode. The waveforms for the measures within the load circuitry for rectifier operation mode (transfer AC → DC) are shown in Figure 13.12.

13.3 Grid Interface Converters

Based on the previous discussion, someone can reverse the logic and start from the voltage source e_A previously introduced within the load circuit. This voltage source can now be considered an AC grid or supply voltage. Furthermore, the convention for the direction of the circuit current (now grid current) can be reversed. These premises are used as a basis for building an interface for supplying various circuits from the AC grid.

The energy flow can now be observed in Figure 13.13. The update for the circuitry is shown in Figure 13.14 to consider the voltage source as

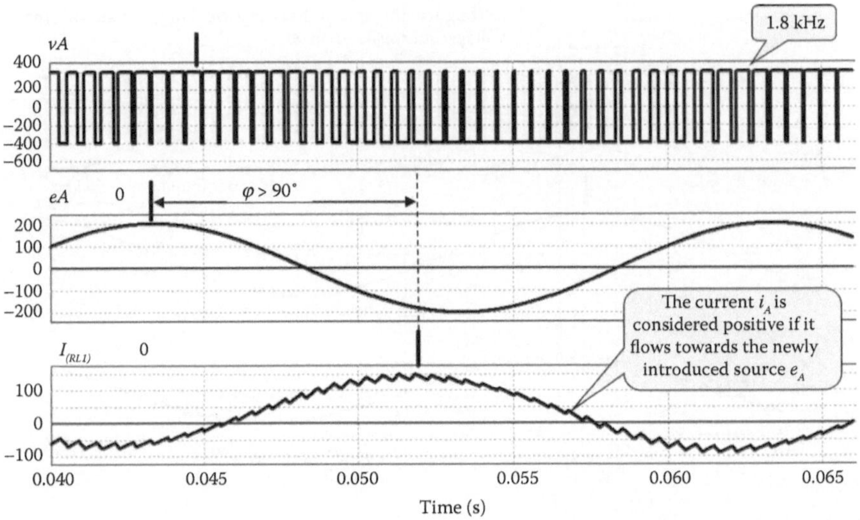

FIGURE 13.12
Waveforms for the rectifier operation mode of an inverter.

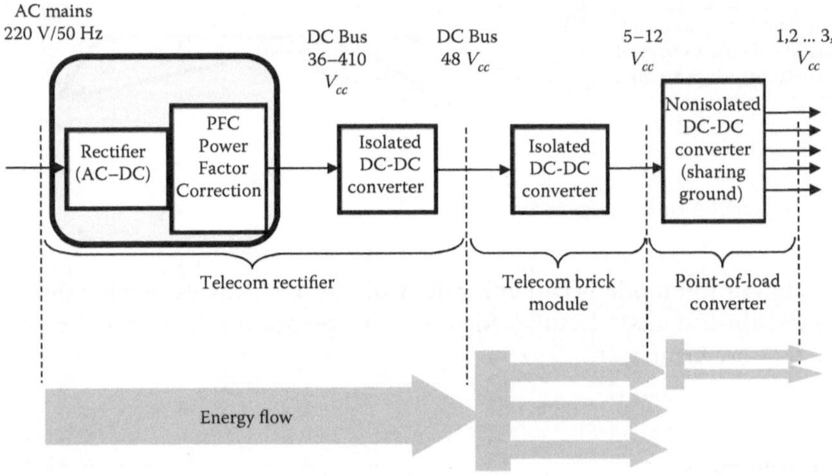

FIGURE 13.13
Energy flow for a grid interface converter.

only the AC grid. The same definition for the bridge converter is shown in Figure 13.15.

Furthermore, the DC voltage can be replaced with a capacitor whose voltage is subject to our control system with charging from the AC grid. This operation mode can be achieved without a real DC voltage source, and an interface for the AC grid can therefore be built that is able to take energy from the grid and transfer it towards the DC load.

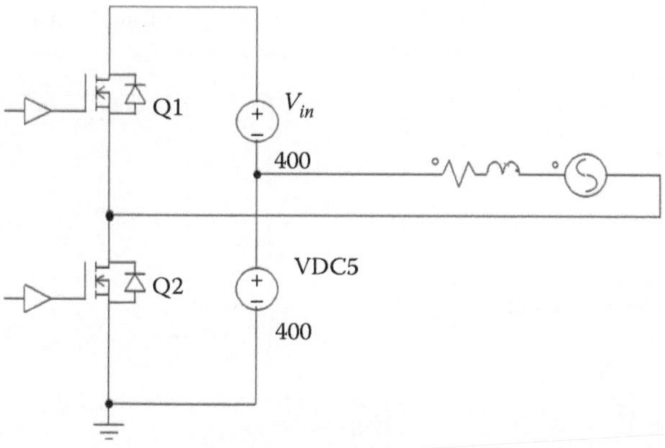

FIGURE 13.14
Circuitry for the supply from AC grid.

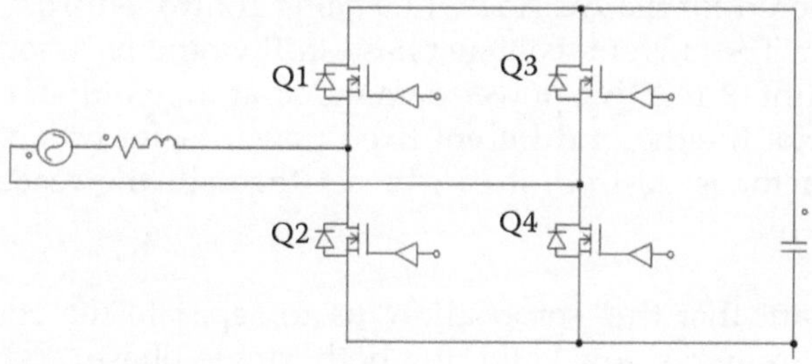

FIGURE 13.15
Circuitry for bridge converter from AC grid.

This implies two requirements for the design of the control system (Figure 13.16):

- Maintain a constant DC voltage on the DC capacitor for any load connected after this capacitor. This requirement corresponds to a transfer of active power (useful load effect) between the AC grid and the load. When the load asks for more active power, the DC-side capacitor would decrease, and the control system would see the error and demand more grid current to charge the DC-side capacitor back. Similarly, for the case when the load needs less active power, the DC voltage would increase across the capacitor, and the control system would sense the error and demand less current from the grid.
- AC grid energy extraction should be made at a power factor close to unity, with a current in phase with voltage. This requirement corresponds to reactive power (provided from the reactive current–voltage phase shift). Sometimes a power factor different from unity

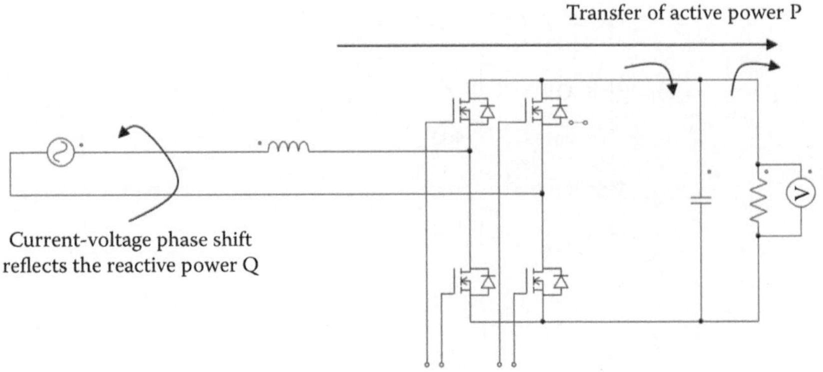

FIGURE 13.16
Definition of active and reactive power components for a grid interface carried out with a single-phase converter.

is considered for the AC side while being controlled from inside the converter. The current–voltage phase shift would be associated with the amount of reactive power demanded at the point of connection to the grid. If either a different fixed power factor or a controllable power factor is desired, it can be set through the reactive power component.

It is important that this setup allow us to separate the control actions. The previous remarks are valid for both single-phase and three-phase systems. The two requirements can be shown with a block scheme like that in Figure 13.17, where the unity power factor case yields $i_q = 0$.

After prescribing the components of the grid current, the control of the converter to produce this current has to be designed. The following steps are considered in the design of the control algorithm:

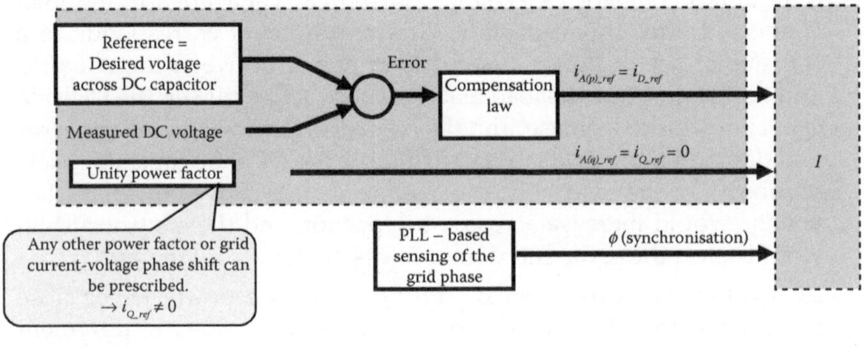

FIGURE 13.17
Control scheme for the two components of power.

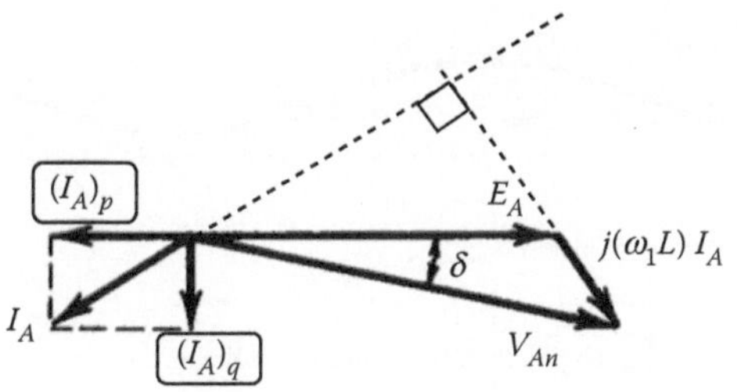

FIGURE 13.18
Phasor diagram used for controller design.

- Start off with known (i_{Ap}, i_{Aq}) current components.
- Determine the grid parameters (E, ϕ) through measurement.
  - Where ϕ is the phase after an arbitrary reference
- Calculate $(j \cdot 2 \cdot \pi \cdot 50 \cdot L) \cdot I$ at 90° off the total current I_A.
- Graphically or analytically calculate the voltage V_A produced by the inverter (see Figure 13.18 for the phasor diagram).

Since the design is made at the unity power factor and the reactive component of the current is zero, the entire grid current comes from the active component. The DC-side voltage regulator is included within a gain block K and provides a reference for the grid current as $I_{d,ref}$. The active component of the current becomes the magnitude of the sinusoidal waveform, which also needs to be synchronised with the grid voltage through the angular coordinate.

Prescribing the current reference is followed by a current controller. This is shown schematically in Figure 13.19, with results outlined with waveforms

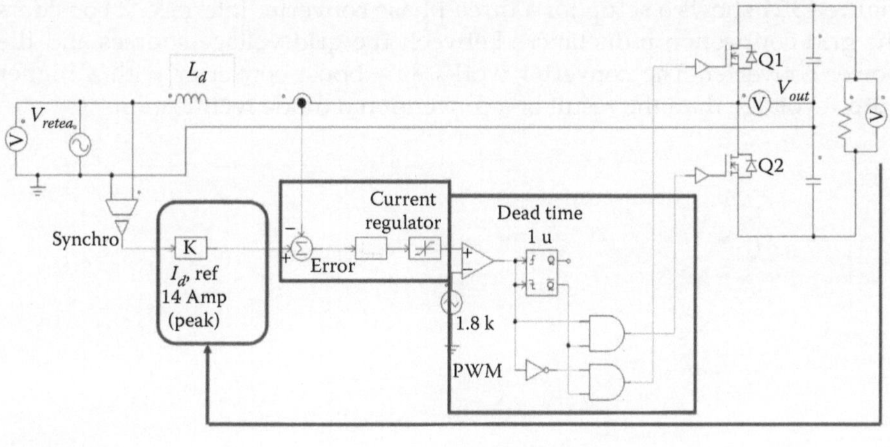

FIGURE 13.19
Controller design for the grid interface.

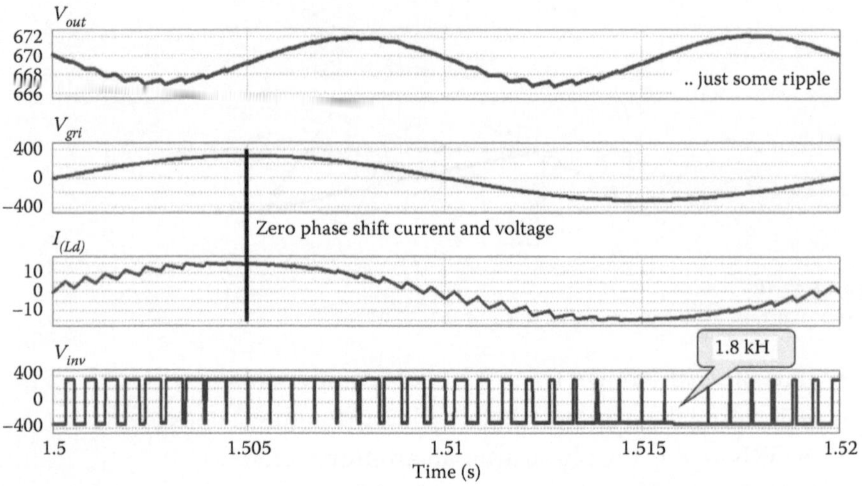

FIGURE 13.20
Waveforms for the operation of the controller and converter as a grid interface.

from Figure 13.20. The current waveform can be improved with a higher switching frequency.

The current regulator has to be designed for the case of tracking a sinusoidal waveform. This is a particular control system case which will be studied in a separate section of this chapter.

13.4 Three-Phase Converters as Grid Interface

Figure 13.21 shows a setup for a three-phase converter interface. It considers the grid connection inductances between the grid voltage sources and the power converter. The converter works as a boost converter with a higher output voltage than the result of a conventional diode rectification.

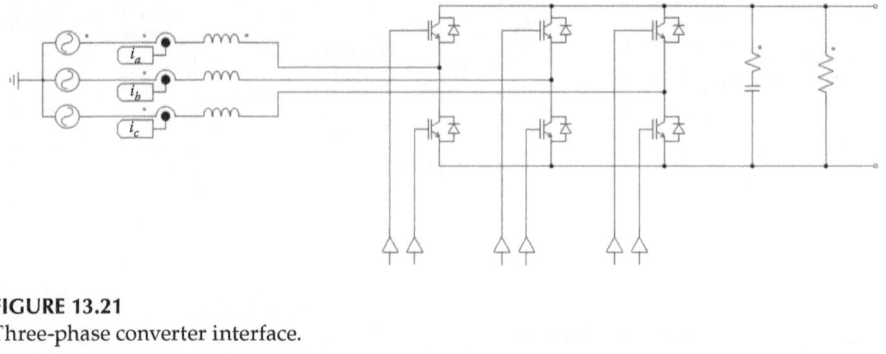

FIGURE 13.21
Three-phase converter interface.

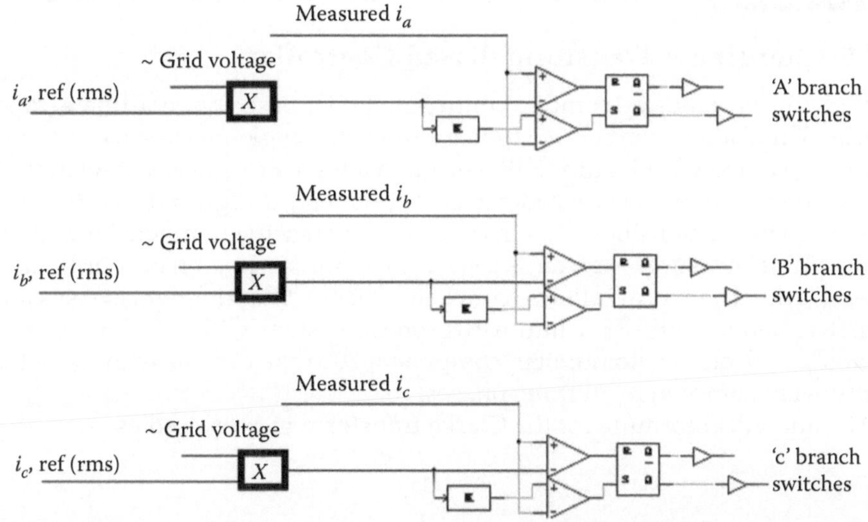

FIGURE 13.22
Hysteresis control of currents.

The simplest control circuitry can be realised after the definition of a phase-current reference, synchronisation with grid voltage and a hysteresis controller. The schematic is shown in Figure 13.22, while the resulting waveforms are provided in Figure 13.23. It can be seen that the switching frequency is not constant due to the hysteresis controller.

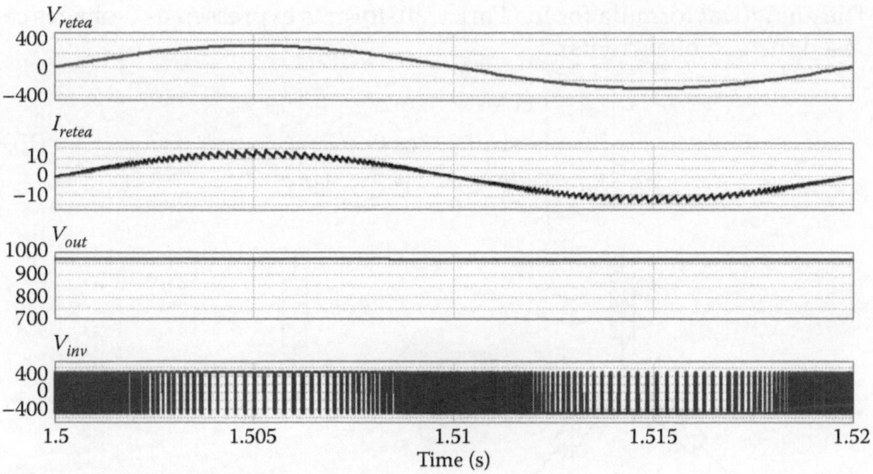

FIGURE 13.23
Waveforms for steady-state operation of the hysteresis current controller.

13.5 Coordinate Transform–Based Controller

A high-performance yet complex control method is based on coordinate transforms. This follows the control strategy from the single-phase system, previously explained with Figure 13.19. The main advantage consists of separation of the control action on two coordinates, previously designated as active and reactive. The relationship between the active and reactive components and the three-phase measures is given as a successive transformation of coordinates.

A Clarke transform allows conversion from the three-phase system, denoted herein with RST, into a two-phase system with AC components denoted α,β plus a homopolar component '0' that can be seen as a DC component common to all three phases.

The analytical formula for the Clarke transform is expressed as:

$$\begin{bmatrix} I_\alpha \\ I_\beta \\ I_0 \end{bmatrix} = \frac{2}{3} \cdot \begin{bmatrix} 1 & -\frac{1}{2} & -\frac{1}{2} \\ 0 & \frac{\sqrt{3}}{2} & \frac{-\sqrt{3}}{2} \\ \frac{1}{2} & \frac{1}{2} & \frac{1}{2} \end{bmatrix} \cdot \begin{bmatrix} i_R \\ i_S \\ i_T \end{bmatrix} \tag{13.7}$$

and can be observed in Figure 13.24.

The Park transform eliminates the AC form of signals with a coordinate change able to directly provide the active and reactive components as quasi-DC measures. This transform requires knowledge of the instantaneous angular coordinate (phase) of the grid.

The analytical formula for the Park transform is expressed as (voltages can be used instead of currents):

$$\begin{bmatrix} I_d \\ I_q \\ I_0 \end{bmatrix} = \begin{bmatrix} \cos\theta & \sin\theta & 0 \\ -\sin\theta & \cos\theta & 0 \\ 0 & 0 & 1 \end{bmatrix} \cdot \begin{bmatrix} I_\alpha \\ I_\beta \\ I_0 \end{bmatrix} \tag{13.8}$$

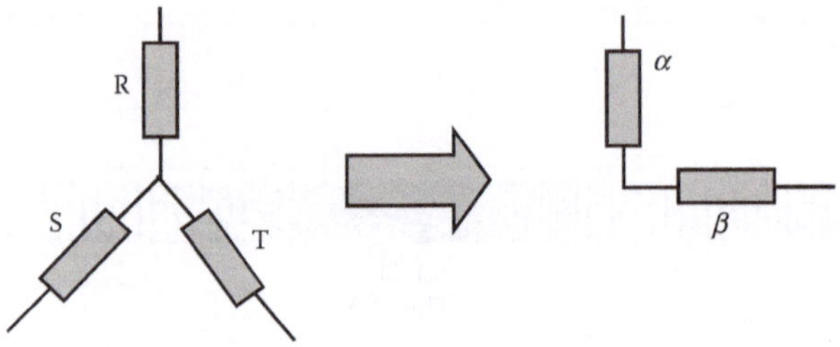

FIGURE 13.24
Schematic representation of the Clarke transform.

Both transforms can be grouped within a compact analytical form:

$$\begin{bmatrix} I_d \\ I_q \\ I_0 \end{bmatrix} = \frac{2}{3} \cdot \begin{bmatrix} \cos\theta & \cos\left[\theta - \frac{2\cdot\pi}{3}\right] & \cos\left[\theta - \frac{4\cdot\pi}{3}\right] \\ \sin\theta & \sin\left[\theta - \frac{2\cdot\pi}{3}\right] & \sin\left[\theta - \frac{4\cdot\pi}{3}\right] \\ \frac{1}{2} & \frac{1}{2} & \frac{1}{2} \end{bmatrix} \cdot \begin{bmatrix} i_R \\ i_S \\ i_T \end{bmatrix} \qquad (13.9)$$

Use of these transforms in a generic three-phase voltage system (220 V AC rms/phase at 50 Hz or 120 V AC rms/phase at 60 Hz) provides the waveforms shown in Figure 13.25.

Similarly, one may use reverse-transform relationships.

The reverse Park transform requires knowledge of the grid's angular coordinate (phase), and it is expressed with the analytical formula:

$$\begin{bmatrix} I_\alpha \\ I_\beta \\ I_0 \end{bmatrix} = \begin{bmatrix} \cos\theta & -\sin\theta & 0 \\ \sin\theta & \cos\theta & 0 \\ 0 & 0 & 1 \end{bmatrix} \cdot \begin{bmatrix} I_d \\ I_q \\ I_0 \end{bmatrix} \qquad (13.10)$$

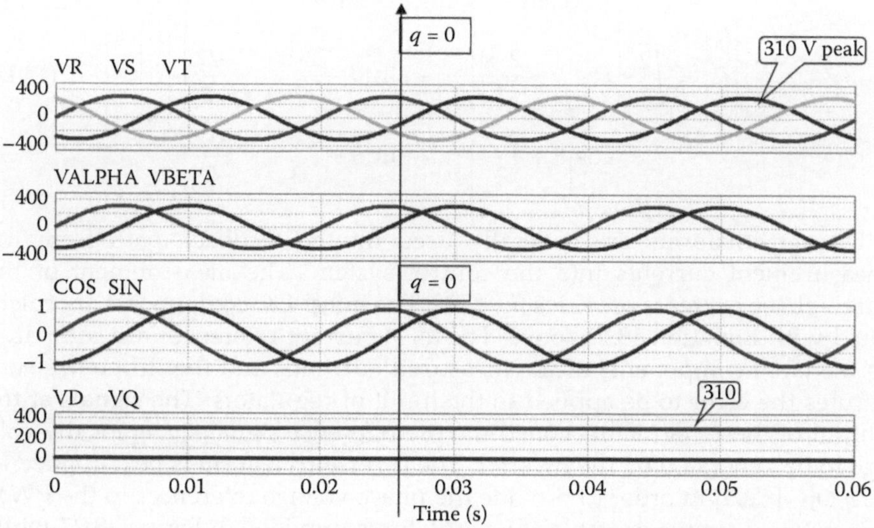

FIGURE 13.25
Waveforms to illustrate the transforms applied to a three-phase voltage system with 220 V rms and 50 Hz.

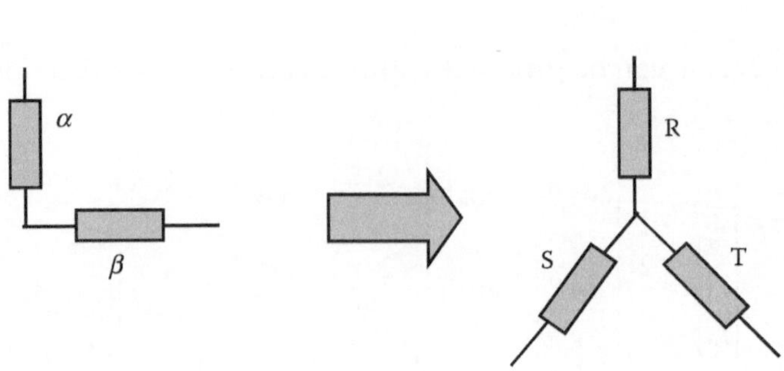

FIGURE 13.26
Graphical representation for the Clarke transform.

The reverse Clarke transform uses the equation:

$$\begin{bmatrix} i_R \\ i_S \\ i_T \end{bmatrix} = \begin{bmatrix} 1 & 0 & 0 \\ -\frac{1}{2} & \frac{\sqrt{3}}{2} & 0 \\ -\frac{1}{2} & \frac{\sqrt{3}}{2} & 0 \end{bmatrix} \cdot \begin{bmatrix} I_\alpha \\ I_\beta \\ I_0 \end{bmatrix} \tag{13.11}$$

and it can be illustrated with the phasor diagram from Figure 13.26.

Both transforms can be merged within the following form:

$$\begin{bmatrix} i_R \\ i_S \\ i_T \end{bmatrix} = \sqrt{\frac{2}{3}} \cdot \begin{bmatrix} \cos\theta & -\sin\theta & \frac{\sqrt{2}}{2} \\ \cos\left[\theta - \frac{2\cdot\pi}{3}\right] & -\sin\left[\theta - \frac{2\cdot\pi}{3}\right] & \frac{\sqrt{2}}{2} \\ \cos\left[\theta + \frac{2\cdot\pi}{3}\right] & -\sin\left[\theta + \frac{2\cdot\pi}{3}\right] & \frac{\sqrt{2}}{2} \end{bmatrix} \cdot \begin{bmatrix} i_d \\ i_q \\ i_0 \end{bmatrix} \tag{13.12}$$

Direct transforms are generally used on the feedback path from the measurement currents into the control system. The measurement of the three-phase currents provides $[i_R, i_S, i_T]$. The quasi-DC components are calculated with Equation 13.12 above. The $[d, q]$ current references are compared to the $[d, q]$ components of the measured currents, and the difference constitutes the error to be applied to the input of regulators. The signals at the output of these regulators constitute the reference signals $[d, q]$ for the voltage to be generated by the inverter. The reverse transform is next applied as $[d, q, 0] \rightarrow [r, s, t]$ in order to provide the phase voltage references to the PWM generator. A simple example of a regulator is provided in Figure 13.27, while a more complete diagram showing both the power and the control circuits is shown in Figure 13.28. It can also serve for understanding the information flux within a power converter controller.

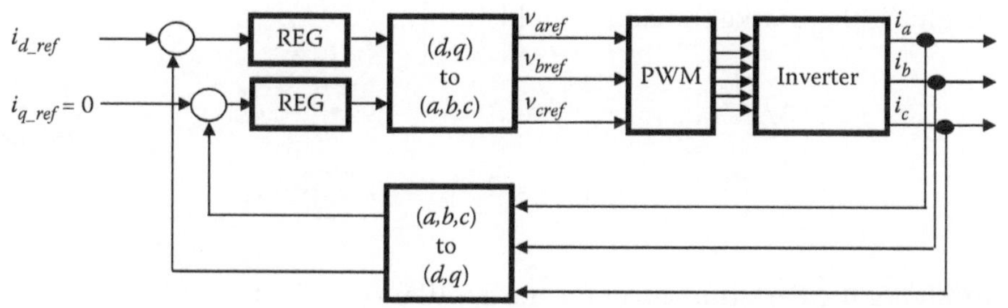

FIGURE 13.27
Control schematic for a three-phase system.

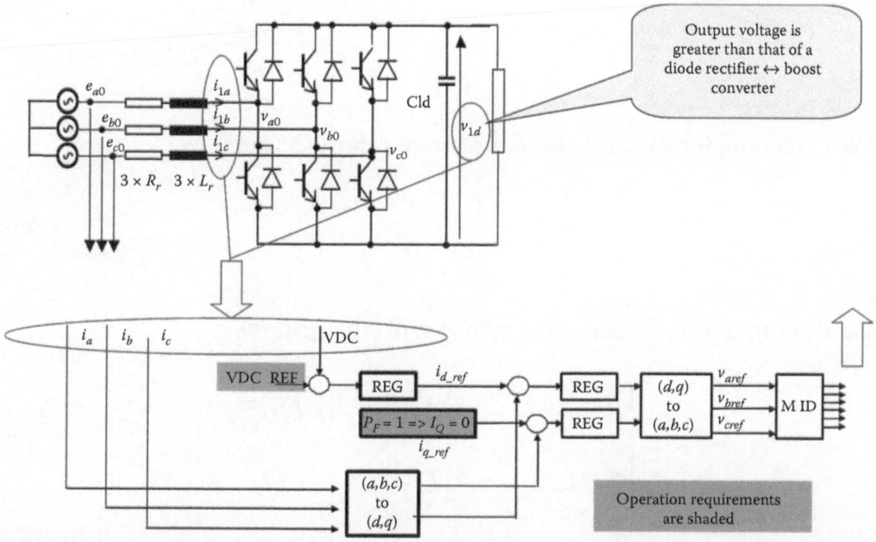

FIGURE 13.28
Complete schematic for the information flow within a three-phase controlled converter.

13.6 Circuit Equations

The circuit equations for the case of a grid supply can be written for each phase individually.

$$
\begin{aligned}
e_{R0} &= R_r \cdot i_{1R} + L_r \cdot \frac{di_{1R}}{dt} + v_{R0} \\
e_{S0} &= R_r \cdot i_{1S} + L_r \cdot \frac{di_{1S}}{dt} + v_{S0} \\
e_{T0} &= R_r \cdot i_{1T} + L_r \cdot \frac{di_{1T}}{dt} + v_{T0}
\end{aligned}
\tag{13.13}
$$

Applying the Clarke transform yields:

$$\underset{\Rightarrow}{3/2\ \text{Transform}} \begin{cases} e_{\alpha 0} = Rr \cdot i_{1\alpha} + L_r \cdot \dfrac{di_{1\alpha}}{dt} + v_{\alpha 0} \\ e_{\beta 0} = Rr \cdot i_{1\beta} + L_r \cdot \dfrac{di_{1\beta}}{dt} + v_{\beta 0} \end{cases} \tag{13.14}$$

Applying the Park transform for the angular coordinate yields:

$$\begin{cases} v_d = v_{\alpha 0} \cdot \cos\theta + v_{\beta 0} \cdot \sin\theta \\ v_q = -v_{\alpha 0} \cdot \sin\theta + v_{\beta 0} \cdot \cos\theta \end{cases} \Rightarrow \begin{cases} e_d = R_r \cdot i_d + v_d + L_r \cdot \dfrac{di_d}{dt} - \omega \cdot L_r \cdot i_q \\ e_q = R_r \cdot i_q + v_q + L_r \cdot \dfrac{di_q}{dt} + \omega \cdot L_r \cdot i_d \end{cases} \tag{13.15}$$

We can consider the angular coordinate reference, such as:

$$\begin{cases} e_d = E \\ e_q = 0 \end{cases} \tag{13.16}$$

Further on, consider the same phase with the grid (θ).

$$\begin{cases} E = R_r \cdot i_d + v_d + L_r \cdot \dfrac{di_d}{dt} - \omega \cdot L_r \cdot i_q \\ 0 = R_r \cdot i_q + v_q + L_r \cdot \dfrac{di_q}{dt} + \omega \cdot L_r \cdot i_d \end{cases} \tag{13.17}$$

This equivalent circuit can be schematically shown, as in Figure 13.29. Furthermore, consider the operation with the unity power factor (PF = 1), which provides $i_q = 0$.

13.7 Phase Current Tracking Methods

13.7.1 Proportional-Integrative-Sinusoidal Controller

After setting up the control on active and reactive power components to provide the references of the two components of the grid current, controllers need to be established for the current. If Park/Clarke transforms are used, the current components can be controlled directly on quasi-DC (d, q) components with PI controllers.

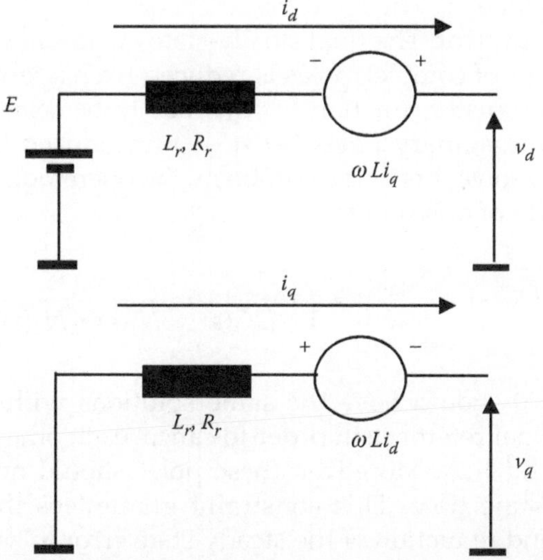

FIGURE 13.29
Simplified equivalent circuits for the two phases (active and reactive circuits).

Both single-phase and three-phase grid-connected power converters can also have the currents controlled directly as sinusoidal currents. The current control system would be composed of one PI controller for each phase and have a sinusoidal reference. Unfortunately, applying a sinusoidal reference to the input of a PI control law has been proven to produce a nonzero steady-state error. Several solutions can be used instead of the conventional PI controller.

A control law derived from the internal model principle [1] is presented herein to work with a sinusoidal reference:

$$i_{ref} = I_{REF} \cdot \cos(\omega t) \tag{13.18}$$

with the Laplace transform:

$$I_{ref}(s) = \frac{I_{REF} \cdot s}{s^2 + \omega_0^2} \tag{13.19}$$

Considering a conventional closed-loop system around an open-loop transfer function $D \cdot G(s)$ with the previous sinusoidal reference yields a Laplace transform for the error signal as:

$$E(s) = \frac{I_{ref}(s)}{1 + DG(s)} = \frac{I_{REF} \cdot s}{s^2 + \omega_0^2} \cdot \frac{1}{1 + DG(s)} \tag{13.20}$$

This Laplace function has a pole pair on the imaginary axis, and the Final Value Theorem does not apply. It can be used only when all poles are in the left half of the s-plane [2]. The Final Value Theorem is usually used to calculate the steady-state error. The final steady-state error will not be zero in this case unless the pair of complex poles is reduced from a term within $D \cdot G(s)$.

The open-loop transfer function $D \cdot G(s)$ has to be selected to cancel the effect of the two imaginary poles $(+/-j \cdot \omega_0)$ introduced by the sinusoidal reference term. To solve this design problem, the open-loop transfer function is written as a ratio of polynomials:

$$DG(s) = \frac{N_1(s)}{N_2(s)} \Rightarrow \frac{1}{1+DG(s)} = \frac{N_2(s)}{N_1(s)+N_2(s)} \tag{13.21}$$

Herein, $N_2(s) = 0$ should have the same solutions with the poles as the model for sinusoidal reference in order to cancel each other out. This means including a term $s^2 + \omega_0^2$. Moreover, these poles should not be solutions of $N_1(s) = 0$ at the same time. This constraint guarantees the usage of Final Value Theorem and reduction of the steady-state error to zero.

Acknowledging the advantages of using a PI control law, such terms will be maintained in DG(s), while a resonant term $s^2 + \omega_0^2$ is added therein. This yields the controller in Figure 13.30. The conventional PI regulator is enhanced by a term corresponding to the two imaginary solutions $(+/-j \cdot \omega_0)$. The equivalent transfer function of the regulator becomes:

$$G_{PIS}(s) = K_p + \frac{K_i}{s} + \frac{K_s \cdot s}{s^2 + \omega_0^2} = \frac{K_p \cdot s \cdot \left[s^2 + \omega_0^2\right] + K_i \cdot \left[s^2 + \omega_0^2\right] + K_s \cdot s^2}{s \cdot \left[s^2 + \omega_0^2\right]} \tag{13.22}$$

The open-loop transfer function is calculated as:

$$DG(s) = \frac{F_1(s)}{\left[s^2 + \omega_0^2\right] \cdot F_2(s)} \tag{13.23}$$

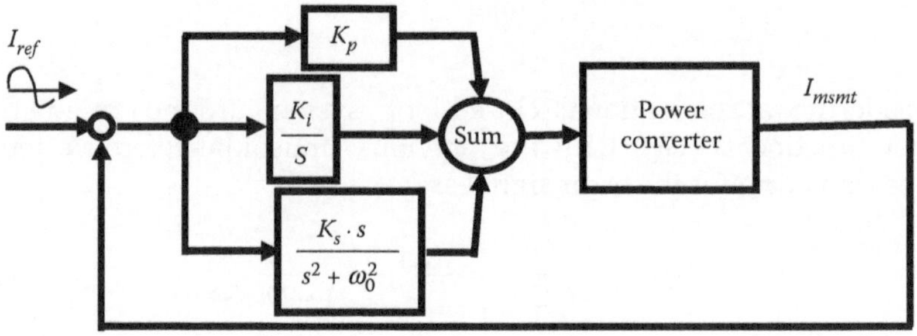

FIGURE 13.30
Block diagram of the compensated controller.

and Equation (13.20) is modified as:

$$E(s) = \frac{I_{REF} \cdot s}{s^2 + \omega_0^2} \cdot \frac{1}{1 + DG(s)} = \frac{I_{REF} \cdot s}{\left[s^2 + \omega_0^2\right] \cdot F_2(s) + F_1(s)} \tag{13.24}$$

The Final Value Theorem can now be applied and yields:

$$e_{ss} = \lim_{s \to 0}[s \cdot E(s)] = \lim_{s \to 0}\left[s \cdot \frac{I_R \cdot s}{\left[s^2 + \omega_0^2\right] \cdot F_2(s) + F_1(s)}\right] = 0 \tag{13.25}$$

13.7.2 Feed-Forward Controller

Proportional-integrative-sinusoidal control is currently very often used for controlling grid currents. Improved forms are proposed in [3,4]. A variation of these controllers is used for higher harmonic cancellation [5]. The main drawback of this method consists of its dependency on accurate knowledge of the grid frequency (ω_o), since this is part of the control law. Any slight change in grid frequency produces a steady-state error of the tracking system (Figure 13.31).

The current controller acts upon an inverter leg that is composed of two IGBT devices. The operation can be separated on the positive and negative polarities of this AC current: positive polarity can be achieved with control of the top-side IGBT, while negative polarity can result from control of the low-side IGBT (Figure 13.32).

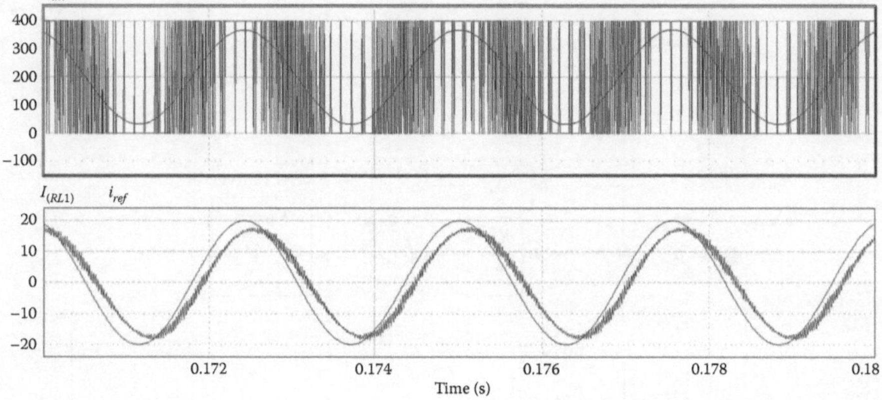

FIGURE 13.31
P-I-S control applied to the system operated out of synchronisation, with the grid at 390 Hz instead of the resonant controller frequency of 400 Hz. Top waveform = phase current reference and measurement. Bottom waveform = inverter pole voltage and grid voltage.

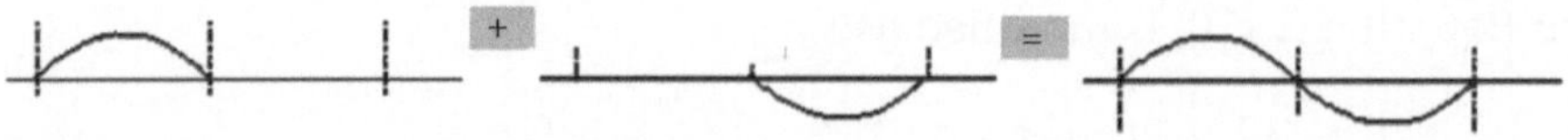

FIGURE 13.32
Principle of the control system.

The control problem changes from tracking a periodic sinusoidal waveform into tracking a varying waveform [6]. Having two independent control systems allows the intervals with a null reference to reset the steady-state error, if any. The reference approximates a piecewise waveform, and the design of the control system is similar to the case of a ramp input (also called *velocity* input) [7].

This approximation of the reference input is:

$$i_{ref} = \omega_0 \cdot I_{REF} \cdot t \cdot 1(t) \tag{13.26}$$

with the Laplace transform:

$$I_{ref}(s) = \frac{\omega_0 \cdot I_{REF}}{s^2} \tag{13.27}$$

A feed-forward component K_n is added to compensate for nonideal behaviour of the control loop under varying input waveforms (Figure 13.33).

The controller output is:

$$\begin{aligned} D_{PIFF}(s) &= \left(K_p + \frac{K_i}{s}\right) \cdot (I_{ref}(s) - I_{msmt}(s)) + K_n \cdot I_{ref}(s) \\ I_m(s) &= D_{PIFF}(s) \cdot G_C(s) \end{aligned} \tag{13.28}$$

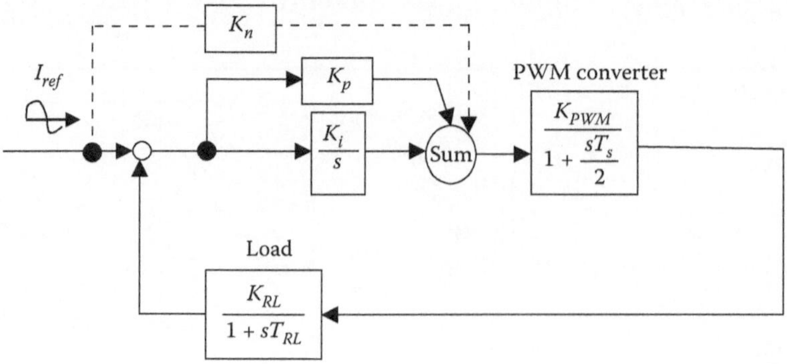

FIGURE 13.33
Feed-forward controller.

This further yields:

$$I_{msmt}(s) = \frac{\left(K_p + \dfrac{K_i}{s} + K_n\right)}{\left(\dfrac{1}{G_C(s)} + K_p + \dfrac{K_i}{s}\right)} \cdot (I_{ref}(s)) \tag{13.29}$$

And finally:

$$E(s) = I_{ref}(s) - I_{msmt}(s) = \left[\frac{1 - K_n \cdot G_C(s)}{1 + \left(K_p + \dfrac{K_i}{s}\right) \cdot G_C(s)}\right] \cdot I_{ref}(s) \tag{13.30}$$

The Final Value Theorem is applicable herein since there is no right-half pole, and, accordingly, the steady-state error can be calculated with:

$$\begin{aligned} e_{ss} &= \lim_{t\to\infty} e(t) = \lim_{s\to 0} s \cdot E(s) \\ &= \lim_{s\to 0}\left\{ s \cdot \left[\frac{1 - K_n \cdot G_C(s)}{1 + \left(K_p + \dfrac{K_i}{s}\right) \cdot G_C(s)}\right] \cdot \frac{I_{ref}}{s^2}\right\} \\ &= \lim_{s\to 0}\left\{\left[1 - K_n \cdot \frac{K1}{s + K2}\right] \cdot \left[\frac{s}{s + GK_0}\right] \cdot \frac{I_{ref}}{s}\right\} = 0 \end{aligned} \tag{13.31}$$

Sample results are shown in Figure 13.34 and prove the advantage over the P-I-S controller when the grid frequency varies. Otherwise, the performance of the feed-forward controller is comparable with the performance of the proportional-integral-sinusoidal controller. When analogue power management ICs are used for implementation, the system is simpler than the P-I-S controller.

Another advantage consists of improved power efficiency and faster transient response since the two IGBTs composing an inverter leg are operated for 180° only. This means reduction of power loss, especially with driver circuits.

13.8 Grid Synchronisation

Grid-connected power converters show a large impedance when connected to the grid [8]. This requires synchronisation of the PWM pattern with the

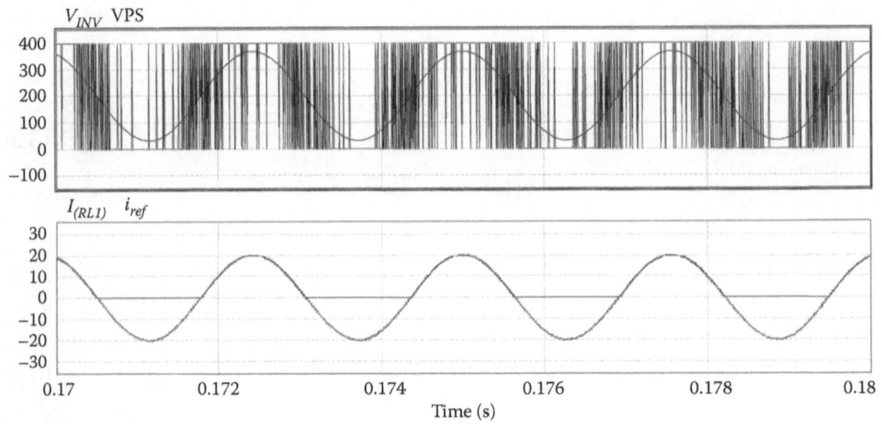

FIGURE 13.34
The use of the feedforward component with the grid at 390 Hz instead of 400 Hz. Top waveform = phase current reference and measurement. Bottom waveform = inverter pole voltage and grid voltage.

phase of the grid in order to control the power factor. There are different methods possible for grid synchronisation. The simplest solution consists of a zero crossing detector and a counter circuit (Figure 13.35 for single phase grid and Figure 13.36 for three phase grid).

The grid voltage is sensed before any filter associated with the power converter and converted to a lower level with a signal transformer or resistive divider. The resulting signal is compared against a threshold device, such as the base–emitter junction of a bipolar transistor or an IC comparator, to provide a logic signal corresponding to the polarity of the grid voltage. The detected logic signal resets a digital counter whose up-counting provides the phase coordinate of the PWM pattern.

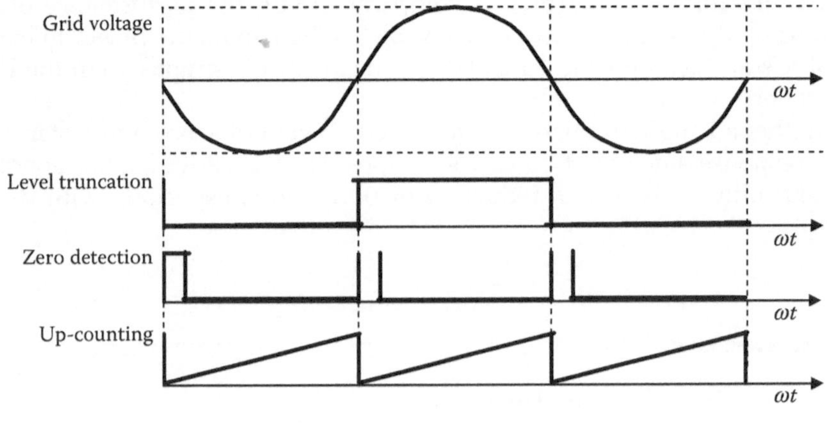

FIGURE 13.35
Principle of a simple synchronisation circuit.

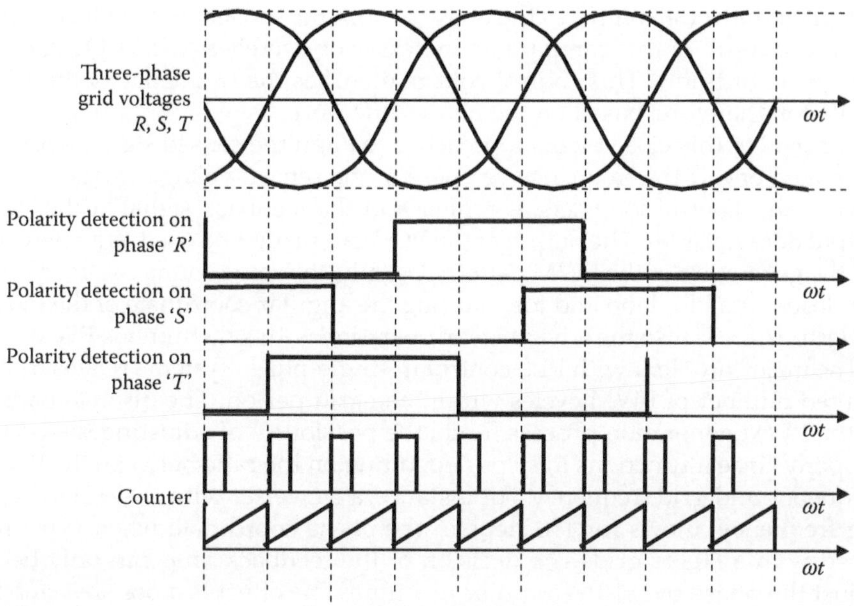

FIGURE 13.36
Principle of a synchronisation circuit for a three-phase grid.

There are sources of errors in this detection, such as noise in the line voltage due to unfiltered switching, dispersion and temperature variations of the threshold level and delay in detecting the exact zero-crossing moment. The sum of these errors should be kept smaller than the quantisation step of the digital system. Improved systems include a phase-locked loop configuration able to filter unwanted small variations in the detecting process. This helps eliminate the jitter around the zero-crossing moments.

A generic PLL circuit schematic is shown in Figure 13.37, and it is based on a voltage-controlled oscillator (VCO) working at a frequency multiple of N times the detected frequency. In power converters, it makes sense to select the VCO frequency equal to the PWM sampling/switching frequency to induce a synchronous PWM operation.

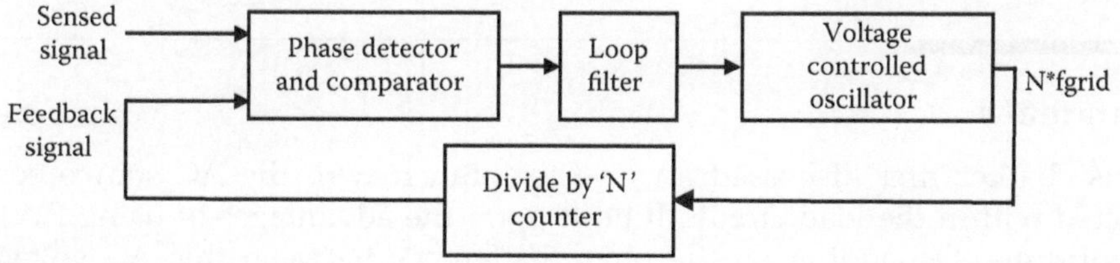

FIGURE 13.37
Principle of a PLL circuit.

A phase detector compares the phase of both the sensed and feedback voltages. The result of this comparison increases or decreases the VCO's control voltage accordingly. This control voltage modifies the frequency of the VCO oscillator, therefore, based on the phase difference.

The goal of this closed-loop approach is to align the sensed signal with the generated one. If the frequency or phase of the sensed voltage varies for any reason, this control loop acts as a filter and the feedback signal at the VCO output does not jitter. The output of the VCO can further be used as a reference for the generation of the PWM pattern. Usually, the same counter is used both for closing the PLL loop and for counting the angular coordinate of the PWM pattern. There is also the advantage of operation with synchronous PWM.

The major problem with PLL control in single-phase systems relates to the limited number of PWM cycles within one grid period. The discrete nature of the PWM generation process limits the possibility of adjusting the phase properly since it functions in steps depending on the ratio between the PWM frequency and grid frequency. For instance, a converter with a 9-kHz switching frequency allows for 150 steps in the phase coordinate when synchronised with a 60 Hz grid. The decision of the feedback loop can only be to adjust the phase by ± 1 PWM pulse at a time. The effect is a possible step in the filtered voltage at each zero crossing when the phase is abruptly reset to zero.

A solution in the single-phase case consists of using asynchronous PWM generators, but they do not guarantee the best harmonic content of the grid current. However, evolved VCO systems can work with asynchronous PWM with a noninteger ratio of switching and grid frequencies.

Grid synchronisation for three-phase converters benefits from multiple zero crossings and the symmetry of a grid-generated three-phase system. The detection module includes zero-crossing detection for all three phases, and this results in a signal with a frequency six times larger than the grid frequency. The PLL circuit can therefore provide more resolution with much finer tuning of the detected frequency.

PLL controllers can be implemented with dedicated analogue or digital ICs or a proper software routine in the microcontroller or digital signal processor (DSP) control system.

Summary

This chapter first discussed an inverter branch with the AC source connected within the load circuit. It built upon the advantages of using PWM algorithms. Depending on the phase difference between this AC voltage source and the inverter reference, the operation has an energy transfer from inverter to source or the reverse, from source to inverter.

The phasor diagram shows decomposition into two components:

- Active current = The current component and the grid voltage are in phase.
- Reactive current = The current component is orthogonal to the grid voltage.

It is important that the system can be decoupled and the two components can be controlled independently. Depending on the active current direction, the active energy transfer can be from the inverter to the source or from the source to the inverter. Operation of the converter can be defined as an interface for the AC grid while featuring low harmonics of the supply current.

Finally, different control methods were presented. The use of coordinate transforms (Park/Clarke) is recommended in order to establish equivalence between the three-phase AC measures and DC quantities. Control methods based on PWM were demonstrated.

References

1. Fukuda, S., Yoda, T., 2001, "A Novel Current-Tracking Method for Active Filters Based on a Sinusoidal Internal Model", *IEEE Transactions on Industry Applications*, 37, 888–895.
2. Franklin, G.F., Powell, J.D., Emami-Naeini, A., 2015, *Feedback Control of Dynamic Systems*, 7th edition, Prentice-Hall, Englewood Cliffs, NJ.
3. Timbus, A., Liserre, M., Teodorescu, R., Rodriguez, P., Blaabjerg, F., 2009, "Evaluation of Current Controllers for Distributed Power Generation Systems", *IEEE Transactions on Power Electronics*, 24(3), 654–664.
4. Lisserre, M., Teodorescu, R., Blaabjerg, F., 2006, "Multiple Harmonics Control for Three-Phase Grid Converter Systems with the Use of PI-RES Current Controller in a Rotating Fame", *IEEE Transactions on Power Electronics*, 21(3), 836–841.
5. Bojoi, R.I., Griva, G., Bostan, V., Guerrero, M., Farina, F., Profumo, F., 2005, "Current control strategy for power conditioners using sinusoidal signal integrators in synchronous reference frame", *IEEE Transactions on Power Electronics*, 20(6), 1402–1412.
6. Franklin, G., Powell, D., Emami-Naeimi, A., 2002, *Feedback Control of Dynamic Systems*, Prentice-Hall, Englewood Cliffs, NJ.
7. Neacsu, D.O., 2010, "Analytical Investigation of a Novel Solution to AC Waveform Tracking Control", *IEEE International Symposium in Industrial Electronics*, Bari, Italy, July 2010, pp. 2684–2689.
8. Neacsu, D.O., 2013, *Switching Power Converters—Medium and High Power*, 2nd edition, CRC Press, Taylor & Francis, Boca Raton, FL.

14

Uninterruptible Power Supplies

14.1 Role of Uninterruptible Power Supply Systems

The conventional role of uninterruptible power supply systems is to deliver energy when the main AC grid voltage fails. However, modern requirements add up to this goal:

- Voltage conditioning for waveform quality
- System functions like intelligent power management (peak shaving, demand management and so on)

This chapter provides an introduction to modern UPS systems, combining the traditional task of reserve power supply with modern functions aimed at energy savings.

Since 2006, electricity outages have been estimated to cost the United States approximately $79 billion annually, and two-thirds of those costs were from outages lasting less than 5 minutes [1]. Energy storage devices together with electronic processing and transfer of energy can provide short-term solutions to power outages to minimise their costly impact.

Most voltage failure situations are very short in duration, yet they have important negative effects. For instance, a grid voltage fail of 0.25 sec can interrupt work within a computing or communications centre for over 15 minutes due to required operating system rebooting and other startup operations. Statistics collected about the U.S. national power system show at least nine hours of AC mains failure per year, which is unacceptable for telecom equipment. The reliability constraints for the AC supply of computer or communication systems are more stringent than for the national power system or for most other applications. This is illustrated with data in Table 14.1.

Due to the importance of reliable energy sources for the telecom industry, global telecommunications network provider spending on distributed generation and energy storage purchases is expected to grow from $2.4 billion in 2015 to $3.4 billion in 2024 (according to Navigant Research [2]).

Standard waveform quality requirements or supply rules for national electrical grid performance are under the requirements for performance and

TABLE 14.1
Reliability Requirements for Various Applications

Reliability (%)	Down Time per Year	Example
99	88 hours	
99.9	8.8 hours	Statistic corresponding to U.S. national grid (120 V AC)
99.99	53 minutes	
99.999	5.3 minutes	Typical requirements for telecom systems (phone centres, networks of communications)
99.9999	32 second	
99.99999	3.2 second	

quality within a data or communications centre. It is recommended to seek local improvement of performance through harmonic filtering and elimination of other incidental variation of voltage.

Furthermore, it is recommended to transfer energy with the electrical grid at the unity power factor without any reactive power component. Electrical power supply systems are rated depending on apparent power, which includes the active power (that produces the active power in the load) as well as the reactive power (useful simply to facilitate good operation). The power balance is shown in Figure 14.1. These principles are standardised with various standards dedicated to uninterruptible power supplies, like the IEC62040-3. Due to their decisive importance, the cost of UPS systems is an important component within a data or communications centre budget.

The UPS system secures an alternative source of energy and – preferably – also conditions interaction with the electrical grid. Industrial standards require switching the power supply to a reserve source in less than 25 msec in order to allow seamless and uninterrupted operation of the computing or communications systems. This time constraint comes from design requirements for point-of-load power supplies (Chapter 4), which are supposed to deliver power for another 25 msec after their input supply disappears. Figure 14.2 considers a practical example and shows the input and output voltages and currents for a power supply before and after the AC grid voltage

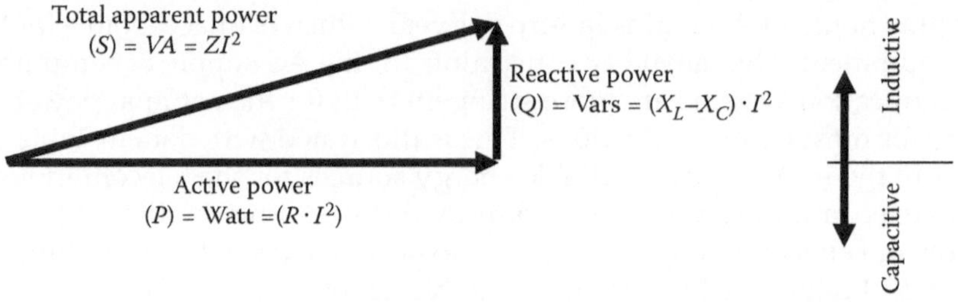

FIGURE 14.1
Relationship between power components.

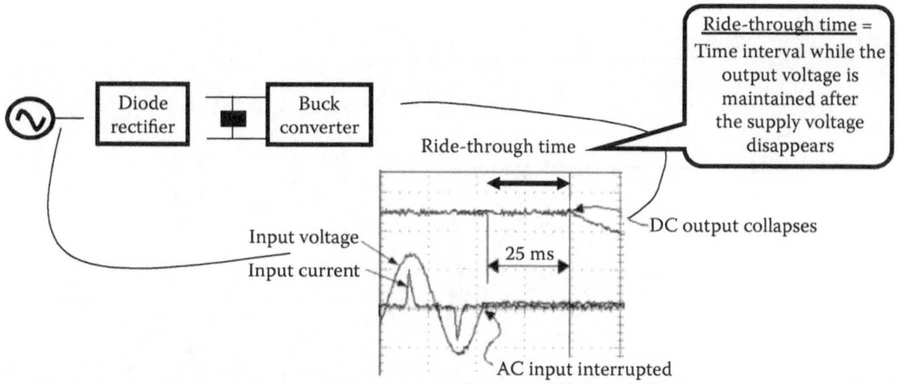

FIGURE 14.2
Ride-through operation of a local power supply.

collapses. The UPS system needs to take over the supply voltage faster than this *ride-through time* considered within the design of point-of-load power supplies.

A series of possible situations and definitions are identified when the quality of AC energy supply is jeopardised. This includes the cases shown in Figure 14.3. Along with the supply of energy through the UPS system, electronic conditioning of the input power source can be used to eliminate all these negative effects. Electronic conditioning of the voltage at the input power source can be achieved with either passive or active solutions.

A series of passive solutions can be considered at first without the involvement of switch-mode power supplies in order to minimise the negative effects shown in Figure 14.3. Passive components act more quickly than switched-mode power supplies and do not need fault detection circuitry before acting. Their operation also does not depend on the actual presence of a reliable power source.

- *Metal-oxide varistor* (MOV) devices limit possible and occasional overvoltage.
- *EMI filters* eliminate the high-frequency components from the harmonic spectra of the PWM signals.
- *Isolation transformers* offer an electrostatic barrier and also eliminate the common mode voltages.
- *Ferro-resonant transformers* allow voltage regulation and overvoltage mitigation.

In addition to these passive solutions, electronic power converters can have a role in power conditioning with the task of mitigating the previously shown effects. Two major classes of applications emerge:

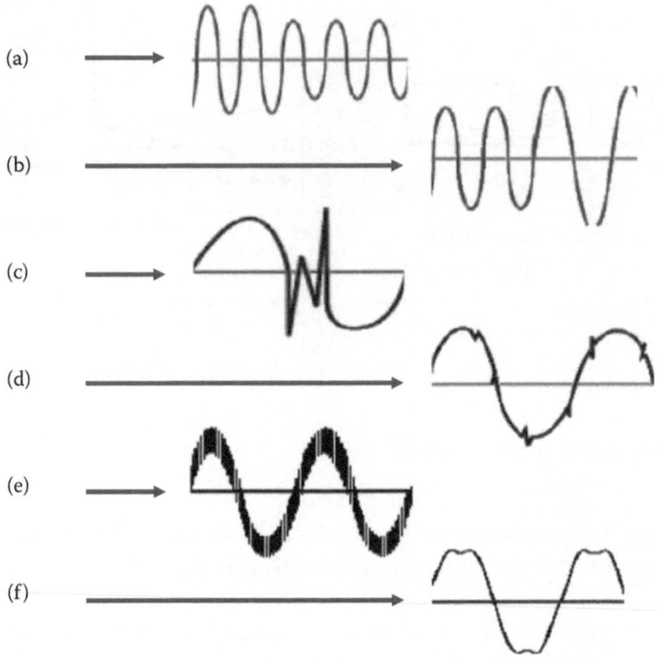

FIGURE 14.3
Quality problems in grid energy: (a) temporary decrease in voltage (sags), (b) sudden increase in voltage (swells), (c) transitory spike (transient, spike, impulse), (d) fast transient with reversed peak (notch), (e) high-frequency noise (noise), (f) distortion.

- Filtering the load current with a *parallel active filter* (Figure 14.4). The power converter is connected in the same node as the load, and its current is controlled to compensate for the harmonics and nonlinearity of the load current. However, the grid current remains clear of harmonics as a pure sinusoidal waveform. The power converter is actually an inverter and works with voltage to control a current produced by applying its output voltage against a connecting inductor.
- Filtering the load voltage with a *series filter* (Figure 14.5). The electronic power converter is connected through a series transformer with the goal of taking on the voltage harmonics.

14.2 Classification of Uninterruptible Power Supply Topologies

UPS systems have the primary role of delivering energy to loads in the case of grid voltage failure. This primary function is analysed first herein, while special usage modes for UPS are described at the end of this chapter.

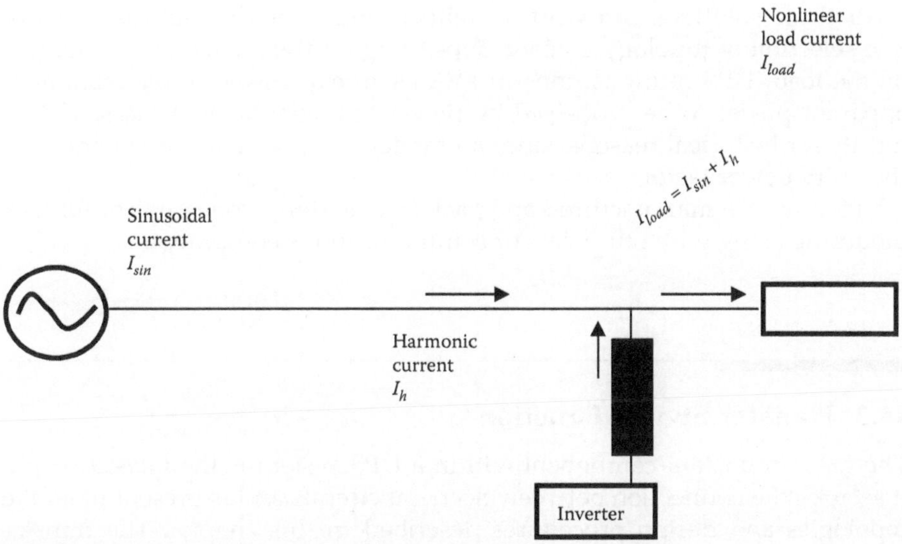

FIGURE 14.4
Power converter used as parallel-connected filter.

There are several UPS topologies used currently by residential or commercial customers [3–5], and the most known design principles are:

- Uninterruptible power supply in standby state (*standby UPS*)
- Uninterruptible power supply in direct connection with the grid (*line-interactive UPS*)
- Uninterruptible power supply with ferro-transformer (*standby-ferro UPS*)
- Uninterruptible power supply with double conversion (*double conversion online UPS*)
- Uninterruptible power supply with delta conversion [6] (*delta conversion online UPS*)

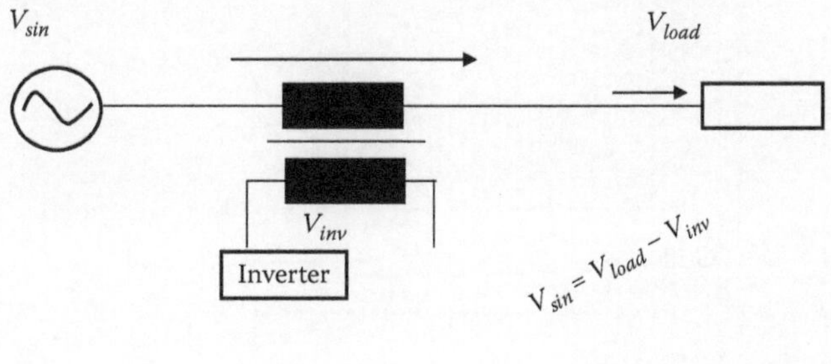

FIGURE 14.5
Series-connected filter.

All these solutions can work for either single- or three-phase systems. The selection of topology is made depending on the power level requested by the load. UPS rating is done in kVA as an expression of the maximum apparent power to be processed by power semiconductor devices. This is mostly for historical reasons, since all modern UPS systems are operated at the unity power factor.

UPS units are manufactured and packaged as standalone units or for rack mounting along with other data or communications equipment.

14.3 Transfer Switch Function

The most important component within a UPS system is the *transfer switch*. It secures the connection between electrical circuits and is present in all the topologies and design procedures described in this chapter. The transfer switch is shown schematically in Figure 14.6.

The transfer switch can be implemented with controlled power semiconductor devices, where it is also called a static or solid-state switch. For

FIGURE 14.6
Transfer switch.

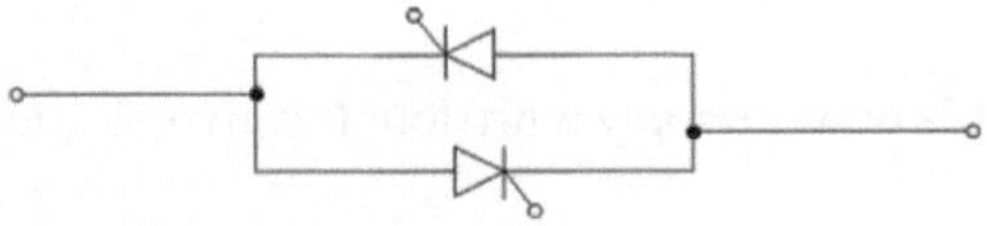

FIGURE 14.7
Transfer switch implemented with thyristors.

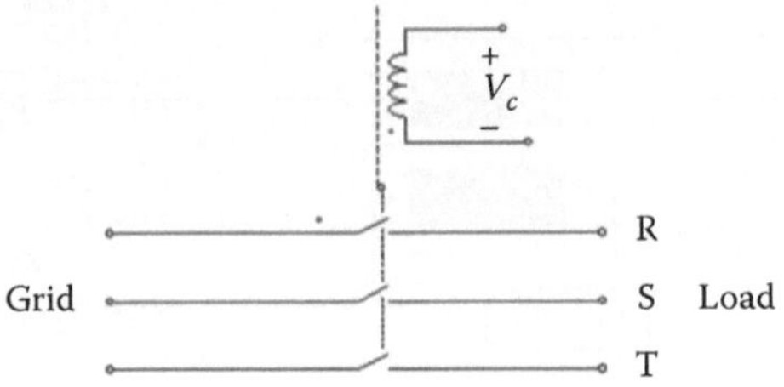

FIGURE 14.8
Transfer switch using electromagnetic devices.

instance, a switch implemented with thyristors is shown in Figure 14.7 for a single-phase system.

The transfer switch can also be implemented with electromagnetic devices that have electronic control of the electrical switch. An example is shown for a three-phase system in Figure 14.8.

14.4 Standby Uninterruptible Power Supply

Standby UPS systems are very popular, and they are used to supply residential or commercial computer systems with low installed power. Figure 14.9 illustrates the principle of operation. The transfer switch herein has a role in energy delivery to the load either from the grid or from a battery. The grid is considered the primary energy source. At grid failure, the load is supplied from a battery through an inverter able to generate the appropriate AC voltage. The inverter starts to operate only for a grid voltage failure – hence the name standby system.

The control system for the inverter is shown in Figure 14.10. It has to operate independently from the transfer switch position, even if the output voltage is taken directly from the grid. The angular coordinate (phase) of the grid system is continuously monitored and tracked through a phase lock loop (PLL) system (see Section 13.8). The PWM pulses produced by a single-phase or three-phase inverter are aligned with the angular coordinate.

The RMS of the voltage produced by the inverter is calculated with a numerical software module, and the feedback control loop adjusts the PWM pulses for the generation of the desired RMS voltage.

Finally, the train of PWM pulses produced by the inverter is filtered with a low-pass filter for a smooth waveform produced in the connection node. The output of this system is a sinusoidal voltage waveform.

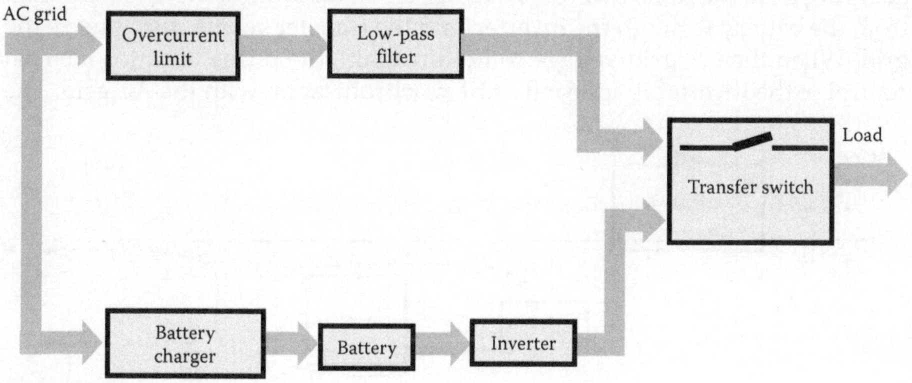

FIGURE 14.9
Principle of standby UPS.

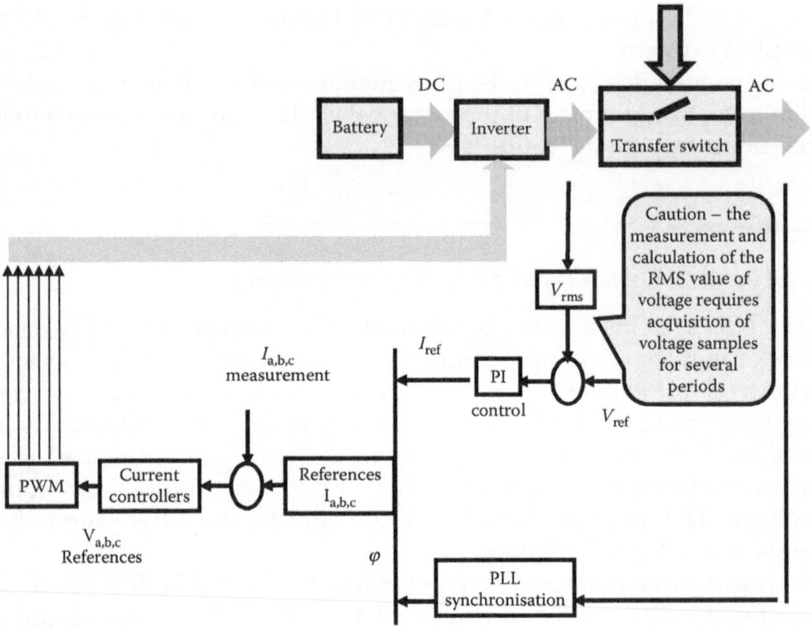

FIGURE 14.10
Principle of control system for a standby UPS system.

14.5 Line-Interactive Uninterruptible Power Supply

Line-interactive UPS systems are very commonly used by small corporations, internet suppliers or computer networks at the usual power level within the range 0.5–5 kVA. The principle is illustrated in Figure 14.11. The inverter is always connected to the load in the same node as the grid. This allows battery charging from the same inverter. At AC grid voltage failure, energy is supplied from the battery through the inverter, and the transfer switch disconnects the grid. When the AC grid voltage comes up again, the phase used for internal control of the inverter is adjusted until synchronisation with the AC grid. The

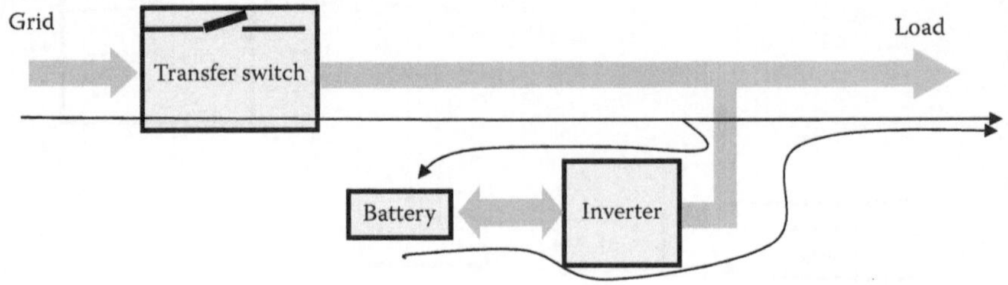

FIGURE 14.11
Principle of line-interactive UPS.

voltages at both ends of the transfer switch have to be quasi-identical, and afterwards, the switch connects the AC grid power. The load current transfers from the inverter to the AC grid source until the inverter current is zero.

14.6 Standby-Ferro Uninterruptible Power Supply

The standby-ferro UPS was used in the past for load power in the range 3–15 kVA. The principle schematic is shown in Figure 14.12. The construction is based on a transformer working with core saturation, featuring three windings. The actual connection to the load is done through this magnetic device, also ensuring galvanic isolation.

However, this solution has been abandoned since the efficiency through the transformer has been very low, and the transformer would heat up. Another major drawback consists of possible instability due to the resonance between the transformer inductance and the rectifier capacitance.

This transition from magnetic components to more semiconductor-based converters constitutes a more general current trend inspired by the impressive performance of modern semiconductor devices.

During operation, the energy flows from the AC grid through the transfer switch towards the load. When the AC grid voltage fails, the switch is opened, and the inverter generates voltage out of the battery. The transformer herein also has a role in overload protection.

14.7 Double Conversion Online Uninterruptible Power Supply

This is the most-used solution at power levels above 10 kVA, and the principle illustrating this topology is shown in Figure 14.13. The power converter is

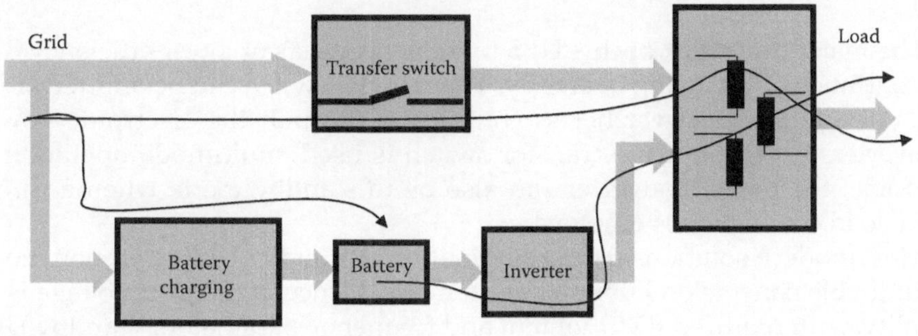

FIGURE 14.12
Principle of a standby-ferro UPS.

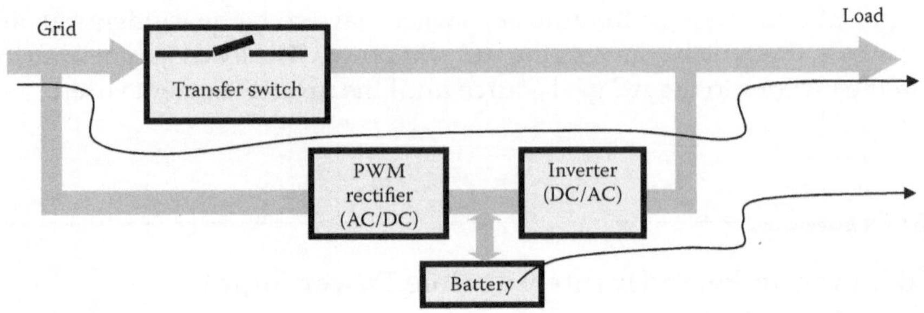

FIGURE 14.13
Principle of a double conversion online UPS.

composed of a fully controlled AC/DC converter and a DC/AC converter. This centrepiece is also called a back-to-back converter. The roles of the components are:

- The PWM rectifier takes energy from the AC grid and maintains a constant voltage across the capacitor bank used for setting up the DC voltage.
- The PWM inverter is supplied from the DC bus and used for AC voltage generation. The battery is connected in parallel with the DC bus. Contrary to the standby UPS system, the AC/DC/AC converter is now the main path for the energy flow, and it is operated continuously.

The major advantage of this solution consists of an instantaneous load transfer without transfer delay time or transition time. A second advantage consists of the superior quality of the voltage generated by a controlled inverter:

- In terms of reduced voltage and current harmonics
- Frequency and RMS voltage content stability

The major drawback of this UPS topology consists of lower efficiency due to continuous processing of energy. The transfer switch can be omitted since the energy flow is nevertheless continuous through the electronic power converter. However, if the transfer switch is used, multimode operation is possible: the power converter can also be in standby mode when a direct grid–load connection is employed.

Most modern solutions use a combination of standby UPS operation mode and double conversion UPS operation mode. Hence, if the grid voltage is of quality, with required RMS content and frequency parameters, standby UPS operation mode is used. Efficiency and reliability are improved since the electronic power converter does not process energy. When the grid voltage performance is jeopardised and not up to par with required performance, possibly with parameters out of specification, operation in double conversion UPS operation mode is desirable. The electronic power converter can thus generate AC voltage with the desired RMS content and frequency. Overall, this multimode operation provides an ideal balance between efficiency and protection.

Since the double conversion UPS system is a solution frequently used in modern power systems, the problem of reliability comes into discussion. Processing energy continuously through an electronic converter with semiconductor devices may introduce failure risks and a somewhat limited lifetime of the equipment. A semiconductor failure would stop operation of all the equipment. This risk is mostly not accepted within a telecom or computer system, as is shown with Table 14.1. The risks associated with double conversion UPS systems are alleviated by introducing redundancy, either as a cascaded or parallel connection of several UPS systems. Redundant operation of UPS systems is independent of the actual UPS topology, and it can be used with any of the previous topologies.

First, Figure 14.14 shows the redundant operation of two cascaded UPS systems able to provide increased reliability. If equipment composed of a double conversion rectifier-inverter system stops its operation due to an internal fault, then it is disconnected from the grid by the closing of the bypass switch. The energy flows from the AC grid through the bypass transfer switch and the second UPS system into the load. This allows the operator to replace the damaged unit and provide load supply continuously. This can be done with the hot-swap feature, which is connection–disconnection under voltage.

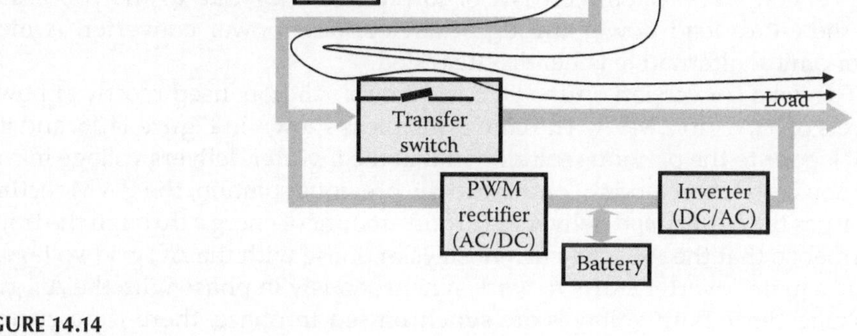

FIGURE 14.14
Redundant system with cascaded connection of UPS system.

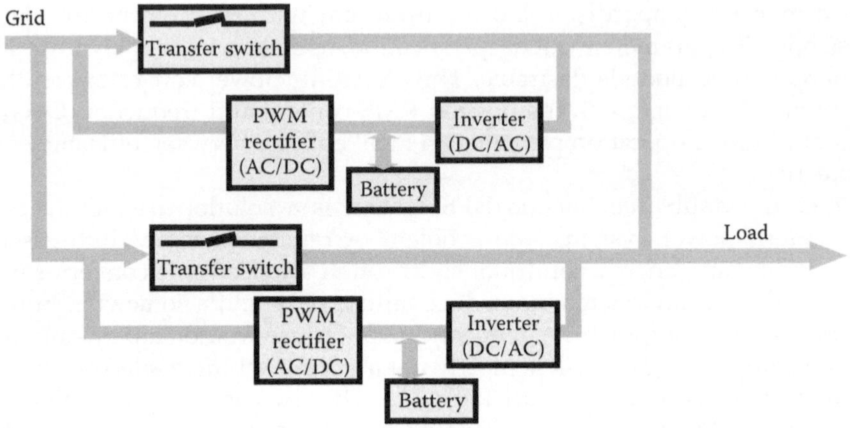

FIGURE 14.15
Redundant system with parallel connection of UPS systems.

A second solution involves the redundant operation of two parallel-connected UPS systems, aiming to provide increased reliability. An example of the complete system is shown in Figure 14.15. If equipment composed of a double conversion rectifier-inverter system stops its operation due to an internal failure, then it is disconnected through the bypass transfer switch.

Special control algorithms have to secure synchronisation of operation, since the parallel connection on the AC grid is not easy. Similar to the cascaded UPS systems, modern control systems add the hot-swap operation, which is connection–disconnection under voltage.

14.8 Delta Conversion Online Uninterruptible Power Supply

Despite the great waveform quality, the main drawback of the double conversion UPS system consists of lower efficiency due to the processing of the entire load power through the electronic power converter. A more economical alternative is herein suggested.

The delta conversion online UPS is a newer solution, used mostly at power levels of 5 kVA to 2 MVA. The core principle is shown in Figure 14.16, and it is analogous to the previous solution in that the inverter delivers voltage in continuous operation mode. Contrary to the previous solution, the PWM rectifier charges the battery *and* delivers a certain amount of energy through the transformer so that the AC grid current stays in phase with the AC grid voltage.

The main inverter starts to work synchronously in phase with the AC grid voltage. Since both voltages are synchronised in phase, there is no current circulation through the transformer, and the entire load current circulates

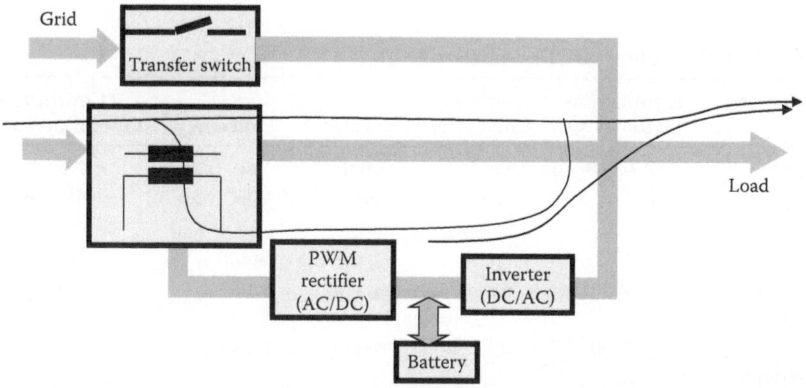

FIGURE 14.16
Principle of the delta conversion UPS system.

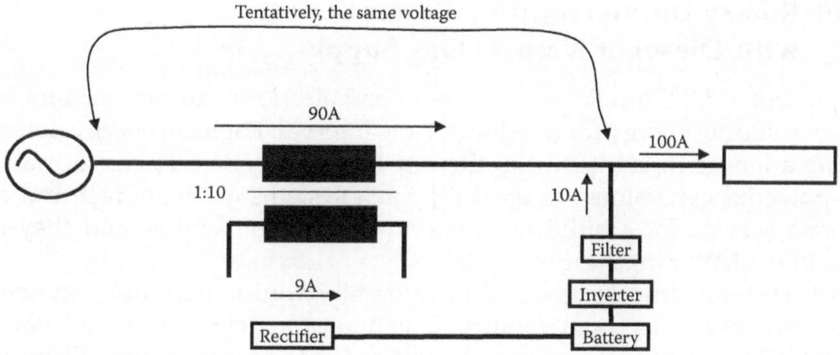

FIGURE 14.17
Numerical example of the principle of a delta-conversion UPS.

from the battery and through the inverter. The PWM rectifier is controlled as a current-controlled inverter so that it delivers current in the secondary winding of the transformer. This current is reflected in the primary winding and circulates from the grid to the load.

For a numerical example, if the load current is 100 A, the PWM rectifier regulates a 90-A current through a direct path within the transformer into the load. Differently from the required load current, the current through the inverter stays at 10 A. This numerical example is further illustrated graphically in Figure 14.17.

14.9 Comparison of Uninterruptible Power Supply Solutions

A quick review of solutions is shown in Table 14.2.

TABLE 14.2
Comparative Overview of UPS Solutions

	Application Field (kVA)	Voltage Correction	Cost	Efficiency	Continuous Operation Inverter
Standby	0–0.5	No	Small	Very high	No
Line interactive	0.5–5	Depends	Average	Very high	Depends
Standby-ferro	3–15	Yes	Large	Small	No
Double Conversion	5–5000	Yes	Average	Small	Yes
Delta Conversion	5–5000	Yes	Average	Large	Yes

14.10 Rotary Uninterruptible Power Supply with Diesel or Natural Gas Supply

All presented UPS topologies are based on batteries or ultracapacitors, and they can supply energy for a reduced time interval. For local energy delivery during a longer time interval in the day-long range, diesel-, gas- or natural gas–powered generators are used [7]. Such systems are manufactured and put into service for a building, company or whole campus, and they can work in the MW range (Figure 14.18).

Such systems are not based mainly on electronics: they use a synchronous generator (electrical motor) to generate a three-phase AC voltage system. The generator is engaged with a diesel or gas engine. Generator operation is, however, conditioned with electronic circuitry for improved performance. Since the generator presents a small impedance on the AC side, it can be used close to an ideal voltage source for any load with any current harmonics. Different configurations are possible, according to the

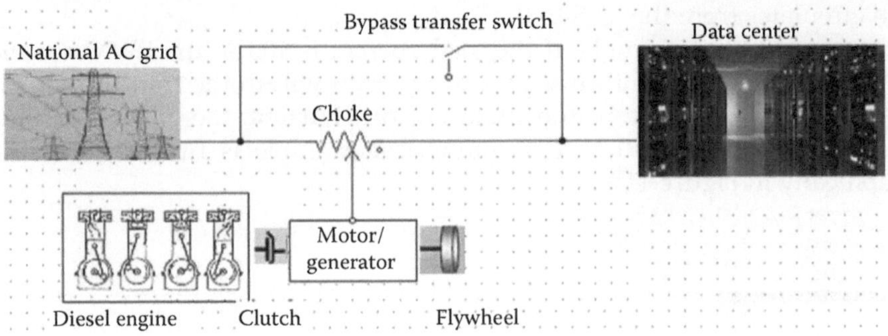

FIGURE 14.18
Principle of the rotary UPS system.

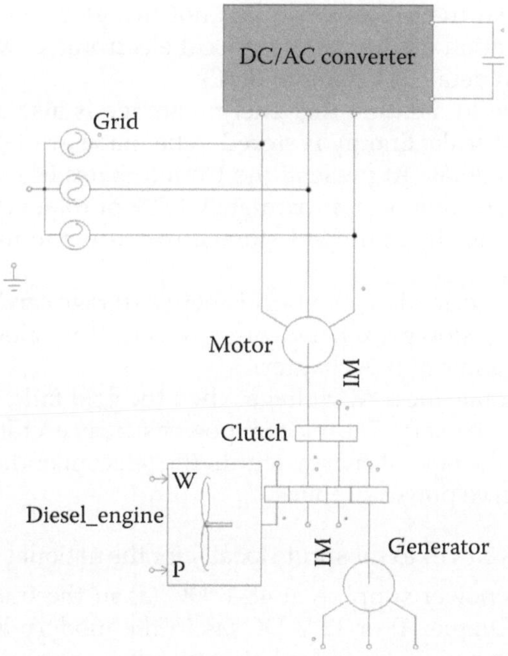

FIGURE 14.19
Details of the circuitry for a rotary UPS considered as an example.

previous classification (standby power, grid interactive, double conversion and so on).

An example is next considered to illustrate details of operation, and it follows Figure 14.19.

At normal operation, the AC voltage source supplies the asynchronous machine **M**, which rotates the generator **G** in order to deliver the AC output voltage. During this time, the converter and diesel machine do not operate (are off). When the AC grid fails, the DC/AC converter works in reverse and supplies the asynchronous machine **M** until the diesel engine reaches the desired speed. Then the electronic converter ends its operation and a clutch couples the diesel engine to the asynchronous machine's axle. The rotation of the asynchronous machine working as generator produces sinusoidal voltage(s).

14.11 Energy Storage

Previous sections have described system architecture for building up reserve AC power supplies using electronic power converters. They are intended to deliver energy near the telecom load and not as a global energy source. The circuits and operation modes for these local electronic power supplies have been presented in detail in Chapters 10–13.

It is worthwhile to mention that energy storage is also a very important issue at the global scale. Energy is stored to be made available when needed for large regional areas. At present, the United States has about 24.6 GW of grid energy storage; that is, approximately 2.3% of total electric production capacity [8]. Among this, pumped storage hydro is the most efficient and accounts for 95%.

While current information about grid energy storage can be collected from http://www.energystorageexchange.org/projects, the global grid energy storage is not the topic of this chapter.

In order to generate local AC voltage when the grid fails, another primary energy source is necessary. The telecom power supply architecture presented in Chapter 1 and the flow of energy within the telecom or data centre suggest two levels of reserve power supplies:

- AC voltage sources to substitute locally for the national AC grid utility
- DC reserve power supplies at 48 V DC (as in the traditional architecture in Chapter 1) or 12 V DC (as in the modern architecture in Chapter 1) to supply lower-end electronics

Two cases of primary energy sources are distinguished depending on supply need:

- For a long-term or permanent alternative energy supply, diesel or natural gas generators are used. However, the need to reduce maintenance and fuel costs of generators has driven manufacturers of telecom equipment to tentatively opt more and more for batteries for power storage and therefore achieve economies of scale.
- For temporary energy supply, batteries or capacitors are used.

Since short-term grid failure covers most cases when UPS systems are needed, this section focuses on batteries and capacitors [1].

14.11.1 Batteries

14.11.1.1 Historical Perspective

The first commercially available battery was made available in 1799 [9] by Volta and showed that by combining different metals that are separated by a salt or acidic solution, it is possible to generate electricity. The chemical energy is converted to electrical energy within a battery through an oxidation reaction at the anode terminal and a reduction reaction at the cathode terminal (Figure 14.20).

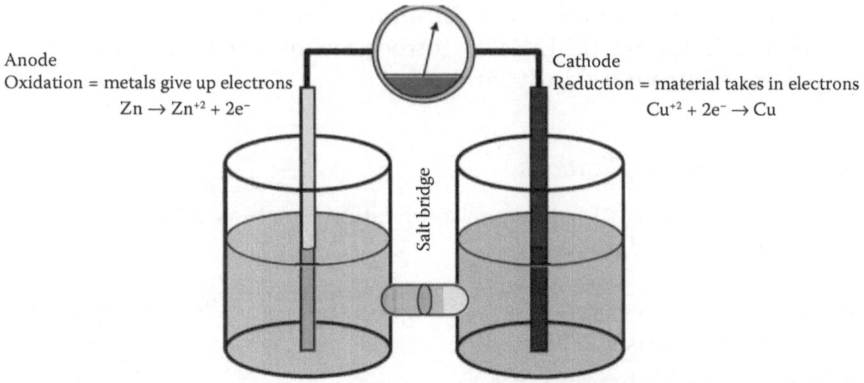

FIGURE 14.20
Historical perspective on Volta pile.

The oxidation reaction determines a loss of electrons from a metal plate like zinc while the ions dissolve into a salt solution. Contrarily, the reduction reaction determines a gain of electrons into a metal plate like copper while the ions are attracted to and deposited from the same salt solution. The ions travel through the salt bridge and the electrons through the supplied circuit. This was called a *Volta pile*. Within the original experiment, the voltage of the cell was related to the change in free energy from the two selected metals (zinc and copper):

$$V = volt(Zn) + volt(Cu) = 0.763 + 0.337 = 1.1V \quad (14.1)$$

Selection of other metals for the two metal plates in Figure 14.20 would produce other voltage levels. For any selection, the amount of produced voltage is very low, in the range of several volts. Hence, a multitude of cells need to be built and set up in cascade to provide a higher useable level of voltage.

Starting from this principle, the industry packaged and improved battery products over the years. Due to the need for a salt or acidic solution, each constituent cell was called a *wet cell*. This introduced a major problem in mobility of wet cells.

The next technological step forward was the introduction of an enclosed cell by Leclanche (zinc anode and chloride electrolyte, 1866), Edison (alkaline electrolyte, 1911) and Gassner (dry cell, 1888) [9], with a special mention of the introduction of the lead-acid (Pb-PbSO4) battery in 1859. All these solutions had the support of the emerging telegraph industry.

Battery technology has not changed much since the 1800s. Other modern and very commonly used technologies for batteries include:

- Nickel-based batteries like nickel-cadmium (Cd-NiOOH, introduced in 1899) and nickel-metal-hydride (MH-NiOOH, introduced in 1990)

- Lithium-based batteries like lithium-iodine (Li-I2, introduced in 1968), lithium-ion (Li-$LiCoO_2$, introduced in 1991) and Li-plastic/polymer (introduced in 1995)

14.11.1.2 Telecom Applications

Backup energy systems for telecom applications such as wireless, cable and cellular networks or data centres:

- Need to operate reliably, with good resistance to electrical or mechanical abuse
- Need minimal maintenance
- Should be able to sustain large variations in temperature or humidity
- Need to be versatile to either urban or remote operation
- Need to offer large energy/power available in a small space
- Need to offer good cycling capability and a high charging efficiency

As mentioned before, two different applications take advantage of batteries for backup power:

- 12/48 V DC systems for uninterruptible supply of the telecom intermediary bus
- High-voltage battery banks (300–400 V DC) for AC generation through any of the previously presented UPS systems

Lead-acid batteries are the most-used batteries for either automotive or telecommunications (including older telegraph systems) applications [10,11] since they are cheap, readily available and easy to integrate and have high power capacity. They are also the world's most recycled product. These contemporary batteries use a method based on acid and electrodes and can be classified as:

- *Sealed acid batteries* = also called valve-regulated lead acid (VRLA)
- Open or *flooded batteries* = vented lead acid (VLA) batteries

Lead-acid batteries are used at low voltage as DC bus backup power (12 V DC or 48 V DC), as well as for AC voltage generation through UPS topologies. The usability of lead-acid batteries has been proven in national utility power systems with a very large amount of energy stored. For instance, the largest installation based on VLA technology is a 10-MW/40-MWh system in Chino, California, that is used for load levelling by the Southern California Edison utility, while the largest installation based on VRLA technology is a 15- MWh facility in Hawaii built in 1993 that is used for frequency regulation, peak shaving and load shifting [11].

Obviously, a data centre requires less stored energy, and lead acid batteries are a viable solution.

While lead-acid batteries are the most used in existing systems, new developments are relying more on Li-ion technology despite the higher cost [10]. This technology is less robust: when in a fully charged state, the battery is sensitive to overtemperature, overcharge and internal pressure buildup, necessitating advanced monitoring equipment and safety precautions. Modern Li-ion backup battery systems include electronic and thermal controls packaged within the same unit as the battery itself.

Large-scale energy storage projects show that there is no limitation in usage of Li-ion batteries for data or telecom centre storage systems. Concerning the production of AC energy through UPS systems, a 12-MW lithium-ion battery system was installed in 2011 by U.S. corporation 123Systems in Chile for grid stabilisation services, with plans for an additional 20-MW system in Chile. A similar 8-MW/32-MWh lithium-ion battery system was considered in 2012 to improve grid performance and integration with large-scale wind-powered electricity generation in Tehachapi, California.

In terms of the energy storage for the intermediary DC bus architecture, an important performance metric is the energy density, currently around 180 Wh/L for Li-ion batteries, which is double that of conventional lead-acid batteries. With a stored energy of 2–4 kWh per rack mount 48 V DC storage unit, these battery systems can be used to supply energy for 3–30 minutes to local electronic equipment, with the possibility of increasing power through paralleling.

Batteries based on technologies like NiCd and NiMH are produced to be used as direct replacements for low-voltage VRLA batteries due to their maintenance-free packaging able to offer an energy density of about 100 Wh/L [7], with less performance dependency on temperature than their lead-acid counterparts, and they are 30% lighter than lead-acid batteries. As a bonus, NiCd batteries can be fully discharged, which means there is no need for oversizing.

14.11.1.3 Powering the Cell Tower

Battery technologies are most preferred for remote locations housing a *base transceiver station* (BTS) or a *basic station controller* (Figure 14.21). A BTS represents the global system for mobile communication (GSM) equipment required to secure wireless communication between the user's equipment, like a cell phone, and the communications network. Even though a typical BTS tower holds just the antenna, the tower is quite widely misinterpreted as the BTS itself. Hence, powering the cell power actually refers to the energy system for the BTS equipment. Furthermore, the BTS equipment is housed in a shelter at the base of the antenna. The temperature inside the telecom shelter during operation exceeds 55°C, which is a bad operating condition for BTS electronic equipment. Hence, NiCd or NIMH batteries are preferred to a lead-acid

FIGURE 14.21
Use of a fuel cell for cell tower supply.

battery since their performance under operation and their lifetimes are less sensitive to temperature. They are packaged in 3.6 V to 12 V DC units, with solutions for 12 V DC and 48 V DC as well.

14.11.1.4 Future Battery Development

The R&D in battery technology is currently impressive. Other battery technologies include flow batteries, vanadium redox batteries, zinc-bromine flow batteries, hydrogen bromine flow batteries, polysulfide bromide flow batteries, sodium sulphur batteries, sodium-metal chloride batteries, iron-air rechargeable batteries, metal-air batteries, Na-ion batteries including Na-halide chemistries, new types of NaS cells (e.g., flat, bipolar, low temperature, high-power), new Li-ion chemistries that improve performance and safety characteristics, advanced lead-carbon batteries, ultrabatteries (a hybrid energy storage device that combines a VRLA battery with an electrochemical capacitor) and new flow battery couples including iron-chrome and zinc/chlorine (Zn/Cl).

14.11.1.5 Lead-Acid Battery Charging Systems

While all types of batteries are rechargeable, the focus is herein on lead-acid batteries given their wider spread in use and the particularities of the

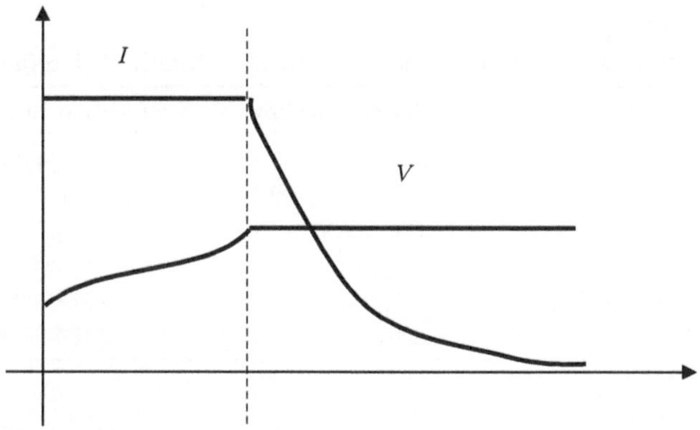

FIGURE 14.22
Graphical representation of the charging process.

charging process. After the load is supplied from a lead-acid battery, the battery is discharged. The discharge has to be stopped before a certain minimum voltage level; otherwise, the battery may lose its properties and decrease its lifetime. Thus, a battery of 10 Ah may be discharged at 1 A for 10 hours. After the grid comes back, battery recharge can be achieved with a charging circuit.

This is done in two steps:

- A large constant current is applied until a previously established voltage. This moment is called *trickle charge.*
- The voltage is kept constant and the current is decreased down to a minimum value which is characteristic to the waiting state.

The graphical representation of this principle is shown in Figure 14.22.

14.11.2 Ultra- or Supercapacitors

Other compatible solutions include supercapacitors or ultracapacitors [12]. The first ultracapacitor was patented in 1957 using porous carbon electrodes. Ultracapacitors feature a novel structure of capacitor, different from ceramic or electrolytic capacitors and from a conventional battery [13,14]. While a conventional capacitor uses electronically conductive electrons and a battery uses ionic conduction in liquid, the ultracapacitor (sometimes called supercapacitor) uses a two-layer structure with a solid-state electrolyte, hence also the name *double-layer capacitor* (DLC). The structure uses electrodes and electrolytes not unlike a battery and is able to deliver a capacitance thousands of times larger than an electrolytic capacitor. The capacitance value of a supercapacitor is proportional to the surface area of its electrodes.

TABLE 14.3

Comparative Overview of Ultracapacitors and Modern Li-Ion Batteries

Function	Supercapacitor	Lithium-Ion Battery
Charging time	1–10 sec	10–60 minutes
Lifetime (cycles)	1 mil or 30,000 h	>500
Cell voltage	2.3–2.75 V	3.6–3.7 V
Specific energy (Wh/kg)	5 (typ)	100–200
Specific power (W/kg)	<10,000	1000–3000
Cost per Wh	$20	$0.50–$1.00

An ultracapacitor has properties of both a battery and a capacitor: it has high power density and can be cycled like a capacitor, but it also has a significant energy density, like a battery. A comparative and detailed overview of performance is shown in Table 14.3, while more data is provided later in Figure 14.27.

The advantages of a supercapacitor over a battery system consist of a greater lifetime in operation, better reversibility, lower capacity fade, higher power ratings and rapid charging and discharging processes. The advantages of a battery over an ultracapacitor system consist of higher energy density, greater shelf lifetime (during storage at zero current), lower self-discharge and more affordable cost.

Similar to a battery cell, supercapacitor cells are rated in the range of 2.5–3.3 V at room temperature. This voltage rating drops at higher temperatures and should be decreased when a longer lifetime is desired. Since the cell voltage rating is typically under 3.3 V DC and telecom loads often require higher supplies to accomplish a bus at 12 V c, 48 V DC or even higher voltages, the options for supercapacitor cell configuration and switching regulator are:

- To use a single cell with a boost converter
- To use multiple cells in series and a buck or buck-boost regulator

It is preferable to set up an ultracapacitor string with the highest voltage for more energy storage [15]. Electronic circuits may need to be used for voltage balancing across individual ultracapacitors. An example of a circuit is shown in Figure 14.23. If $I_{C2} > I_{C1}$, V_{OUT} drops until M_1 is turned on and M_2 is turned off. This brings I_{OUT2} to zero, and $I_{OUT1} + I_{C1} = I_{C2}$.

This balancing precaution significantly increases the cost of the ultracapacitor bank.

For generation of high voltage at the AC grid level, an ultracapacitor around 400 V DC is required. This can be produced as 42 F capacitance at 400 V DC when using 286 cells of 2.7 V DC×3000 F each (Figure 14.24), connected as two parallel strings of 143 cells. Note a huge 3360-kW power storage capacity. In practical application with this ultracapacitor, the charging current can

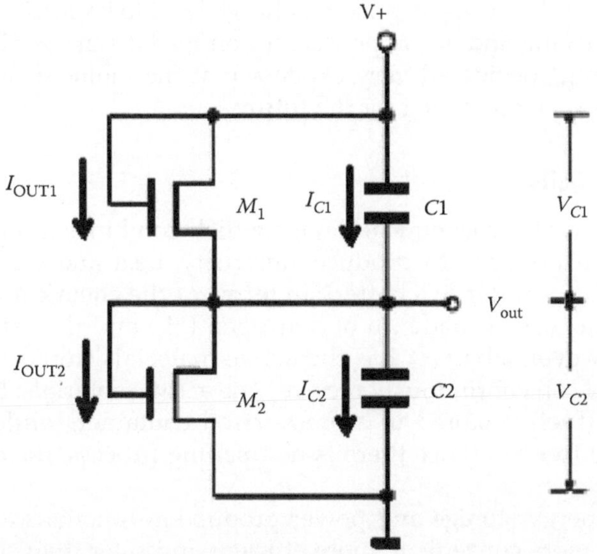

FIGURE 14.23
Possible balancing circuit with FET transistors.

go up to 500 A; that is, much more than with any battery. Figure 14.24 also shows the wire used to short the capacitor's terminals during transportation. This is done in order to avoid the deposit of high voltage.

A power converter configured as a constant current source can be used for charging the entire supercapacitor bank. The voltage linearly increases up to a target voltage, at which time the constant voltage loop becomes active and

FIGURE 14.24
Cell used in building up a high voltage storage unit.

accurately controls the supercapacitor charge level to be constant in order to avoid overcharging and to compensate for on-load discharge [15].

Despite being considered very expensive at the moment, supercapacitor technology is very promising for the future.

14.11.3 Fuel Cells

A *fuel cell* is an electrochemical device which combines a liquid combustible (fuel) with oxygen to produce electricity, heat and water [16]. Fuel-cell systems are similar to a battery in terms of the chemical reaction. The anode of a fuel cell is made up of hydrogen (H_2) and the cathode of oxygen (O_2). However, a battery has the active materials stored in electrodes, while the fuel cell continuously receives the active materials: the hydrogen combustible (fuel) is stored in a pressurised container, while the oxygen is taken from free air. Since there is no burning process, the only waste is water.

Since the energy storage and power production functions are separated, fuel cells are more convenient, more efficient and safer than storage batteries. They can also be 30% cheaper than batteries. Finally, fuel cells are safer because the short-circuit of a fuel cell harmlessly dissipates only the energy associated with the small amount of hydrogen present in the cell, while a short circuit at a battery's terminals dissipates all of its stored energy.

Fuel-cell systems need 5 seconds to start; thus, we need another battery or ultracapacitor within the same system to supply energy for the short term. After start, safe operation is limited to certain conditions (Figure 14.25).

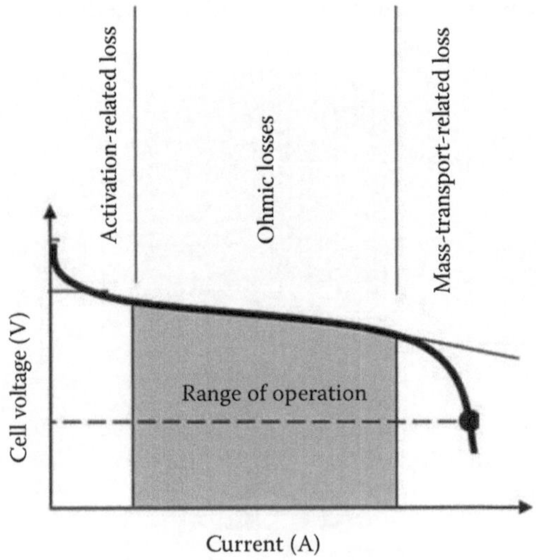

FIGURE 14.25
V–I S-curve characteristics of a fuel cell and safe operation range.

Two important applications take advantage of fuel-cell technology to enhance telecom systems:

- Power a base transceiver station, as described in Section 14.11.1.3 (Figure 14.21)
- Form 48–56 V DC storage to serve the 48 V DC intermediary bus at levels of tens of kW

14.11.4 Electromechanical Storage (Flywheel Systems)

Due to the drawbacks linked to manipulation of toxic substances, the usage of batteries can be completed with alternative solutions for energy storage such as flywheel systems. A flywheel represents a mechanical system of high inertia which stores kinetic energy in a spinning rotor that is charged and discharged through an electric generator [11,17]. Flywheel systems have been used for energy storage for a long time, with applications in both power quality and uninterruptible power supplies. The conversion from kinetic energy to electrical energy is done towards the AC voltage in grid-type applications. Current flywheel energy storage systems can operate at low or high speed [18]:

- Low speed = Most low-speed flywheels are designed for 10,000 revolutions per minute (rpm) or less, and are typically made of extremely heavy and compact steel discs. The shaft is either vertical or horizontal, and may have mechanical or magnetic bearings.
- High Speed = High-speed designs operate above 10,000 rpm, some at even more than 100,000 rpm. Operation at high speed brings associated fatigue failure risks. This imposes the use of stronger materials, like composites of graphite or fibreglass, and requires magnetic bearings and a vertical shaft.

14.11.5 Selection of the Alternate Power Source

An optimal energy transfer between the source and load occurs when the impedance of the source equals the impedance of the load. However, this criterion cannot be used herein since the internal resistance of most energy sources (like the battery) is very small, and it would require a very large current.

Hence, the selection is sought considering the energy availability from the source. This is strongly dependent on the source's electrochemistry, and it is defined with the *Ragone chart* (Figure 14.26).

The Ragone chart was first introduced to compare batteries and extended afterwards for various energy sources. The vertical axis describes how much energy is available, while the horizontal axis shows how quickly that energy can be delivered; that is, power. Both measures are expressed per unit mass. A point in a Ragone chart represents the amount of time during which the

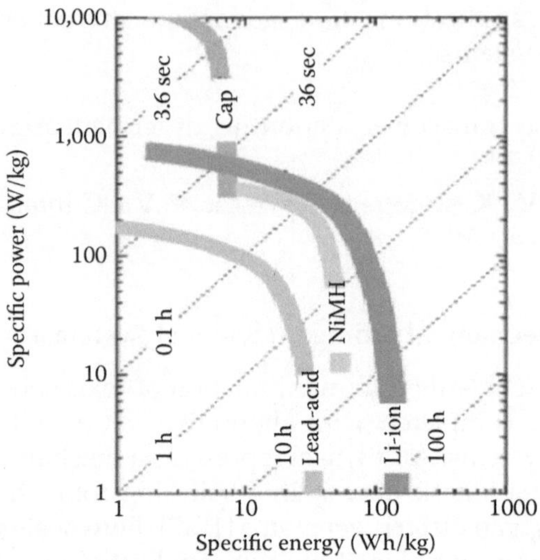

FIGURE 14.26
Ragone chart for definition of source's availability for a given chemistry.

energy-per-mass can be delivered at the *power-per-mass*. Such an amount of time is expressed in hours and is equal to the ratio between the energy and the power densities. The isolines for a Ragone chart are straight lines with the unity slope at distances in time units like hours.

Analogously, performance can be defined in volume units instead of mass (Figure 14.27).

Other definitions for availability relate to the rate at which the energy is needed since this involves the rate of ion movement across the electrolyte as

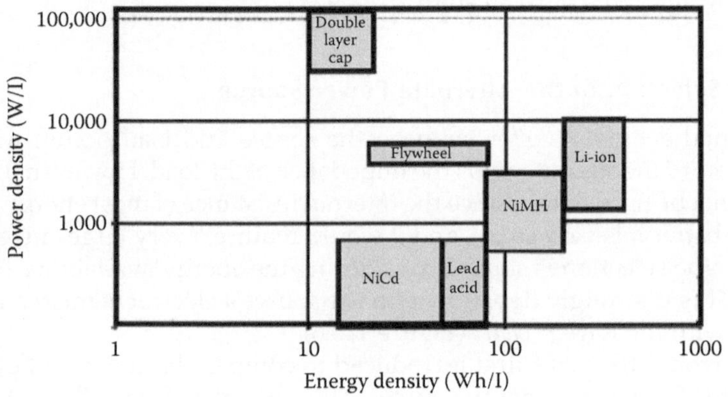

FIGURE 14.27
Comparison of power density and energy density in relation to volume [18], shown on logarithmic scales.

a main limiting factor. As a consequence, the available voltage and capacity decrease as demand current is increased. This is expressed with Peukert's law, presented by the German scientist Wilhelm Peukert in 1897. It expresses the approximate change in capacity of rechargeable lead-acid batteries at different rates of discharge.

For a 1-ampere discharge rate, Peukert's law is:

$$C_p = I^k \cdot t \tag{14.2}$$

where C_p is the capacity at a 1-ampere discharge rate, expressed in ampere hours; I is the actual discharge current in amperes; t is the actual discharge time and k is the dimensionless Peukert constant.

This relationship can be enhanced with a dependency on a discharge rate different from a 1-ampere discharge:

$$t = H \cdot \left(\frac{C}{I \cdot H} \right)^k \tag{14.3}$$

where H equals the rated discharge time in hours and C is the rated capacity at the discharge rate in ampere-hours.

Let us consider an example with a battery rated at 100 Ah and having a Peukert constant of 1.2 that is being discharged at a rate of 10 A for a rated time of 10 hours. However, according to Peukert's law, it is fully discharged in:

$$t = 20 \cdot \left(\frac{100}{10 \cdot 10} \right)^{1.2} \approx 8.7 \text{ hours} \tag{14.4}$$

instead of the rated 10 hours.

14.12 Energy Management

14.12.1 Communication Requirements

UPS units feature connectivity interfaces such as RS232 or USB for low-level transport data, used for easier management at the level of computing or communications centres. The time intervals for operation or shutdown can thus be decided for any computing equipment as well as for energy zoning for various equipment within the data or telecom centre.

In order to generalise all solutions within the same standard for communications and management, the majority of manufacturers offer high-level communications like the *simple network management protocol* (SNMP) and *hypertext transmission protocol* (HTTP). This allows each UPS unit to be connected with a TCP/IP (that is Transmission Control Protocol – TCP and the Internet Protocol – IP) address and access through Ethernet local-area network (LAN). The control and monitoring of the UPS system can be allowed through the internet protocol.

Solutions based on SNMP require a dedicated computer to run the network management program, while the HTTP solution allows management at a distance through the internet.

14.12.2 Energy Peak Shaving

The importance of providing reliable and continuous energy to the data or telecom centre requires managers to set up energy storage facilities. These systems are sized to the capacity of the entire data centre: a 1-MW data centre will have a storage system with a power rating of 1 MW with several minutes of energy capacity. As shown at the beginning of this chapter, this is considered enough to survive most short-term grid failures. Alternatively, energy storage and a UPS system are implemented at each equipment rack.

Furthermore, the energy storage facility (either for the entire centre or at the rack level) is solely used for incidental grid failure and is rarely involved in power system performance. This is mainly because the operator does not want to jeopardise the battery system lifetime with additional tasks. For instance, a telecom centre may have planned to replace batteries every four years, considering occasional failures. Using this storage equipment for other means would reduce lifetime and require replacement sooner [19].

Once such a large energy storage facility is in place, the operator may, however, use the surplus energy to compensate for energy peaks or system deregulation, with energy savings advantages. Cost can thus be optimised from either

- Reduction of energy bill expenditure
- Reduction of installed capacity

An example of operation of a power-shaving algorithm is set forth in Figure 14.28 for a single server rack with 16 kW installed power. The rack also houses a local UPS able to take over the entire power supply task, hence 16 kW.

The consumption of the entire data centre or that of a single rack is averaged each 15 minutes [19,20]. Data are collected for a month to establish daily patterns in consumption. A power threshold limit is set continuously based on energy cost, schedule and availability. In our example, this was set at a fixed 10 kW, but the threshold limit can vary during the day.

If the consumption is above the threshold limit, the battery-based UPS is used to take over the difference between the consumption and limit. In our example, this is the case for hours 5–9 of the day, or occasionally at

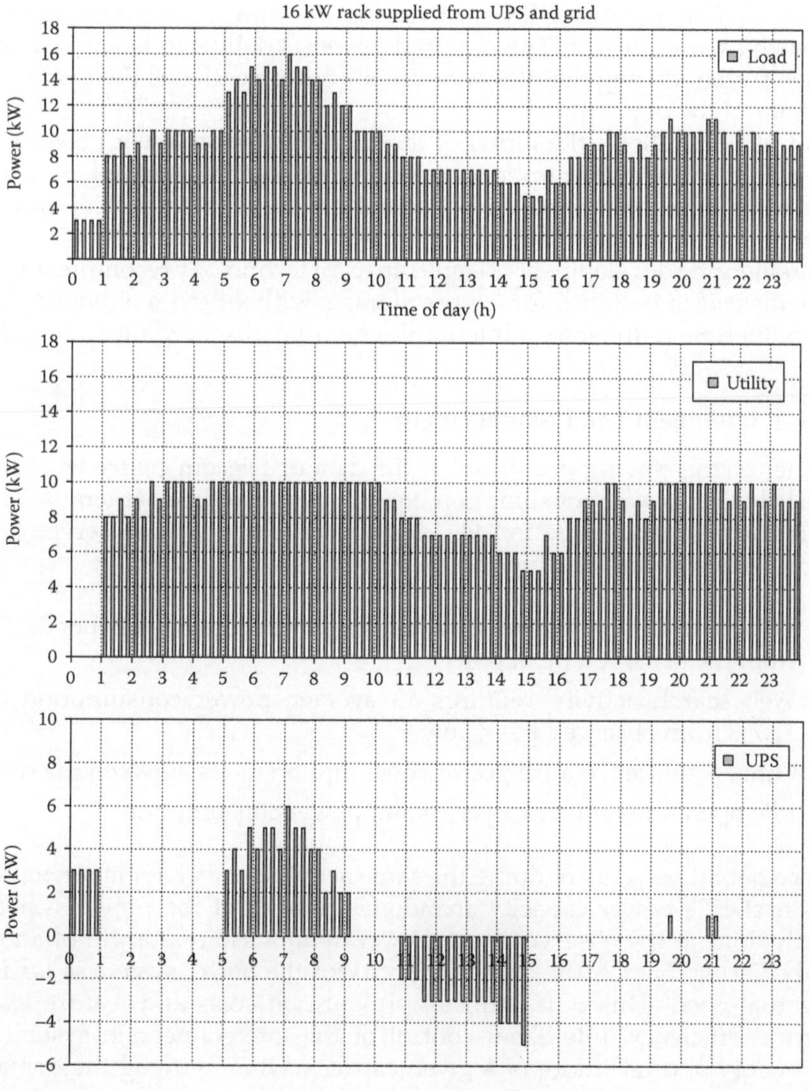

FIGURE 14.28
Example for power shaving: when power consumption is above the threshold, the storage battery supplies the difference, while a level under the threshold allows battery charging if need be.

hours 20 and 22 of the day. This obviously discharges the batteries somewhat. When the power consumption is under the limit, the batteries can be recharged with a power less than the difference between the set limit and the smaller actual level required by load. In our example, this is the case for hours 11–15 of the day.

This planning can go further, and a downtime of one hour or so can be planned each day for scheduled maintenance operations to the entire power system. Preferably, this will be set during the lowest consumption of the day, such as at midnight. The required power is delivered from the energy storage system during maintenance. In our example, this is the case for the first hour of the day.

Peak shaving allows the rating of the entire system at lower power levels than the absolute consumption peak. Various optimisation plans can be adopted to save money based on actual consumption patterns and cost variation during the day [19,20].

The emerging development of supercapacitor technology recommends their usage instead of batteries for energy storage, without the mentioned drawback of lifetime reduction with multiple charging–discharging cycles [14].

14.12.3 Intelligent Load Management

Another energy-saving procedure at the data or telecom centre level refers to distributing power consumption equally over multiple servers. A recent report from Google – cited in [19] – demonstrates that various server activities produce different loadings of the power system:

- Web mail activity requires an average power consumption of 89.9% from the observed peak power.
- Web search activity requires an average power consumption of 72.7% from observed peak power.
- Other activities require power consumption ratios between the two.
- Idle operation requires a very small power consumption.

Since not all servers are doing the same activity, at a given moment, there is at least 17% power capacity abandoned. Rating of the power system is usually done at the observed peak power, with a safety margin of another 40% buffering. Hence, the supply capacity of the entire system is far from the actual need. This puts artificial pressure on costs and feature loss in electrical efficiency. Intelligent control of the interconnection system can save energy and use more of a given capacity while moving the demands from one outlet to another.

Summary

The topic of uninterruptible power supplies was addressed. Most industrial standards require a fast connection of the AC grid supply in order to allow uninterrupted operation of computing and communications equipment. A series of possible situations aiming at jeopardizing the quality of the energy supply were identified and explained in this chapter. The conditioning of the energy supply can be done with passive or active solutions.

Uninterruptible power supply systems have the role of delivering energy to the load when the grid voltage fails. Several topologies can be used:

- Uninterruptible power supply in standby state (standby UPS)
- Uninterruptible power supply in direct connection with the grid (line-interactive UPS)
- Uninterruptible power supply with ferro-transformer (standby-ferro UPS)
- Uninterruptible power supply with double conversion (double conversion online UPS)
- Uninterruptible power supply with delta conversion (delta conversion online UPS)

The energy is stored in rechargeable batteries, high-capacity ultracapacitors or by other storage methods. The topic of energy storage is very important, and a brief information was provided herein.

The status of UPS systems is monitored with special systems for communication and management, such as SNMP and HTTP. Using this information, peak power shaving and intelligent control of energy are considered in data centres or in order to supply remote equipment like a BTS station.

Modern systems expand the traditional use of UPS systems to power conditioning or power-shaving functions. Their meaning was explained in this chapter.

References

1. LaCommare, K.H., Eto, J.H., 2006, "Cost of Power Interruptions to Electricity Consumers in the United States (U.S.)", Report LBNL-58164, Berkeley: LBNL, February.
2. Anon, 2015, "Distributed Generation and Energy Storage in Telecom Networks", Navigant Market report, pp. 1–63.
3. Rasmussen, N., 2003, "The Different Types of UPS Systems", APC Corporation White Paper #1, version 2004-5.
4. Anon, 2000, "High Availability Power Systems, Part I: UPS Internal Topology", Liebert Corporation White Paper, November 2000.
5. Loeffler, C., Spears, E., 2011, "UPS Basics", Eaton Corporation Application Note WP153005EN, pp. 1–14.
6. Anon, 2004, "Understanding Delta Conversion Online™ 'The Difference'—Part 1", APC Corporation Application Note #39.

7. Anunciada, V., Santana, J., 1996, "A New Configuration of Low Cost Rotative Diesel UPS System", *IEEE PESC 1996*, Baveno, Italy, pp. 420–427.
8. Anon, 2013, "Grid Energy Storage", U.S. Department of Energy Report, pp. 1–66, December 2013.
9. Arnold, C.B., 2000, "Batteries and Energy Storage", Lecture Notes, Department of Mechanical and Aerospace Engineering, Princeton University.
10. Anon, 2009, "Innovative Energy Storage Systems—Solutions for Dynamic Telecom Needs", Saft Corporation Marketing Note, pp. 1–9, June.
11. Hausheer, R., Heinze, K., Katai, S., Kaufman, K., Litten, J., Madaiah, G., 2012, "Evaluating Energy Storage Options—A Case Study at Los Angeles Harbor College", Internal report, University of California at Santa Barbara.
12. Grbovic, P., 2013, *Ultra-Capacitors in Power Conversion Systems: Applications, Analysis, Design, from Theory to Practice*, IEEE – Wiley, Chichester, West Sussex, United Kingdom.
13. Anon, 2000, "Murata Super-Capacitor", Murata Technical Note No. C2M1CXS-053L, pp. 1–39.
14. Zheng, W., Ma, K., Wang, X., 2017, "Hybrid Energy Storage with Super-Capacitor for Cost-Efficient Data Center Power Shaving and Capping", *IEEE Transactions on Parallel and Distributed Systems*, 28(4), 1105–1117.
15. Anon, 2014, "How to Charge Super-Capacitor Banks for Energy Storage", Intersil Corporation Technical Note, pp. 1–8.
16. Miller, A.R., 2011, *Fuel Cell Tutorial*, Internet Tutorial, Supersonic Institute.
17. Anon, 2011, "Electrical Energy Storage, International Electro-Technical Commission", White Paper IEC WP EES:2011-12(en), pp. 1–78.
18. Bradbury, K., 2010, Energy Storage Technology Review, internet source at http://www.kylebradbury.org, accessed on 3 October 2017, pp. 1–33.
19. Shi, Y., Xu, B., Wang, D., Zhang, B., 2017, "Using Battery Storage for Peak Shaving and Frequency Regulation: Joint Optimization for Superlinear Gains", *IEEE Transactions on Power Systems*, accepted for publication, available on www.ieee.org, accessed on 3 October 2017.
20. Adams, M., 2016, "Solving Power Capacity Challenge with Software Defined Power", *Bodo's Power Systems*, pp. 24–28, September 2016 issue.

Index

C

L